TITANIUM ALLOYS – TOWARDS ACHIEVING ENHANCED PROPERTIES FOR DIVERSIFIED APPLICATIONS

Edited by **A.K.M. Nurul Amin**

Titanium Alloys – Towards Achieving Enhanced Properties for Diversified Applications
A.K.M. Nurul Amin

Published by InTech
Janeza Trdine 9, 51000 Rijeka, Croatia

Publishing Process Manager Ana Skalamera

Technical Editor Teodora Smiljanic

Cover Designer InTech Design Team

Additional hard copies can be obtained from orders@intechopen.com

Titanium Alloys – Towards Achieving Enhanced Properties for Diversified Applications,
Edited by A.K.M. Nurul Amin
p. cm.
ISBN 978-953-51-0354-7

For EU product safety concerns:
IN TECH d.o.o., Prolaz Marije Krucifikse Kozulić 3, 51000 Rijeka, Croatia,
info@intechopen.com or visit our website at intechopen.com.

Meet the editor

Professor Amin had completed his Master of Manufacturing Engineering and his Ph.D. in Technical Sciences from Volgograd Poly Technique Institute of the former USSR in 1979 and 1982. He started his teaching career at Bangladesh University of Engineering and Technology (BUET) in 1993. After 1997 he had moved to his current job at the International Islamic University Malaysia (IIUM) where he has been working as a Head of Department for the past 7 years. He currently holds position as a senior Professor at the Manufacturing and Materials Engineering Department. His current research interests are in: heat assisted machining of difficult-to-cut metals and alloys including titanium alloys, hard machining, high speed machining of metals and ceramics and chatter control during metal cutting.

Contents

Preface

Though titanium and its alloys are relatively new engineering materials, they have found wide application in the aerospace, shipbuilding, automotive, sports, chemical and food processing industries due to their extreme lightness, high specific strength and good corrosion resistance at temperatures below 500°C. They are also considered suitable materials for biomedical application due to their biological passivity and biocompatibility. However, besides these positive properties titanium alloys have a number of adverse favorable properties which are related to their processing, machinability and long time use in open and corrosive environment. Chemical reactivity of titanium with other materials at elevated temperature is high, which necessitates the development of non conventional melting, refining and casting techniques, making this material very expensive. Numerous research is directed towards addressing these issues in order to ease their processing and further applications.

This nine diverse chapters of this book are distributed under three sections and address problems related to the processing and application of this precious metal and its alloys. The book chapters are contributed by researchers who devoted long periods of their research career working on titanium and its alloys looking for solutions to some of these specific problems.. From this perspective this book will serve as an excellent reference material for researchers whose works is in anyway related to titanium and its alloys - from processing to applications. The chapters are designed to address the issues that arise at the material development, processing and the application stages. For instance, the α case formation defects that arise at the investment casting stage and the optimization aspects of the additive manufacturing (AM) processes through numerical modelling and simulation has been addressed in two different chapters. Similarly, the metallurgical defects resulting from entrapped gases during casting processes are common to titanium parts. The morphology of formation of these defects in the production of Ti–6Al–4V alloy is presented in another chapter of the book with the objective of addressing effectively the problem at the casting stage. Titanium parts are sometimes designed to work as die components or as projectiles and as such are subjected to high pressures and temperatures. However, high pressure raises a number of scientific and engineering issues, mainly because under such pressures the relatively ductile α phase may get transformed into a fairly brittle ω phase, which may significantly limit the use of titanium alloys in high pressure applications. One of the chapters of the book deals with these issues and indicates how the formation of ω phase may be avoided through adoption of proper processing techniques.

In the Chapter 'Hot Plasticity of alpha Beta alloys' the authors have shared their invaluable experimental results explaining different aspects of hot plasticity of two-phase titanium alloys and have indicated techniques for developing appropriate microstructure yielding optimum plastic flow stresses under elevated temperatures. The phenomenon of super plasticity is also addressed in the same chapter. Furthermore, application of titanium alloys under corrosive environment and friction requires additional strengthening through effective surface treatment. One chapter of the book addresses different aspects of a common chemico-thermal method - nitriding to apply an effective coating on titanium parts. Specific issues related to the intricacies of the nitriding process suitable for application on titanium parts have been elaborated in the same chapter with excellent illustrations. Titanium alloys are common options in biomedical and dental applications and the ternary alloys Ti-6Al-7Nb are widely used for these purposes due to their unique mechanical and chemical properties, excellent corrosion resistance and biocompatibility. Development of nanotube anodic layers for medical applications on these materials are addressed in one of the chapters while the mechanism of electrochemical deposition of calcium phosphate on titanium substrate and the related process parameters and optimization techniques are presented in another. Titanium and its alloys are well known for their poor machinability properties. Useful experimental materials are presented in one chapter dealing with machining of titanium alloys, specifically drilling, a common machining operation.

We hope the research materials presented in the different chapters of this book will contribute to the ongoing research works on titanium and its alloys and help further improvement in the properties and application of titanium alloys.

Acknowledgements

The Editor would like congratulate the publishing team of INTECH for taking up this vital project and successfully steering it through its various reviewing, editing and publishing stages. Deep appreciation is extended to all the authors of the book chapters for their contribution in composing this valuable book. He would also like to acknowledge his deep appreciation to the Publishing Process Managers of the book for their sincere cooperation in rendering Editor's duties during the entire period of the editing and compilation process. Finally, he would like to express his gratefulness to the publisher for choosing him as the Editor of this book.

Prof. Dr. A.K.M. Nurul Amin
Department of Manufacturing and Materials Engineering,
Faculty of Engineering, International Islamic University of Malaysia,
Malaysia

Part 1

Manufacturing Processes and Inherent Defects in Titanium Parts

1

Numerical Modeling of the Additive Manufacturing (AM) Processes of Titanium Alloy

Zhiqiang Fan and Frank Liou
Missouri University of Science and Technology
USA

1. Introduction

It is easy to understand why industry and, especially, aerospace engineers love titanium. Titanium parts weigh roughly half as much as steel parts, but its strength is far greater than the strength of many alloy steels giving it an extremely high strength-to-weight ratio. Most titanium alloys are poor thermal conductors, thus heat generated during cutting does not dissipate through the part and machine structure, but concentrates in the cutting area. The high temperature generated during the cutting process also causes a work hardening phenomenon that affects the surface integrity of titanium, and could lead to geometric inaccuracies in the part and severe reduction in its fatigue strength [Benes, 2007]. On the contrary, additive manufacturing (AM) is an effective way to process titanium alloys as AM is principally thermal based, the effectiveness of AM processes depends on the material's thermal properties and its absorption of laser energy rather than on its mechanical properties. Therefore, brittle and hard materials can be processed easily if their thermal properties (e.g., conductivity, heat of fusion, etc.) are favorable, such as titanium. Cost effectiveness is also an important consideration for using additive manufacturing for titanium processing. Parts or products cast and/or machined from titanium and its alloys are very expensive, due to the processing difficulties and complexities during machining and casting. AM processes however, have been found to be very cost effective because they can produce near-net shape parts from these high performance metals with little or no machining [Liou & Kinsella, 2009]. In the aerospace industry, titanium and its alloys are used for many large structural components. When traditional machining/cast routines are adopted, conversion costs for these heavy section components can be prohibitive due to long lead time and low-yield material utilization [Eylon & Froes, 1984]. AM processes have the potential to shorten lead time and increase material utilization in these applications. The following sections 1.1, 1.2 and 1.3 summarize the fundamental knowledge for the modeling of additive manufacturing processes.

1.1 Additive manufacturing

Additive manufacturing can be achieved by powder-based spray (e.g., thermal spray or cold spray), sintering (e.g., selective laser sintering), or fusion-based processes (or direct metal deposition) which use a laser beam, an electron beam, a plasma beam, or an electric arc as an energy source and either metallic powder or wire as feedstock [Kobryn et al., 2006]. For the aerospace industry which is the biggest titanium market in the U.S. [Yu & Imam, 2007], fusion-based AM processes are more advantageous since they can produce 100% dense functional metal parts. This chapter will focus on fusion-based AM processes with application to titanium.

Numerical modeling and simulation is a very useful tool for assessing the impact of process parameters and predicting optimized conditions in AM processes. AM processes involve many process parameters, including total power and power intensity distribution of the energy source, travel speed, translation path, material feed rate and shielding gas pressure. These process parameters not only vary from part to part, but also frequently vary locally within a single part to attain the desired deposit shape [Kobryn et al., 2006]. Physical phenomena associated with AM processes are complex, including melting/solidification and vaporization phase changes, surface tension-dominated free-surface flow, heat and mass transfer, and moving heat source. The variable process parameters together with the interacting physical phenomena involved in AM complicate the development of process-property relationships and appropriate process control. Thus, an effective numerical modeling of the processing is very useful for assessing the impact of process parameters and predicting optimized conditions.

Currently process-scale modeling mainly addresses transport phenomena such as heat transfer and fluid dynamics, which are closely related to the mechanical properties of the final structure. For example, the buoyancy-driven flow due to temperature and species gradients in the melt pool strongly influences the microstructure and thus the mechanical properties of the final products. The surface tension-driven free-surface flow determines the shape and smoothness of the clad. In this chapter, numerical modeling of transport phenomena in fusion-based AM processes will be presented, using the laser metal deposition process as an example. Coaxial laser deposition systems with blown powder as shown in Fig. 1 are considered for simulations and experiments. The material studied is Ti-6Al-4V for both the substrate and powder. As the main challenges in modeling of fusion-based AM processes are related to melting/solidification phase change and free-surface flow in the melt pool, modeling approaches for these physical phenomena will be introduced in Sections 1.2 and 1.3.

1.2 Modeling of melting/solidification phase change

Fusion-based AM processes involve a melting/solidification phase change. Numerical modeling of the solidification of metal alloys is very challenging because a general solidification of metal alloys involves a so-called "mushy region" over which both solid and liquid coexist and the transport phenomena occur across a wide range of time and length scales [Voller, 2006]. A rapidly developing approach that tries to resolve the smallest scales of the solid-liquid interface can be thought of as direct microstructure simulation. In order to simulate the microstructure development directly, the evolution of the interface between different phases or different microstructure constituents has to be calculated, coupled with the physical fields such as temperature and concentration [Pavlyk & Dilthey, 2004]. To this approach belong phase-field [Beckermann et al., 1999; Boettinger et al., 2002; Caginalp, 1989; Karma & Rappel, 1996,1998; Kobayashi,1993; Provatas et al., 1998; Steinbach et al., 1996; Warren & Boettinger, 1995; Wheeler et al., 1992], cellular-automaton [Boettinger et al., 2000;

Fan et al., 2007a; Gandin & Rappaz, 1994; Grujicic et al. 2001; Rappaz & Gandin, 1993; Zhu et al., 2004], front tracking [Juric & Tryggvason, 1996; Sullivan et al., 1987; Tryggvason et al., 2001], immersed boundary [Udaykumar et al., 1999, 2003] and level set [Gibou et al., 2003; Kim et al. 2000] methods. Due to the limits of current computing power, the above methods only apply to small domains on a continuum scale from about 0.1 μm to 10 mm.

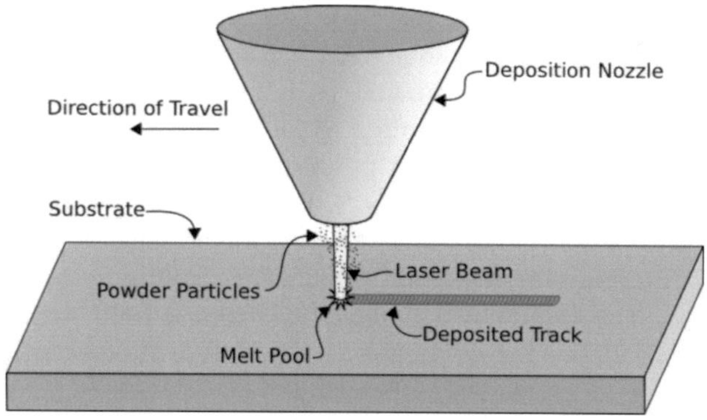

Fig. 1. Schematics of a coaxial laser metal deposition system with powder injection

To treat the effects of transport phenomena at the process-scale (~ 1 m), a macroscopic model needs to be adopted, where a representative elementary volume (REV) is selected to include a representative and uniform sampling of the mushy region such that local scale solidification processes can be described by variables averaged over the REV [Voller et al., 2004]. Based on the REV concept, governing equations for the mass, momentum, energy and species conservation at the process scale are developed and solved. Two main approaches have been used for the derivation and solution of the macroscopic conservation equations. One approach is the two-phase model [Beckermann & Viskanta, 1988; Ganesan & Poirier, 1990; Ni & Beckermann, 1991], in which the two phases are treated as separate and separate volume-averaged conservation equations are derived for solid and liquid phases using a volume averaging technique. This approach gives the complete mathematical models for solidification developed today, which have the potential to build a strong linkage between physical phenomena occurring on macroscopic and microscopic scales [Ni & Incropera, 1995]. However, the numerical procedures of this model are fairly involved since two separate sets of conservation equations need to be solved and the interface between the two phases must be determined for each time step [Jaluria, 2006]. This places a great demand on computational capabilities. In addition, the lack of information about the microscopic configuration at the solid-liquid interface is still a serious obstacle in the implementation of this model for practical applications [Stefanescu, 2002]. An alternative approach to the development of macroscopic conservation equations is the continuum model [Bennon & Incropera, 1987; Hills et al., 1983; Prantil & Dawson, 1983; Prescott et al., 1991; Voller & Prakash, 1987; Voller et al., 1989]. This model uses the classical mixture theory [Muller, 1968] to develop a single set of mass, momentum, energy and species conservation equations, which concurrently apply to the solid, liquid and mushy regions. The numerical procedures for this model are much simpler since the same equations are employed over the entire computational domain, thereby facilitating use of standard, single-phase CFD procedures. In this study, the continuum model is adopted to develop the governing equations.

1.3 Modeling of free-surface flow

In fusion-based AM processes, the melt pool created by the energy source on the substrate is usually modelled as a free-surface flow, in which the pressure of the lighter fluid is not dependent on space, and viscous stresses in the lighter fluid is negligible. The techniques to find the shape of the free surface can be classified into two major groups: Lagrangian (or moving grid) methods and Eulerian (or fixed grid) methods. In Lagrangian methods [Hansbo, 2000; Idelsohn et al., 2001; Ramaswany& Kahawara, 1987; Takizawa et al., 1992], every point of the liquid domain is moved with the liquid velocity. A continuous re-meshing of the domain or part of it is required at each time step so as to follow the interface movement. A special procedure is needed to enforce volume conservation in the moving cells. All of this can lead to complex algorithms. They are mainly used if the deformation of the interface is small, for example, in fluid-structure interactions or small amplitude waves [Caboussat, 2005]. In Eulerian methods, the interface is moving within a fixed grid, and no re-meshing is needed. The interface is determined from a field variable, for example, a volume fraction [DeBar, 1974; Hirt & Nichols, 1981; Noh & Woodward, 1976], a level-set [Sethian, 1996, 1999] or a phase-field [Boettinger et al., 2002; Jacqmin, 1999]. While Lagrangian techniques are superior for small deformations of the interfaces, Eulerian techniques are usually preferred for highly distorted, complex interfaces, which is the case for fusion-based additive manufacturing processes. For example, in AM processes with metallic powder as feedstock, powder injection causes intermittent mergers and breakups at the interface of the melt pool, which needs a robust Eulerian technique to handle.

Among the Eulerian methods, VOF (for Volume-Of-Fluid) [Hirt & Nichols, 1981] is probably the most widely used. It has been adopted by many in-house codes and built into commercial codes (SOLA-VOF [Nichols et al, 1980], NASA-VOF2D [Torrey et al 1985], NASA-VOF3D [Torrey et al 1987], RIPPLE [Kothe & Mjolsness 1991], and FLOW3D [Hirt & Nichols 1988], ANSYS Fluent, to name a few). In this method a scalar indicator function, F, is defined on the grid to indicate the liquid-volume fraction in each computational cell. Volume fraction values between zero and unity indicate the presence of the interface. The VOF method consists of an interface reconstruction algorithm and a volume fraction advection scheme. The features of these two steps are used to distinguish different VOF versions. For modeling of AM processes, an advantage of VOF is that it can be readily integrated with the techniques for simulation of the melting /solidification phase change. VOF methods have gone through a continuous process of development and improvement. Reviews of the historical development of VOF can be found in [Benson, 2002; Rider & Kothe, 1998; Rudman, 1997; Tang et al., 2004]. In earlier versions of VOF [Chorin, 1980; Debar, 1974; Hirt & Nichols, 1981; Noh & Woodward, 1976], reconstruction algorithms are based on a piecewise-constant or "stair-stepped" representation of the interface and advection schemes are at best first-order accurate. These first-order VOF methods are numerically unstable in the absence of surface tension, leading to the deterioration of the interface in the form of flotsam and jetsam [Scardovelli & Zaleski, 1999]. The current generation of VOF methods approximate the interface as a plane within a computational cell, and are commonly referred to as piecewise linear interface construction (PLIC) methods [Gueyffier et al., 1999; Rider & Kothe, 1998; Youngs, 1982, 1984]. PLIC-VOF is more accurate and avoids the numerical instability [Scardovelli & Zaleski, 1999].

2. Mathematical model

2.1 Governing equations

In this study the calculation domain for a laser deposition system includes the substrate, melt pool, remelted zone, deposited layer and part of the gas region, as shown in Fig.2. The continuum model [Bennon & Incropera, 1987; Prescott et al., 1991] is adopted to derive the governing equations for melting and solidification with the mushy zone. Some important terms for the melt pool have been added in the momentum equations, including the buoyancy force term and surface tension force term, while some minor terms in the original derivation in [Prescott et al., 1991] have been neglected. The molten metal is assumed to be Newtonian fluid, and the melt pool is assumed to be an incompressible, laminar flow. The laminar flow assumption can be relaxed if turbulence is considered by an appropriate turbulence model, such as a low-Reynolds-number *k*-ε model [Jones & Launder, 1973]. The solid and liquid phases in the mushy zone are assumed to be in local thermal equilibrium.

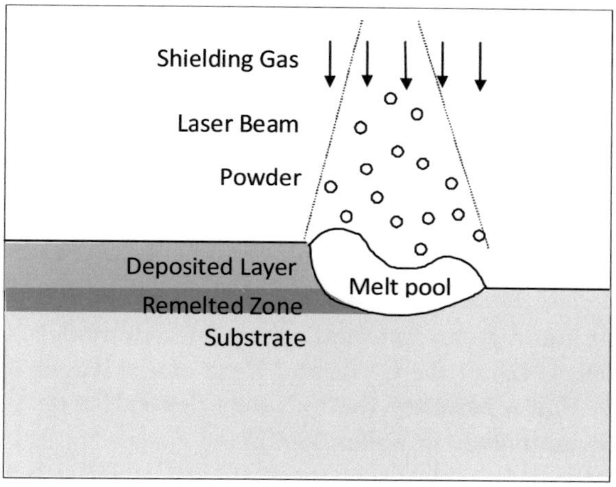

Fig. 2. Schematic diagram of the calculation domain for laser metal deposition process

For the system of interest, the conservation equations are summarized as follows:

Mass conservation:

$$\frac{\partial \rho}{\partial t} + \nabla \cdot (\rho \mathbf{V}) = 0 \tag{1}$$

Momentum conservation:

$$\frac{\partial}{\partial t}(\rho u) + \nabla \cdot (\rho \mathbf{V} u) = \nabla \cdot (\mu_l \frac{\rho}{\rho_l} \nabla u) - \frac{\partial p}{\partial x} - \frac{\mu_l}{K_x}\frac{\rho}{\rho_l}(u - u_s) + \rho \mathbf{g}_x[1 - \alpha(T - T_0)] + \mathbf{F}_{Sx} \tag{2}$$

$$\frac{\partial}{\partial t}(\rho v) + \nabla \cdot (\rho \mathbf{V} v) = \nabla \cdot (\mu_l \frac{\rho}{\rho_l} \nabla v) - \frac{\partial p}{\partial y} - \frac{\mu_l}{K_y}\frac{\rho}{\rho_l}(v - v_s) + \rho \mathbf{g}_y[1 - \alpha(T - T_0)] + \mathbf{F}_{Sy} \tag{3}$$

Energy conservation:

$$\frac{\partial}{\partial t}(\rho h)+\nabla\cdot(\rho\mathbf{V}h)=\nabla\cdot(k\nabla T)-\nabla\cdot[\rho(h_l-h)(\mathbf{V}-\mathbf{V}_s)]+S \tag{4}$$

In equations (1)-(4), the subscripts s and l stand for solid and liquid phase, respectively. t, μ, and T are time, dynamics viscosity and temperature, respectively. u and v are x-direction and y-direction velocity components. The continuum density ρ, vector velocity V, enthalpy h, and thermal conductivity k are defined as follows:

$$\rho = g_s\rho_s + g_l\rho_l \tag{5}$$

$$\mathbf{V} = f_s\mathbf{V}_s + f_l\mathbf{V}_l \tag{6}$$

$$h = f_s h_s + f_l h_l \tag{7}$$

$$k = g_s k_s + g_l k_l \tag{8}$$

Here the subscripts s and l stand for solid and liquid phase, respectively. f_s and f_l refer to mass fractions of solid and liquid phases, and g_s and g_l are volume fractions of solid and liquid phases. To calculate these four quantities, a general practice is that g_l (or g_s) is calculated first and then the other three quantities are obtained according to the following relationships:

$$f_l=\frac{g_l\rho_l}{\rho} \quad f_s=\frac{g_s\rho_s}{\rho} \quad g_s+g_l=1 \quad f_s+f_l=1 \tag{9}$$

The volume fraction of liquid g_l can be found using different models, such as the level rule, the Scheil model [Scheil, 1942], or the Clyne and Kurz model [Clyne & Kurz, 1981]. For the target material Ti-6Al-4V, it is assumed that g_l is only dependent on temperature. The g_l (T) function is given by [Swaminathan & Voller, 1992]:

$$g_l=\begin{cases} 0 & \text{if } T<T_s \\ \dfrac{T-T_s}{T_l-T_s} & \text{if } T_s\le T\le T_l \\ 1 & \text{if } T>T_l \end{cases} \tag{10}$$

The phase enthalpy for the solid and the liquid can be expressed as:

$$h_s=\int_0^T c_s(T)dT \tag{11}$$

$$h_l=\int_0^{T_s} c_s(T)dT+\int_{T_s}^{T} c_l(T)dT+L_m \tag{12}$$

where L_m is the latent heat of melting. c_s and c_l are specific heat of solid and liquid phases.

In Eqs. (2) and (3), the third terms on the right-hand side are the drag interaction terms, and K_x and K_y are the permeability of the two-phase mushy zone in x- and y- directions, which can be calculated from various models [Bhat et al., 1995; Carman, 1937; Drummond & Tahir,

1984; Ganesan et al., 1992; Poirier, 1987; West, 1985]. Here the mushy zone is considered as rigid (i.e. a porous media). If the mushy zone is modeled as a slurry region, these two terms can be treated as in [Ni & Incropera, 1995]. In Eqs. (2) and (3), the fourth terms on the right-hand side are the buoyancy force components due to temperature gradients. Here Boussinesq approximation is applied. α is the thermal expansion coefficient. The fifth terms on the right-hand side of Eqs. (2) and (3) are surface tension force components, which will be described in Section 2.2 below. The term S in Eq. (4) is the heat source.

2.2 Surface tension

The surface tension force, F_S, is given by:

$$\mathbf{F}_S = \gamma\kappa\hat{\mathbf{n}} + \nabla_S\gamma \tag{13}$$

where γ is surface tension coefficient, κ the curvature of the interface, $\hat{\mathbf{n}}$ the unit normal to the local surface, and ∇_S the surface gradient operator. The term $\gamma\kappa\hat{\mathbf{n}}$ is the normal component of the surface tension force. The term $\nabla_S\gamma$ represents the Marangoni effect caused by spatial variations in the surface tension coefficient along the interface due to temperature and/or species gradients. It causes the fluid flow from regions of lower to higher surface tension coefficient.

The conventional approach when dealing with surface tension is to use finite difference schemes to apply a pressure jump at a free-surface discontinuity. More recently, a general practice is to model surface tension as a volume force using a continuum model, either the Continuum Surface Force (CSF) model [Brackbill et al., 1992] or the Continuum Surface Stress (CSS) model [Lafaurie et al., 1994]. The volume force acts everywhere within a finite transition region between the two phases. In this study, the CSF model is adopted, which has been shown to make more accurate use of the free-surface VOF data [Brackbill et al., 1992].

A well-known problem with VOF (and other Eularian methods) modeling of surface tension is so-called "parasitic currents" or "spurious currents", which is a flow induced solely by the discretization and by a lack of convergence with mesh refinement. Under some circumstances, this artificial flow can be strong enough to dominate the solution, and the resulting strong vortices at the interface may lead to catastrophic instability of the interface and may even break-up [Fuster et al., 2009; Gerlach et al. 2006]. Two measures can be taken to relieve or even resolve this problem. One measure is to use a force-balance flow algorithm in which the CSF model is applied in a way that is consistent with the calculation of the pressure gradient field. Thus, imbalance between discrete surface tension and pressure-gradient terms can be avoided. Within a VOF framework, such force-balance flow algorithms can be found in [Francois et al., 2006; Y.Renardy & M. Renardy, 2002; Shirani et al., 2005]. In this study, the algorithm in [Shirani et al., 2005] is followed. The other measure is to get an accurate calculation of surface tension by accurately calculating interface normals and curvatures from volume fractions. For this purpose, many methods have been developed, such as those in [Cummins et al, 2005; Francois et al., 2006; López & Hernández, 2010; Meier et al., 2002; Pilliod Jr. & Puckett, 2004; Y.Renardy & M. Renardy, 2002]. The method we use here is the height function (HF) technique, which has been shown to be second-order accurate, and superior to those based on kernel derivatives of volume fractions or RDF distributions [Cummins et al, 2005; Francois et al., 2006; Liovic et al., 2010]. Specifically, we adopt the HF technique in [López & Hernández, 2010] that has many improvements over earlier versions (such as that in [Torrey et al., 1985]) of HF, including using an error correction procedure to minimize estimation error. Within the HF framework, suppose the absolute value of the y-direction component of the interface normal vector is larger than the x-direction component, interface curvature (in 2D) is given by

$$\kappa = \frac{H_{xx}}{(1+H_x{}^2)^{3/2}} \tag{14}$$

where H is the height function, H_x and H_{xx} are first-order and second-order derivatives of H, respectively. H_x and H_{xx} are obtained by using a finite difference formula. Interface normals are calculated based on the Least-Squares Fit method from [Aulisa et al., 2007].

2.3 Tracking of the free surface

The free surface of the melt pool is tracked using the PLIC-VOF [Gueyffier et al., 1999; Scardovelli & Zaleski, 2000, 2003]. The Volume of Fluid function, F, satisfies the following conservation equation:

$$\frac{\partial F}{\partial t} + (\mathbf{V}\cdot\nabla)F = 0 \tag{15}$$

The PLIC-VOF method consists of two steps: interface reconstruction and interface advection. In 2D calculation, a reconstructed planar surface becomes a straight line which satisfies the following equation:

$$n_x x + n_y y = d \tag{16}$$

where n_x and n_y are x and y components of the interface normal vector. d is a parameter related to the distance between the line and the coordinate origin of the reference cell. In the interface reconstruction step, n_x and n_y of each cell are calculated based on volume fraction data, using the Least-Squares Fit method from [Aulisa et al., 2007]. Then the parameter d is determined to match the given volume fraction. Finally given the velocity field, the reconstructed interface is advected according to the combined Eulerian-Lagrangian scheme in [Aulisa et al., 2007].

2.4 Boundary conditions

Energy balance at the free surface satisfies the following equation:

$$k\frac{\partial T}{\partial \mathbf{n}} = \frac{\eta(P_{laser} - P_{atten})}{\pi R^2} - h_c(T - T_\infty) - \varepsilon\sigma(T^4 - T_\infty^4) - \dot{m}_e L_v \tag{17}$$

where terms on the right-hand side are laser irradiation, convective heat loss, radiation heat loss and evaporation heat loss, respectively. P_{laser} is the power of laser beam, P_{atten} the power attenuated by the powder cloud, R the radius of laser beam spot, η the laser absorption coefficient, $\dot{m}_e$ the evaporation mass flux, L_v the latent heat of evaporation, h_c the heat convective coefficient, ε emissivity, σ the Stefan-Boltzmann constant, and $\mathbf{n}$ the normal vector at the local interface. $\dot{m}_e$ can be evaluated according to the "overall evaporation model" in [Choi et al., 1987], and P_{atten} can be calculated according to [Frenk et al., 1997] with a minor modification.

On the bottom surface and side surfaces, boundary conditions are given by

$$k\frac{\partial T}{\partial \mathbf{n}} + h_c(T - T_\infty) = 0 \tag{18}$$

$$\mathbf{V} = 0 \tag{19}$$

Note that the radiation heat loss at these surfaces is neglected due to the fact that the temperature differences at these surfaces are not large.

2.5 Numerical Implementation

Finite difference and finite volume methods are used for spatial discretization of the governing equations. Staggered grids are employed where the temperatures, pressures and VOF function are located at the cell center and the velocities at the walls. In the numerical implementation, material properties play an important role. The material properties are generally dependent on temperature, concentration, and pressure. For fusion-based additive manufacturing processes, the material experiences a large variation from room temperature to above the melting temperature. For Ti-6Al-4V, many material properties experience large variations over this wide temperature range, as shown in Table 1. For example, the value of specific heat varies from 546 $J \cdot K^{-1} \cdot kg^{-1}$ at room temperature to 831 $J \cdot K^{-1} \cdot kg^{-1}$ at liquidus temperature. Thermal conductivity varies from 7 to 33.4 $W \cdot m^{-1} \cdot K^{-1}$ over the same temperature range. Thus, the temperature dependence of the properties dominates, which necessitates a coupling of the momentum equations with the energy equation and gives rise to strong nonlinearity in the conservation equations.

The variable properties have two effects on the numerical solution procedure [Ferziger & Peric, 2002]. First, although an incompressible flow assumption is made, the thermo-physical properties need to be kept inside the differential operators. Thus, solution methods for incompressible flow can be used. Second, the momentum and energy conservation equations have to be solved in a coupled way. In this study, the coupling between momentum and energy equations is achieved by the following iterative scheme:

1. Eqs. (1) - (3) and the related boundary conditions are solved iteratively using a two-step projection method [Chorin, 1968] to obtain velocities and pressures. Thermo-physical properties used in this step are computed from the old temperature field. At each time step, the discretized momentum equations calculate new velocities in terms of an estimated pressure field. Then the pressure field is iteratively adjusted and velocity changes induced by each pressure correction are added to the previous velocities. This iterative process is repeated until the continuity equation is satisfied under an imposed tolerance by the newly computed velocities. This imposes a requirement for solving a linear system of equations. The preconditioned Bi-CGSTAB (for Bi-Conjugate Gradient Stabilized) method [Barrett et al., 1994] is used to solve the linear system of equations.

2. Eq. (4) is solved by a method [Knoll et al., 1999] based on a finite volume discretization of the enthalpy formulation of Eq. (4). The finite volume approach ensures that the numerical scheme is locally and globally conservative, while the enthalpy formulation can treat phase change in a straightforward and unified manner. Once new temperature field is obtained, the thermo-physical properties are updated.
3. Equation (15) is solved using the PLIC-VOF [Gueyffier et al., 1999; Scardovelli & Zaleski, 2000, 2003] to obtain the updated free surface and geometry of the melt pool.
4. Advance to the next time step and back to step 1 until the desired process time is reached.

Physical Properties	Value	Reference
Liquidus temperature (K)	1923.0	[Mills, 2002]
Solidus temperature (K)	1877.0	[Boyer et al., 1994]
Evaporation temperature (K)	3533.0	[Boyer et al., 1994]
Solid specific heat ($J\ kg^{-1}\ K^{-1}$)	$\begin{cases} 483.04 + 0.215T & T \le 1268K \\ 412.7 + 0.1801T & 1268 < T \le 1923 \end{cases}$	[Mills, 2002]
Liquid specific heat ($J\ kg^{-1}\ K^{-1}$)	831.0	[Mills, 2002]
Thermal conductivity (W m^{-1} K^{-1})	$\begin{cases} 1.2595 + 0.0157T & T \le 1268K \\ 3.5127 + 0.0127T & 1268 < T \le 1923 \\ -12.752 + 0.024T & T > 1923 \end{cases}$	[Mills, 2002]
Solid density (kg m^{-3})	4420 – 0.154 (T – 298 K)	[Mills, 2002]
Liquid density (kg m^{-3})	3920 – 0.68 (T – 1923 K)	[Mills, 2002]
Latent heat of fusion (J kg^{-1})	2.86×10^5	[Mills, 2002]
Latent heat of evaporation (J kg^{-1})	9.83×10^6	[Mills, 2002]
Dynamic viscosity (N m^{-1} s^{-1})	3.25×10^{-3} (1923K) 3.03×10^{-3} (1973K) 2.66×10^{-3} (2073K) 2.36×10^{-3} (2173K)	[Mills, 2002]
Radiation emissivity	$0.1536+1.8377\times10^{-4}\ (T\ -300.0\ K)$	[Lips&Fritsche, 2005]
Surface tension (N m^{-1})	$1.525 - 0.28\times10^{-3}(T - 1941K)^a$	[Mills, 2002]
Thermal expansion coefficient (K^{-1})	1.1×10^{-5}	[Mills, 2002]
Laser absorption coefficient	0.4	
Ambient temperature (K)	300	
Convective coefficient (W m^{-2} K^{-1})	10	

Table 1. Material properties for Ti-6Al-4V and main process parameters used in simulations. [a]Value for commercially pure titanium was used.

The time step is taken at the level of 10^{-6} s initially and adapted subsequently according to the convergence and stability requirements of the Courant–Friedrichs–Lewy (CFL) condition, the explicit differencing of the Newtonian viscous stress tensor, and the explicit treatment of the surface tension force.

3. Simulation results and model validation

The parameters for the simulation were chosen based on the capability of our experimental facilities to compare the simulation results with the experimental measurements. A diode laser deposition system (the LAMP system of Missouri S&T) and a YAG laser deposition system at South Dakota School of Mines and Technology (SDSMT) were used for simulations and experiments. Ti-6Al-V4 plates with a thickness of 0.25 inch were selected as substrates. Ti-6Al-V4 powder particles with a diameter from 40 to 140 μm were used as deposit material. Fig. 3 shows the typical simulation results for temperature, velocity and VOF function.

The numerical model was validated from different aspects. First, it was validated in terms of melt pool peak temperature and melt pool length. The experiments were performed on the LAMP system as shown in Fig. 4. The system consists of a diode laser, powder delivery unit, 5-axis CNC machine, and monitoring subsystem. The laser system used was a Nuvonyx ISL-1000M Diode Laser that is rated for 1 kW of output power. The laser emits at 808 nm and operates in the continuous wave (CW) mode. The laser spot size is 2.5 mm. To protect oxidization of Ti-6Al-V4, the system is covered in an environmental chamber to supply argon gas. The melt pool peak temperature is measured by a non-contact optical pyrometer that is designed for rough conditions, such as high ambient temperatures or electromagnetic interferences. A laser sight within the pyrometer allows for perfect alignment and focal length positioning; the spot size is 2.6 mm which encompasses the melt pool. The pyrometer senses the maximum temperature between 400 and 2500 (degrees C) and correlates the emissivity of the object to the resulting measurement. Temperature measurements are taken in real-time at 500 or 1000 Hz using a National Instruments real-time control system. A 4-20 mA signal is sent to the real-time system which is converted to degrees Celsius, displayed to the user and simultaneously recorded to be analyzed at a later date. Due to the collimator, the pyrometer is mounted to the Z-axis of the CNC at 42 (degrees) and is aligned with the center of the nozzle. Temperature measurements recorded the rise and steady state temperatures and the cooling rates of the melt pool. A complementary metal oxide semiconductor (CMOS) camera was installed right above the nozzle head for a better view in dynamically acquiring the melt pool image. The melt pool dimensions can be calculated from the image by the image process software.

Fig. 5 and Fig. 6 show the measured and predicted melt pool peak temperatures at different laser power levels and at different travel speeds, respectively. It can be seen from the plot that the general trend between simulation and experiment is consistent. At different power intensity level, there is a different error from 10 K (about 0.5%) to 121 K (about 5%). Fig. 7 shows measured and predicted melt pool length at different laser power levels. The biggest disagreement between measured and simulated values is about 7%. It can be seen that the differences between measured and predicted values at higher power intensities (higher power levels or slower travel speeds) are generally bigger than those at lower power intensities. This can be explained by the two-dimensional nature of the numerical model. A 2D model does not consider the heat transfer in the third direction. At a higher power level, heat transfer in the third dimension is more significant.

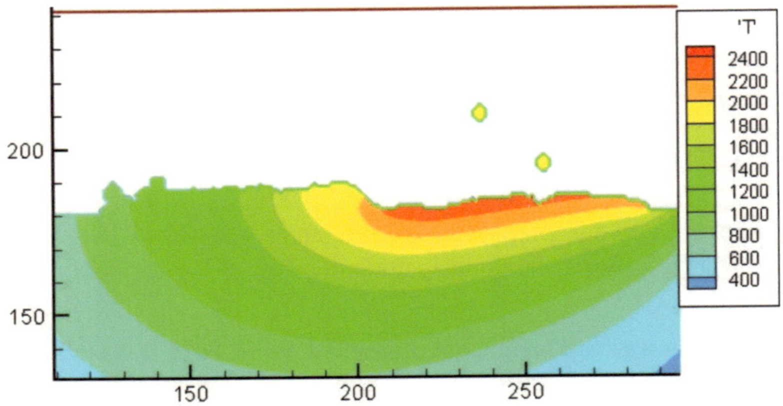

(a) Temperature field of the region around the melt pool

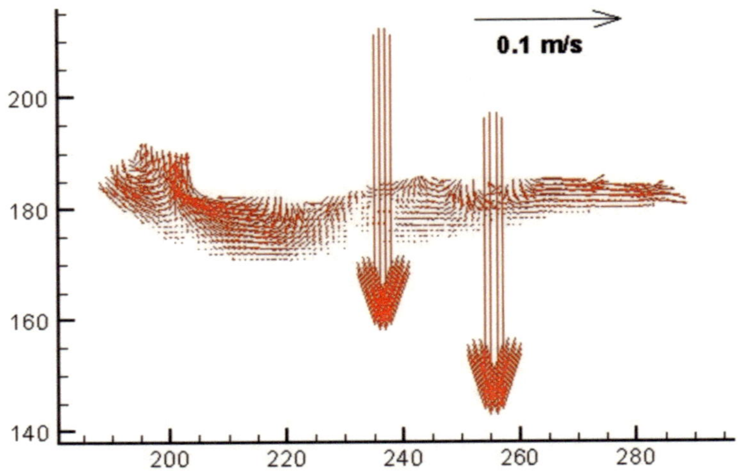

(b) Velocity field of the melt pool and falling powder particles

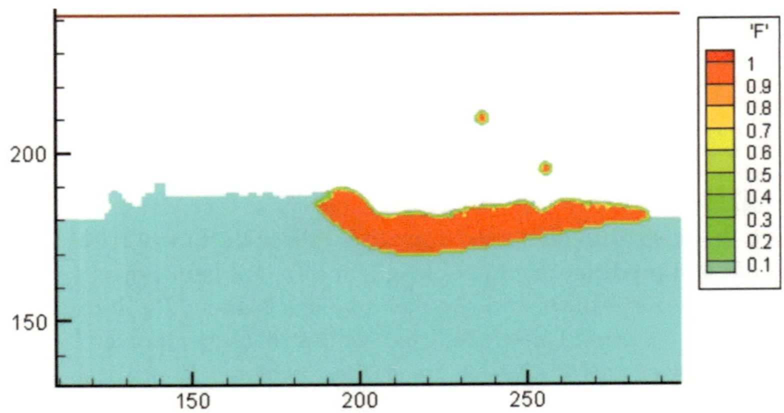

(c) VOF field of part of the region around the melt pool

Fig. 3. Simulation results of laser deposition of Ti-6Al-4V

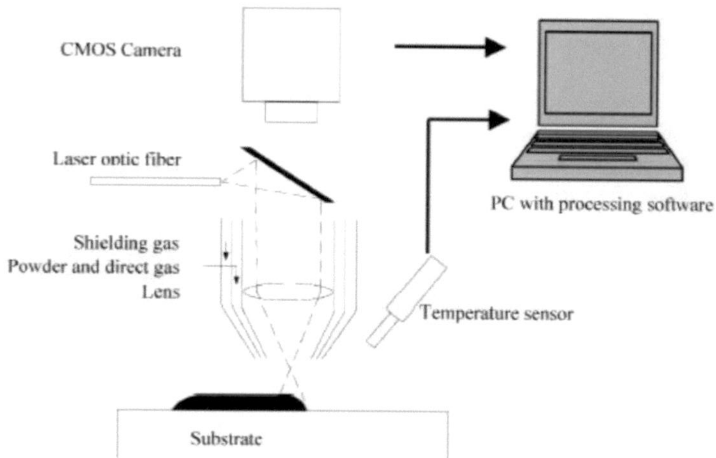

Fig. 4. Schematic of experimental setup

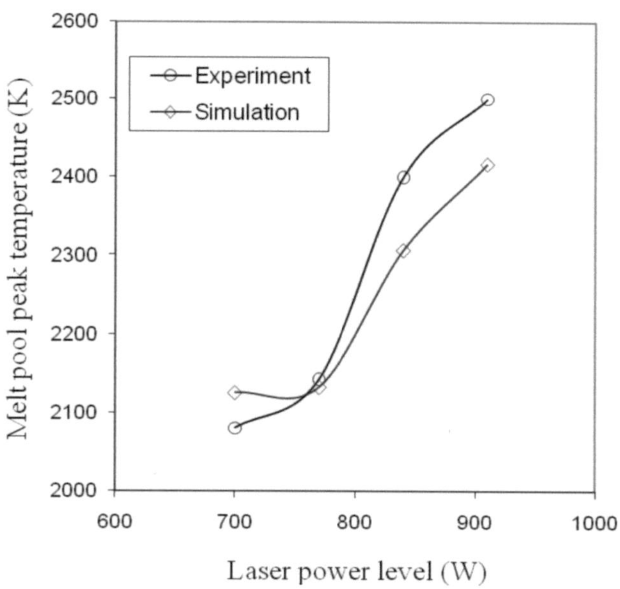

Fig. 5. Melt pool peak temperature comparison between simulation and experiment at different laser power levels (powder mass flow rate = 4.68g/min., travel speed = 20 inch/min.)

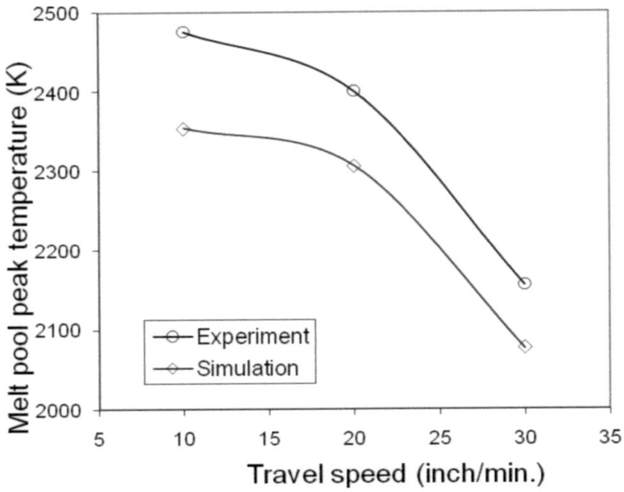

Fig. 6. Melt pool peak temperature comparison between simulation and experiment at different travel speeds (powder mass flow rate = 4.68g/min., laser power = 700 W)

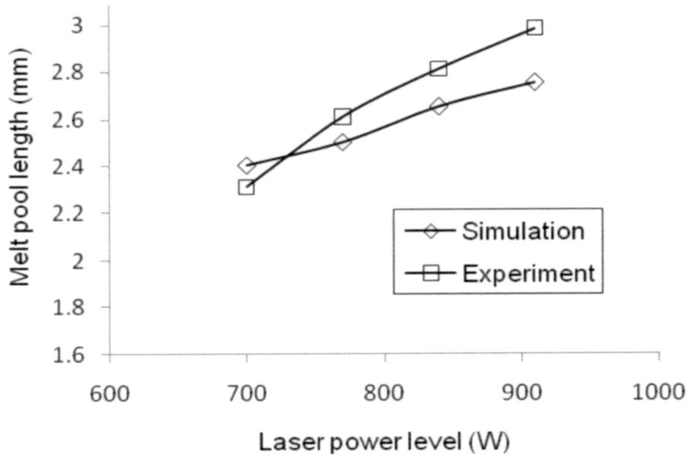

Fig. 7. Melt pool length comparisons between simulation and experiment at different power levels (powder mass flow rate = 4. 68 g/min., travel speed = 20 inch/min.)

The samples were cross-sectioned using a Wire-EDM machine to measure dilution depth. An SEM (Scanning Electron Microscope) line trace was used to determine the dilution of the clad layer. The deposited Ti-6Al-4V is of Widmansttaten structure. The substrate has a rolled equi-axed alpha plus beta structure. Even though these two structures are are easily distinguishable, the HAZ is large and has a martensitic structure that can be associated with it. Hence a small quantity of tool steel in the order of 5% was mixed with Ti-6Al-4V. The small quantity makes sure that it does not drastically change the deposit features of a 100% Ti-6Al-4V deposit. At the same time, the presence of Cr in tool steel makes it easily identifiable by means of EDS scans using SEM. Simulation and experimental results of dilution depth are shown in Figs. 8 - 10.

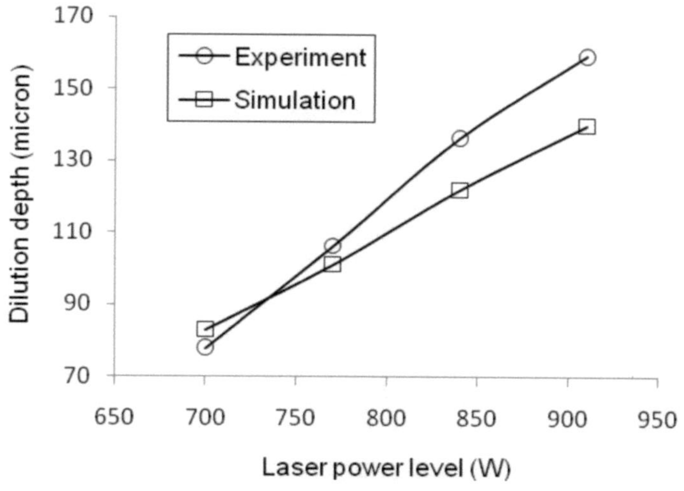

Fig. 8. Comparison of dilution depth between simulation and experiment at different power levels (powder mass flow rate = 4.68 g/min., travel speed = 20 inch/min.)

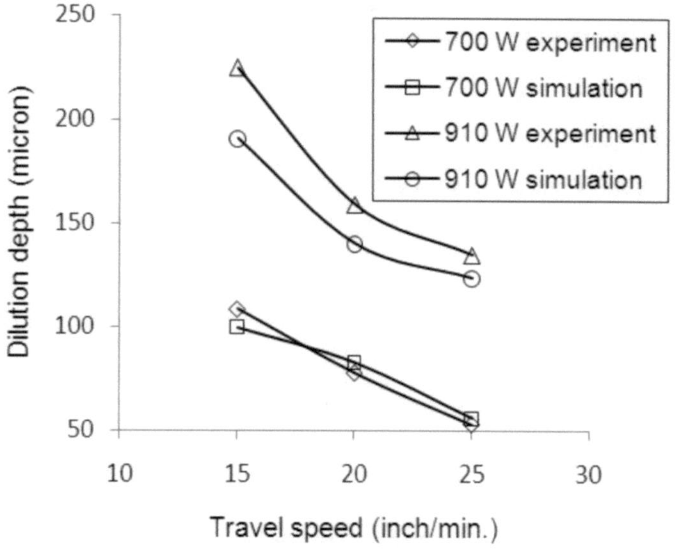

Fig. 9. Comparison of dilution depth between simulation and experiment at different travel speeds and different laser power levels (powder mass flow rate = 4.68 g/min.)

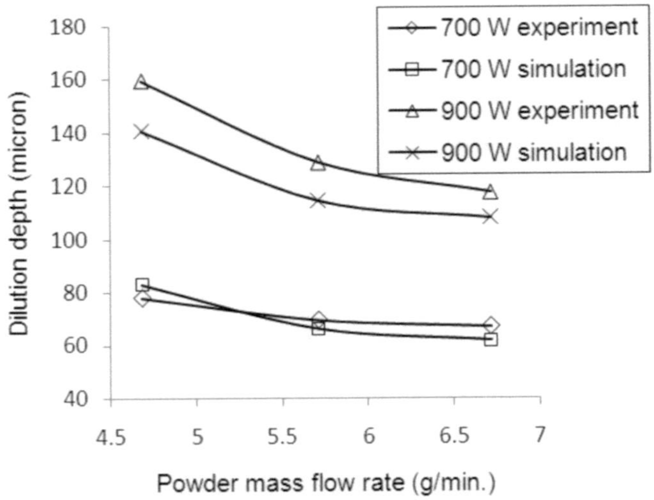

Fig. 10. Comparison of dilution depth between simulation and experiment at different powder mass flow rates and different laser power levels (travel speed = 20 inch/min.)

Good agreements between measured and simulated dilution depths can be found in Figs. 8-10. The differences are from about 4.8% to 15.1%. It can be seen that an increase in the laser power will increase the dilution depth. An increase in the laser travel speed will decrease the dilution depth. It is clear that the dilution depth has a linear dependence on the laser power and the laser travel speed. This is easy to understand. As the laser power increases, more power is available for melting the substrate. As travel speed decreases, the laser material interaction time is extended. From Fig. 10, it can be seen that an increase in powder mass flow rate will decrease the dilution depth. But this effect is more significant at a higher level of laser power. It is likely that at a lower level of laser power, a significant portion of laser energy is consumed to melt the powder. Hence the energy available is barely enough to melt the substrate. Detailed discussion can be found in [Fan et al, 2006, 2007b; Fan, 2007].

Finally, the numerical model was validated in terms of its capability for predicting the lack-of-fusion defect. The test was performed using the YAG laser deposition system at South Dakota School of Mines and Technology (SDSMT). The simulation model determined that 1,200 watts would be the nominal energy level for the test. This means that based on the model, lack of fusion should occur when the laser power is below 1200 W. In accordance with the test matrix, seven energy levels were tested: nominal, nominal ± 10%, nominal ± 20%, and nominal ± 30%. Based on the predicted nominal value of 1,200 watts, the seven energy levels in the test matrix are 840, 960, 1080, 1200, 1320, 1440, and 1540 watts. The deposited Ti-6Al-4V specimens were inspected at Quality Testing Services Co. using ultrasonic and radiographic inspections to determine the extent of lack-of-fusion in the specimens. The determination of whether or not there exists lack of fusion in a deposited specimen can be explained using Figs. 11 - 13. First a substrate without deposit on it was inspected as shown in Fig. 11. Notice that the distance between two peaks are the thickness of the substrate. Then laser deposited specimens were inspected. If there is lack of fusion in a deposited specimen, some form of peaks can be found between the two high peaks in the ultrasonic graph, the distance of which is the height of the deposition and the thickness of the substrate. Fig. 12 shows an ultrasonic graph of a deposited specimen with a very good deposition. The ultrasonic result indicates there is not lack of fusion occurring between layers and the interface. The distance between two peaks is the height of the deposition and the thickness of the substrate. For the deposition as shown in Fig. 13, the lack of fusion occurs as the small peak (in circle) appears between two high peaks. The results revealed that no lack-of-fusion was detected in specimens deposited using 1,200 watts and higher energy levels. However, lack-of-fusion was detected in specimens deposited from lower energy levels (minus 10% up to minus 30% of 1,200 watts.). The test results validated the simulation model.

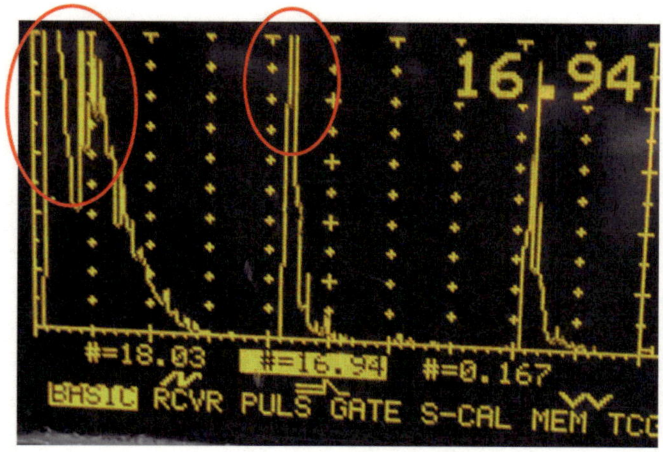

Fig. 11. Ultrasonic graph of a Ti-6Al-4V substrate

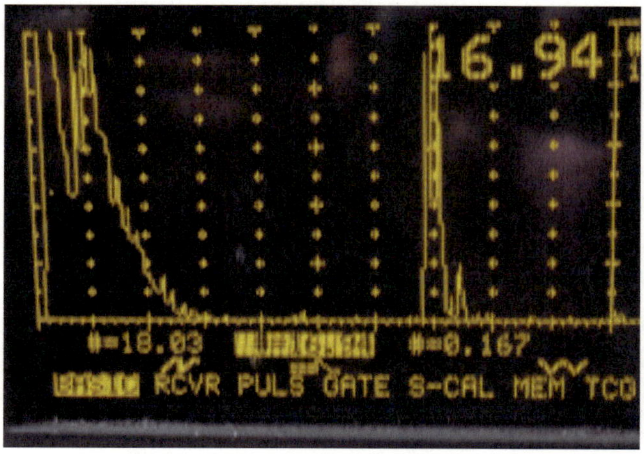

Fig. 12. Ultrasonic graph of a laser deposited Ti-6Al-4V specimen without lack of fusion

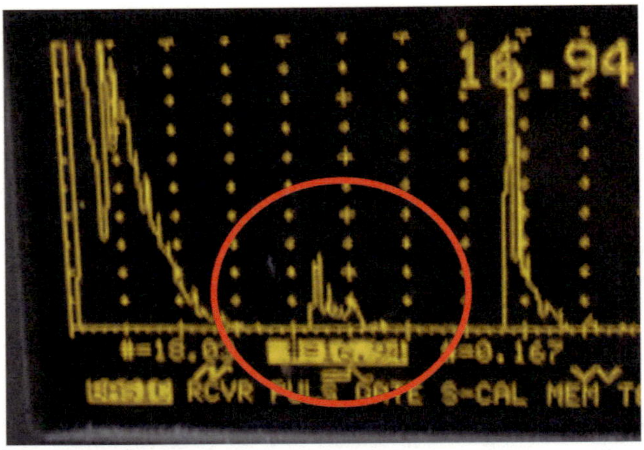

Fig. 13. Ultrasonic graph of a laser deposited Ti-6Al-4V specimen with lack of fusion

4. Conclusion

This chapter has outlined the approach for mathematical and numerical modeling of fusion-based additive manufacturing of titanium. The emphasis is put on modeling of transport phenomena associated with the process, including heat transfer and fluid flow dynamics. Of particular interest are the modeling approaches for solidification and free surface flow with surface tension. The advantages and disadvantages of the main modeling approaches are briefly discussed. Based on the comparisons, the continuum model is adopted for modeling of melting/solidification phase change, and the VOF method for modeling of free-surface flow in the melt pool.

The laser deposition process is selected as an example of fusion-based additive manufacturing processes. The governing equations, auxiliary relationships, and boundary conditions for the solidification system and free-surface flow are presented. The main challenge for modeling of the surface tension-dominant free surface flow is discussed and the measures to overcome the challenge are given. The numerical implementation procedures are outlined, with a focus on the effects of variable material property on the discretization schemes and solution algorithms. Finally the simulation results are presented and compared with experimental measurements. A good agreement has been obtained and thus the numerical model is validated. The modeling approach can be applied to other fusion-based manufacturing processes, such as casting and welding.

5. Acknowledgment

This research was partially supported by the National Aeronautics and Space Administration Grant Number NNX11AI73A, the grant from the U.S. Air Force Research Laboratory, and Missouri S&T's Intelligent Systems Center and Manufacturing Engineering program. Their support is greatly appreciated. The help from Dr. Kevin Slattery and Mr. Hsin Nan Chou at Boeing-St Louis and Dr. James Sears at South Dakota School of Mines and Technology are also acknowledged.

6. References

Aulisa, E.; Manservisi, S.; Scardovelli, R. & Zaleski, S. (2007). Interface reconstruction with least-squares fit and split advection in three-dimensional Cartesian geometry, *Journal of Computational Physics*, Vol. 225, No.2, (August 2007), pp. 2301-2319, ISSN 0021-9991

Barrett, R.; Berry, M.; Chan, T. F.; Demmel, J.; Donato, J.; Dongarra, J.; Eijkhout, V.; Pozo, R.; Romine, C. & Van der Vorst, H. (1994). *Templates for the Solution of Linear Systems: Building Blocks for Iterative Methods* (2nd Edition), SIAM, ISBN 978-0-898713-28-2 Philadelphia, PA

Beckermann C. & Viskanta, R. (1988). Double-diffusive convection during dendritic solidification of a binary mixture. *PCH: Physicochemical Hydrodynamics*, Vol. 10, No.2, pp. 195-213, ISSN 0191-9059

Beckermann, C.; Diepers, H.-J.; Steinbach, I.; Karma, A. & Tong X. (1999). Modeling Melt Convection in Phase-Field Simulations of Solidification, *Journal of Computational Physics*, Vol. 154, No.2, (September 1999), pp. 468–496, ISSN 0021-9991

Benes, J. (2007). Cool Tips for Cutting Titanium, In: *American Mechanist*, 26.09.2011, Available from http://www.americanmachinist.com/304/Issue/Article/False/77297/

Bennon, W. D. & Incropera, F. P. (1987). A continuum model for momentum, heat and species transport in binary solid-liquid phase change systems-I. Model formulation. *International Journal of Heat and Mass Transfer*, Vol. 30, No. 10, (October 1987), pp. 2161-2170, ISSN 0017-9310

Benson, D.J. (2002). Volume of fluid interface reconstruction methods for multi-material problems. *Applied Mechanics Reviews*, Vol. 55, No. 2, (March 2002), pp. 151–165, ISSN 0003-6900

Bhat, M. S.; Poirier, D.R. & Heinrich, J.C. (1995). Permeability for cross flow through columnar-dendritic alloys. *Metallurgical and Materials Transactions B*, Vol. 26, No.5, (October 1995), pp. 1049-1092, ISSN 1073-5615

Boettinger, W.J.; Coriell, S. R.; Greer, A. L.; Karma, A.; Kurz, W.; Rappaz, M. & Trivedi, R. (2000). Solidification microstructures: recent developments, future directions, *Acta Materialia*, Vol. 48, No.1, (January 2000), pp. 43-70, ISSN 1359-6454

Boettinger, W. J.; Warren, J. A.; Beckermann, C. & Karma, A. (2002). Phase-Field Simulation of Solidification, *Annual Review of Materials Research*, Vol. 32, pp. 163-194, ISSN 1531-7331

Boyer, R; Welsch, G. & Collings, E.W. (1994). *Materials properties handbook: titanium alloys*, ASM International, ISBN 978-0871704818, Materials Park, OH

Brackbill, J.U.; Kothe, D.B. & Zemach, C. (1992). A continuum method for modeling surface tension, *Journal of Computational Physics*, Vol. 100, No.2, (June 1992), pp. 335–354, ISSN 0021-9991

Caboussat, A. (2005). Numerical Simulation of Two-Phase Free Surface Flows. *Archives of Computational Methods in Engineering*, Vol. 12, No.2, (June 2005), pp. 165-224, ISSN 1134-3060

Carman, P.C. (1937). Fluid flow through granular beds. *Transactions of the Institution of Chemical Engineers*, Vol. 15, (February 1937), pp. 150-166, ISSN 0046-9858

Caginalp, G. (1989). Stefan and Hele-Shaw type models as asymptotic limits of the phase-field equations. *Physical Review A*, Vol.39, No.11, (June 1989), pp. 5887-5896, ISSN 1050-2947

Choi, M.; Greif, R. & Salcudean, M. (1987). A study of the heat transfer during arc welding with applications to pure metals or alloys and low or high boiling temperature materials, *Numerical Heat Transfer*, Vol. 11, No.4, (April 1987) pp. 477-491, ISSN 0149-5720

Chorin, A.J. (1968). Numerical solution of the Navier-Stokes equations. *Mathematics of Computation*, Vol. 22, No. 104, (October 1968), pp. 745-762, ISSN 0025-5718

Chorin, A.J. (1980). Flame advection and propagation algorithms, *Journal of Computational Physics*, Vol. 35, No. 1, (March 1980), pp. 1 - 11, ISSN 0021-9991

Clyne, T. W. & Kurz, W. (1981). Solute redisribution during solidification with rapid solid state diffusion. *Metallurgical Transactions A*, Vol.12, No.6, (June 1981), pp. 965–971, ISSN 1073-5623

Cummins, S. J.; Francois, M. M. & Kothe, D.B. (2005). Estimating curvature from volume fractions. *Computers & Structures*, Vol.83, No. 6-7, (February 2005), pp. 425 - 434, ISSN 0045-7949

Debar, R. (1974). *Fundamentals of the KRAKEN Code*, Technical Report UCIR-760, Lawrence Livermore National Laboratory, 1974

Drummond, J. E. & Tahir, M.I. (1984). Laminar viscous flow through regular arrays of parallel solid cylinders. *International Journal of Multiphase Flow*, Vol. 10, No. 5, (October 1984), pp. 515–540, ISSN 0301-9322

Eylon, D. & Froes, F.H. (1984). Titanium Net-Shape Technologies. *Journal of Metals*, Vol. 36, No. 6, (June 1984), pp. 36-41, ISSN 1047-4838

Fan,Z.; Sparks, T.E.; Liou, F.; Jambunathan, A.; Bao, Y.; Ruan, J. & Newkirk, J.W. (2007). Numerical Simulation of the Evolution of Solidification Microstructure in Laser Deposition. *Proceedings of the 18th Annual Solid Freeform Fabrication Symposium*, Austin, TX, USA, August 2007

Fan,Z.; Stroble,J.K.; Ruan, J.; Sparks, T.E. & Liou, F. (2007). Numerical and Analytical Modeling of Laser Deposition with Preheating. *Proceedings of ASME 2007 International Manufacturing Science and Engineering Conference*, ISBN 0-7918-4290-8, Atlanta, Georgia, USA, October, 2007, pp. 37-51

Fan,Z. (2007). *Numerical and Experimental Study of Parameter Sensitivity and Dilution in Laser Deposition*, Master Thesis, University of Missouri-Rolla, 2007

Fan,Z.; Jambunathan, A.; Sparks, T.E.; Ruan, J.; Yang, Y.; Bao, Y. & Liou, F. (2006). Numerical simulation and prediction of dilution during laser deposition. *Proceedings of the 17th Annual Solid Freeform Fabrication Symposium*, Austin, TX, USA, August 2006, pp. 532-545

Ferziger, J.H. & Peric, M. (2002). *Computational Methods for Fluid Dynamics* (3rd edition), Springer-Verlag, ISBN 3-540-42074-6, Berlin Heidelberg New York

Francois, M.M.; Cummins, S.J.; Dendy, E.D.; Kothe, D.B.; Sicilian, J.M. & Williams, M.W. (2006). A balanced-force algorithm for continuous and sharp interfacial surface tension models within a volume tracking framework. *Journal of Computational Physics*, Vol. 213, No. 1, (March 2006), pp. 141 - 173, ISSN 0021-9991

Frenk, A.; Vandyoussefi, M.; Wagniere, J.; Zryd, A. & Kurz, W. (1997). Analysis of the laser-cladding process for stellite on steel. *Metallurgical and Materials Transactions B*, Vol. 28, No.3, (June 1997), pp. 501-508, ISSN 1073-5615

Fuster, D.; Agbaglah, G.; Josserand, C.; Popinet, S. & Zaleski, S. (2009). Numerical simulation of droplets, bubbles and waves: state of the art. *Fluid Dynamics Research*, Vol.41, No.6, (December 2009), p. 065001 (24 pp.), ISSN 0169-5983

Gandin, Ch.-A. & Rappaz, M. (1994). A coupled finite element-cellular automaton model for the prediction of dendritic grain structures in solidification processes, *Acta Metallurgica et Materialia*, Vol. 42, No. 7, (July 1994), pp. 2233-2246, ISSN 0956-7151

Ganesan, S. & Poirier, D. R. (1990). Conservation of mass and momentum for the flow of interdendritic liquid during solidification, *Metallurgical Transactions B*, Vol. 21, No.1, (February 1990), pp. 173-181, ISSN 1073-5615

Ganesan, S.; Chan, C.L. & Poirier, D.R. (1992). Permeability for flow parallel to primary dendrite arms. *Materials Science and Engineering: A*, Vol. 151, No. 2, (May 1992), pp. 97-105, ISSN 0921-5093

Gerlach, D.; Tomar, G.; Biswas, G. & Durst, F. (2006). Comparison of volume-of-fluid methods for surface tension-dominant two-phase flows. *International Journal of Heat and Mass Transfer*, Vol. 49, No. 3-4, (February 2006), pp. 740–754, ISSN 0017-9310

Gibou,F.; Fedkiw, R.; Caflisch, R. & Osher, S. (2003). A Level Set Approach for the Numerical Simulation of Dendritic Growth. *Journal of Scientific Computing*, Vol. 19, No.1-3, (December 2003), pp. 183-199, ISSN 0885-7474

Grujicic, M.; Cao, G. & Figliola, R.S. (2001). Computer simulations of the evolution of solidification microstructure in the LENS™ rapid fabrication process, *Applied Surface Science*, Vol. 183, No. 1-2, (November 2001), pp. 43-57, ISSN 0169-4332

Gueyffier, D.; Li, J.; Nadim, A.; Scardovelli, R. & Zaleski, S. (1999). Volume of Fluid interface tracking with smoothed surface stress methods for three-dimensional flows. *Journal of Computational Physics*, Vol. 152, No.2, (July 1999), pp. 423-456, ISSN 0021-9991

Hansbo, P. (2000). A Free-Lagrange Finite Element Method using Space-Time Elements. *Computer Methods in Applied Mechanics and Engineering*, Vol. 188, No. 1-3, (July 2000), pp. 347–361, ISSN 0045-7825

Hills, R.N.; Loper, D.E. & Roberts, P.H. (1983). A thermodynamically consistent model of a mushy zone. *The Quarterly Journal of Mechanics & Applied Mathematics*, Vol. 36, No. 4, (November 1983), pp. 505-536, ISSN 0033-5614

Hirt, C.W. & Nichols, B.D. (1981). Volume-of-fluid (VOF) for the dynamics of free boundaries. *Journal of Computational Physics*, Vol. 39, No.1, (January 1981), pp. 201-225, ISSN 0021-9991

Hirt, C. W. & Nichols, B. D. (1988). *Flow-3D user manual*, Technical Report, Flow Sciences, Inc.

Idelsohn, S. R.; Storti, M. A. & Onate, E. (2001). Lagrangian Formulations to Solve Free

Surface Incompressible Inviscid Fluid Flows. *Computer Methods in Applied Mechanics and Engineering*, Vol. 191, No. 6-7, (December 2001), pp. 583–593, ISSN 0045-7825

Jacqmin, D. (1999). Calculation of two-phase Navier–Stokes flows using phase-field modeling. *Journal of Computational Physics*, Vol. 155, No.1, (October 1999), pp. 96–127, ISSN 0021-9991

Jaluria, Y. (2006). Numerical Modeling of Manufacturing Processes. In: *Handbook of Numerical Heat Transfer* (2 edition), Minkowycz, W.J.; Sparrow, E.M. & Murthy, J.Y., (Ed.), pp. 729-783, Wiley, ISBN 0471348783, Hoboken, New Jersey

Jones, W.P. & Launder, B.E. (1973). The calculation of low-Reynolds-number phenomena with a two-equation model of turbulence. *International Journal of Heat and Mass Transfer*, Vol. 16, No. 6, (June 1973), pp.1119-1130, ISSN 0017-9310

Juric D. & Tryggvason G. (1996). A Front-Tracking Method for Dendritic Solidification, *Journal of Computational Physics*, Vol. 123, No.1, (January 1996) pp. 127–148, ISSN 0021-9991

Karma, A. & Rappel, W.J. (1996). Phase-field method for computationally efficient modeling of solidification with arbitrary interface kinetics, *Physical Review E*, Vol. 53, No.4, (April 1996), pp. R3017-R3020, ISSN 1063-651X

Karma, A. & Rappel, W.J. (1998). Quantitative phase-field modeling of dendritic growth in two and three dimensions. *Physical Review E*, Vol. 57, No.4, (April 1998), pp. 4323-4329, ISSN 1063-651X

Kim, Y.T.; Goldenfeld, N. & Dantzig, J. (2000). Computation of dendritic microstructures using a level set method. *Physical Review E*, Vol. 62, No.2, (August 2000). pp. 2471-2474, ISSN 1063-651X

Knoll, D. A.; Kothe, D. B. & Lally, B. (1999). A new nonlinear solution method for phase change problems. *Numerical Heat Transfer, Part B Fundamentals*, Vol. 35, No. 4, (December 1999), pp. 436–459, ISSN 1040-7790

Kobayashi, R. (1993). Modeling and numerical simulations of dendritic crystal growth. *Physica D: Nonlinear Phenomena*, Vol. 63, No. 3-4, (March 1993), pp. 410-423, ISSN 0167-2789

Kobryn, P.A.; Ontko, N.R.; Perkins, L.P. & Tiley, J.S. (2006) Additive Manufacturing of Aerospace Alloys for Aircraft Structures. In: *Cost Effective Manufacture via Net-Shape Processing*, Meeting Proceedings RTO-MP-AVT-139, Neuilly-sur-Seine, France: RTO, pp. 3-1 – 3-14

Kothe, D. B.; Mjolsness, R. C. & Torrey, M. D. (1991). *RIPPLE: A Computer Program for Incompressible Flows with Free surfaces*, Technical Report, LA-12007-MS, Los Alamos National Lab

Lafaurie, B.; Nardone, C.; Scardovelli, R.; Zaleski, S. & Zanetti, G. (1994). Modelling Merging and Fragmentation in Multiphase Flows with SURFER. *Journal of Computational Physics*, Vol. 113, No.1, (July 1994), pp. 134-147, ISSN 0021-9991

Liou, F. & Kinsella, M. (2009). A Rapid Manufacturing Process for High Performance Precision Metal Parts. *Proceedings of RAPID 2009 Conference & Exposition, Society of Manufacturing Engineers*, Schaumburg, IL, USA, May 2009

Liovic, P.; Francois, M.; Rudman, M. & Manasseh, R. (2010). Efficient simulation of surface tension-dominated flows through enhanced interface geometry interrogation. *Journal of Computational Physics*, Vol. 229, No. 19, (September 2010), pp. 7520-7544, ISSN 0021-9991

Lips,T. & Fritsche, B. (2005). A comparison of commonly used re-entry analysis tools, *Acta Astronautica*, Vol. 57, No.2-8, (July-October 2005), pp. 312-323, ISSN 0094-5765

López, J. & Hernández, J. (2010). On reducing interface curvature computation errors in the height function technique. *Journal of Computational Physics*, Vol.229, No.13, (July 2010), pp. 4855-4868, ISSN 0021-9991

Meier, M; Yadigaroglu,G. & Smith, B.L. (2002). A novel technique for including surface tension in PLIC-VOF methods. *European Journal of Mechanics - B/Fluids*, Vol. 21, No. 1, ISSN 0997-7546, pp. 61-73, ISSN 0997-7546

Mills, K. C. (2002). *Recommended Values of Thermophysical Properties for Selected Commercial Alloys*, Woodhead Publishing Ltd, ISBN 978-1855735699, Cambridge

Muller, I.A. (1968). A Thermodynamic theory of mixtures of fluids, *Archive for Rational Mechanics and Analysis*, Vol. 28, No. 1, (January 1968), pp. 1-39, ISSN 0003-9527

Ni, J. & Beckermann, C. (1991). A volume-averaged two-phase model for transport phenomena during solidification, *Metallurgical Transactions B*, Vol. 22, No.3, (June 1991), pp. 349-361, ISSN 1073-5615

Ni, J. & Incropera, F. P. (1995). Extension of the Continuum Model for Transport Phenomena Occurring during Metal Alloy Solidification, Part I: The Conservation Equations. *International Journal of Heat and Mass Transfer*, Vol. 38, No. 7, (May 1995), pp. 1271-1284, ISSN 0017-9310

Nichols, B. D.; Hirt, C. W. & Hotchkiss, R. S. (1980). *SOLA-VOF: A solution algorithm for transient fluid flow with multiple free boundaries*, Technical Report, LA-8355, Los Alamos National Lab

Noh, W.F. & Woodward, P.R. (1976). SLIC (simple line interface calculation), *Lecture Notes in Physics*, Vol. 59, pp. 330 - 340, ISSN 0075-8450

Pavlyk, V. & Dilthey, U. (2004). Numerical Simulation of Solidification Structures During Fusion Welding. In: *Continuum Scale Simulation of Engineering Materials: Fundamentals - Microstructures - Process Applications*, Raabe, D.; Franz Roters, F.; Barlat, F. & Chen, L.Q., (Ed.), pp. 745-762, Wiley, ISBN 978-3-527-30760-9, Weinheim, Germany

Pilliod Jr., J.E. & Puckett, E.G. (2004). Second-order accurate volume-of-fluid algorithms for tracking material interfaces. *Journal of Computational Physics*, Vol. 199, No. 2, (September 2004), pp. 465 - 502, ISSN 0021-9991

Poirier, D.R. (1987). Permeability for flow of interdendritic liquid in columnar-dendritic alloys. *Metallurgical Transactions B*, Vol. 18, No. 1, (March 1987), pp. 245-255, ISSN 1073-5615

Prantil, V. C. & Dawson, P. R. (1983). Application of a mixture theory to continuous casting. In: *Transport Phenomena in Materials Processing*, Chen, M. M.; Mazumder, J. & Tucker III, C. L., (Ed.), pp. 469-484, ASME, ISBN 978-9994588787, New York, N.Y.

Prescott, P.J.; Incropera, F.P. & Bennon, W.D. (1991). Modeling of Dendritic Solidification Systems: Reassessment of the Continuum Momentum Equation. *International Journal of Heat and Mass Transfer*, Vol. 34, No. 9, (September 1991), pp. 2351-2360, ISSN 0017-9310

Provatas, N.; Goldenfeld, N. & Dantzig, J. (1998). Efficient Computation of Dendritic Microstructures using Adaptive Mesh Refinement. *Physical Review Letters*, Vol. 80, No. 15, (April 1998), pp. 3308-3311, ISSN 0031-9007

Ramaswany, B. & Kahawara, M. (1987). Lagrangian finite element analysis applied to viscous free surface fluid flow. *International Journal for Numerical Methods in Fluids*, Vol. 7, No. 9, (September 1987), pp. 953-984, ISSN 0271-2091

Rappaz, M. & Gandin, Ch.-A. (1993). Probabilistic modelling of microstructure formation in solidification processes. *Acta Metallurgica Et Materialia*, vol. 41, No. 2, (February 1993), pp. 345-360, ISSN 0956-7151

Renardy Y. & Renardy, M. (2002). PROST: A parabolic reconstruction of surface tension for the volume-of-fluid method. *Journal of Computational Physics,* Vol. 183, No. 2, (December 2002), pp. 400 - 421, ISSN 0021-9991

Rider, W.J. & Kothe, D.B. (1998). Reconstructing volume tracking. *Journal of Computational Physics,* Vol. 141, No. 2, (April 1998), pp. 112–152, ISSN 0021-9991

Rudman, M. (1997). Volume-tracking methods for interfacial flow calculations. *International Journal for Numerical Methods in Fluids,* Vol. 24, No.7, (April 1997), pp.671-691, ISSN 0271-2091

Scardovelli, R. & Zaleski, S. (1999). Direct numerical simulation of free-surface and interfacial flow, *Annual Review of Fluid Mechanics,* Vol. 31, (January 1999), pp. 567-603, ISSN 0066-4189

Scardovelli R. & Zaleski S. (2000). Analytical relations connecting linear interfaces and volume fractions in rectangular grids, *Journal of Computational Physics,* Vol. 164, No. 1, (October 2000), pp. 228-237, ISSN 0021-9991

Scardovelli, R. & Zaleski, S. (2003). Interface Reconstruction with Least-Square Fit and Split Eulerian-Lagrangian Advection. *International Journal for Numerical Methods in Fluids,* Vol. 41, No.3, (January 2003), pp. 251-274, ISSN 0271-2091

Scheil, E. (1942). *Z. Metallkd,* Vol.34, pp. 70-72, ISSN 0044-3093

Sethian, J. A. (1996). *Level Set Methods: Evolving Interfaces in Computational Geometry, Fluid Mechanics, Computer Vision, and Materials Science,* Cambridge University Press, ISBN 978-0521572026, Cambridge, UK.

Sethian, J. A. (1999). *Level Set Methods and Fast Marching Methods: Evolving Interfaces in Computational Geometry, Fluid Mechanics, Computer Vision, and Materials Science* (2 edition), Cambridge University Press, ISBN 978-0521645577, Cambridge, UK.

Shirani, E.; Ashgriz, N. & Mostaghimi, J. (2005). Interface pressure calculation based on conservation of momentum for front capturing methods. *Journal of Computational Physics,* Vol. 203, No. 1, (February 2005), pp. 154–175, ISSN 0021-9991

Stefanescu, D.M. (2002). *Science and Engineering of Casting Solidification,* Springer, ISBN 030646750X, New York, NY

Steinbach, I.; Pezzolla, F.; Nestler, B.; Seeszelberg, M.; Prieler, R.; Schmitz G.J. & Rezende J.L.L. (1996). A phase field concept for multiphase systems. *Physica D,* Vol. 94, No.3, (August 1996), pp. 135-147, ISSN 0167-2789

Sullivan Jr., J. M.; Lynch, D.R. & O'Neill, K. (1987) Finite element simulation of planar instabilities during solidification of an undercooled melt. *Journal of Computational Physics,* Vol. 69, No. 1, (March 1987), pp. 81-111, ISSN 0021-9991

Swaminathan, C. R. & Voller, V. R. (1992). A general enthalpy method for modeling solidification processes. *Metallurgical Transactions B,* Vol. 23, No. 5, (October 1992), pp. 651-664, ISSN 1073-5615

Takizawa, A.; Koshizuka, S. & Kondo S. (1992). Generalization of physical component boundary fitted co-ordinate method for the analysis of free surface flow. *International Journal for Numerical Methods in Fluids,* Vol. 15, No. 10, (November 1992), pp. 1213-1237, ISSN 0271-2091

Tang, H.; Wrobel, L.C. & Fan, Z. (2004). Tracking of immiscible interfaces in multiple-material mixing processes, *Computational Materials Science,* Vol. 29, No.1, (January 2004), pp.103 - 118, ISSN 0927-0256

Torrey, M. D.; Cloutman, L. D.; Mjolsness, R. C. & Hirt, C. W. (1985). *NASA-VOF2D: a computer program for incompressible flow with free surfaces*, Technical Report, LA-101612-MS, Los Alamos National Lab

Torrey, M. D.; Mjolsness, R. C. & Stein L. R. (1987). *NASA-VOF3D: a three-dimensional computer program for incompressible flow with free surfaces*, Technical Report, LA-11009-MS, Los Alamos National Lab

Tryggvason, G.; Bunner, B.; Esmaeeli, A.; Juric, D.; Al-Rawahi, N.; Tauber, W.; Han, J.; Nas, S. & Jan, Y.-J. (2001). A Front-Tracking Method for the Computations of Multiphase Flow. *Journal of Computational Physics*, Vol. 169, No. 2, (May 2001), pp. 708–759, ISSN 0021-9991

Udaykumar, H. S.; Mittal, R. & Shyy, W. (1999). Computation of Solid–Liquid Phase Fronts in the Sharp Interface Limit on Fixed Grids. *Journal of Computational Physics*, Vol. 153, No.2, (August 1999), pp. 535–574, ISSN 0021-9991

Udaykumar, H.S.; Marella, S. & Krishnan, S. (2003). Sharp-interface simulation of dendritic growth with convection: benchmarks. *International Journal of Heat and Mass Transfer*, Vol. 46, No.14, (July 2003), pp. 2615-2627, ISSN 0017-9310

Voller, V.R. & Prakash, C. (1987). A Fixed Grid Numerical Modeling Methodology for Convection-Diffusion Mushy Region Phase Change Problems. *International Journal of Heat and Mass Transfer*, Vol. 30, No. 8, (August 1987), pp. 1709-1719, ISSN 0017-9310

Voller, V.R.; Brent, A. & Prakash, C. (1989). The Modeling of Heat, Mass and Solute Transport in Solidification Systems, *International Journal of Heat and Mass Transfer*, Vol. 32, No. 9, (September 1989), pp. 1719-1731, ISSN 0017-9310

Voller, V.R.; Mouchmov, A. & Cross, M. (2004). An explicit scheme for coupling temperature and concentration fields in solidification models. *Applied Mathematical Modelling*, Vol. 28, No. 1, (January 2004), pp. 79 - 94, ISSN 0307-904X

Voller, V.R. (2006). Numerical Methods for Phase-Change Problems. In: *Handbook of Numerical Heat Transfer* (2 edition), Minkowycz, W.J.; Sparrow, E.M. & Murthy, J.Y., (Ed.), pp. 593-622, Wiley, ISBN 978-0-471-34878-8, Hoboken, New Jersey

Warren, J.A. & Boettinger, W.J. (1995) Prediction of dendritic growth and microsegregation patterns in a binary alloy using the phase-field method. *Acta Metallurgica et Materialia*,
Vol. 43, No. 2, (February 1995), pp. 689-703, ISSN 0956-7151

West, R. (1985). On the permeability of the two-phase zone during solidification of alloys. *Metallurgical and Materials Transactions A*, Vol. 16, No. 4, (April 1985), pp. 693, ISSN 1073-5623

Wheeler, A.A.; Boettinger, W.J. & McFadden, G.B. (1992). Phase-field model for isothermal phase transitions in binary alloys. *Physical Review A*, Vol. 45, No. 10, (May 1992), pp. 7424 -7439, ISSN 1050-2947

Youngs, D.L. (1982). Time-dependent multi-material flow with large fluid distortion. In: Numerical Methods for Fluid Dynamics, Morton, K.W. & Baines, M. J. (Ed.), pp. 273–285, Academic Press, ISBN 9780125083607, UK

Youngs, D.L. (1984). *An Interface Tracking Method for a 3D Eulerian Hydrodynamics Code*, Technical Report 44/92/35, AWRE, 1984

Yu, K.O. & Imam, M.A. (2007). Development of Titanium Processing Technology in the USA. *Proceedings of the 11th World Conference on Titanium (Ti-2007)*, Kyoto, Japan, June 2007

Zhu, M. F.; Lee, S. Y. & Hong, C. P. (2004). Modified cellular automaton model for the prediction of dendritic growth with melt convection. *Physical Review E*, Vol. 69, No.6, (June 2004), pp. 061610-1 - 061610-12, ISSN 1063-651X

2

Formation of Alpha Case Mechanism on Titanium Investment Cast Parts

Si-Young Sung[1], Beom-Suck Han[1] and Young-Jig Kim[2]
[1]*KATECH(Korea Automotive Technology Institute),*
[2]*Sungkyunkwan University*
Korea

1. Introduction

At present the practical utilization of titanium (Ti) and its alloys ranges from sports equipment to the aerospace, power generation and chemical processing industries, and automotive, marine and medical engineering [1]. Yet, Ti alloy forming processes are much more laborious than one thinks, which is mainly due to the fact that in solid solution Ti combines with carbon, nitrogen and especially oxygen and additionally Ti is constrained to be melted in inert environments or in a vacuum [2]. For the intended applications, Ti and its alloy forming processes should incorporate forging, powder metallurgy and casting. A center bulkhead of the F-22 Raptor fighter aircraft is a good example of the difficulty in Ti alloy forging. Although the final Ti6Al4V alloy component weighs only about 150 kg, it should be forged initially from a single cast ingot of nearly 3,000 kg, which clearly exhibits extremely high machining losses in the Ti alloy forging process [2]. By the way, the powder metallurgy of Ti alloys offers a practicable way of producing complex components with less machining losses than in forging processes. However, the applications of the powder metallurgy have been limited by the freedom of size and shape. In addition, in order to ensure the soundness of the powder metallurgy products without porosity defects, hot isostatic pressing (HIP) is essentially required, which makes the powder metallurgy even more expensive [3].

Casting is a typical net shape forming technique in which molten metals are poured into a mold to produce an object of desired shape. However, the casting of titanium alloys is considered as only a near net shape forming technique. That is because regardless of mold type, titanium alloys casting has a drawback, called alpha-case (α-case), which makes it difficult to machine them and can lend themselves to crack initiation and propagation, due to their enormous reactivity in molten states [4]. Thus, the depth of α-case must be taken into consideration in the initial design for casting since the brittle α-case must be removed by chemical milling. For the reason, the wide use of titanium alloys casting has been limited, although titanium alloys castings are comparable, and quite often superior, to wrought products in all respects [3].

In order to avoid the α-case problem, the expensive ceramics, such as CaO, ZrO_2 and Y_2O_3 have been adopted as mold materials since the standard free energy changes of the formation of their oxides are more negative than that of TiO_2. Regardless of thermodynamic approaches, the α-case formation reaction still remains to be eliminated in the chemical milling processes. In order to develop the economic net shape forming technique of titanium alloys by casting process, it is necessary to take a much closer look into the α-case formation mechanism. Therefore, the exact α-case formation mechanism must be examined and the development of α-case controlled mold materials is required for practical applications of Ti alloys.

2. Evaluation of the alpha-case formation mechanism

2.1 Conventional alpha-case formation mechanism

The investment casting of titanium has a drawback, called 'α-case', which makes it difficult to machine and can lead to crack initiation and propagation, due to their enormous reactivity in molten states [4]. The α-case is generally known to be developed by the interstitials such as carbon, nitrogen and especially oxygen dissolved from mold materials. In order to avoid this problem, the expensive ceramics, such as ZrO_2, $ZrSiO_4$, $CaZrO_3$, CaO and Y_2O_3 have been adopted for mold materials because their standard free energy changes of the formation of oxides are more negative than that of TiO_2 as shown in figure 1.

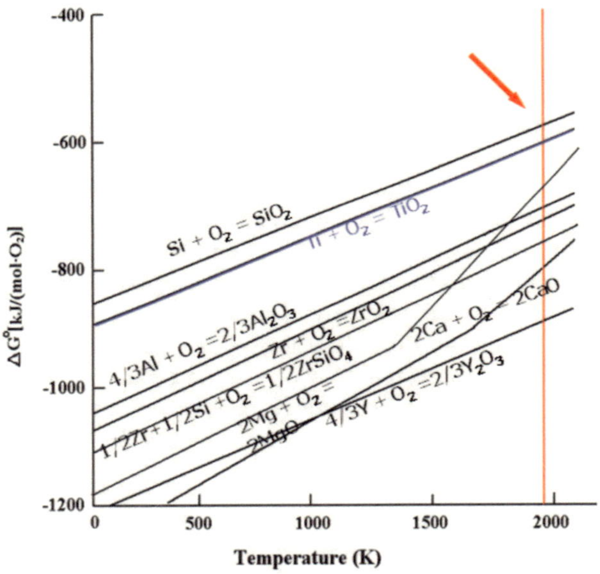

Fig. 1. Standard free energy changes of the formation of various oxides.

Regardless of the thermodynamic approaches, some amount of α-case still remains to be eliminated with the complex chemical milling processes except for Y_2O_3 mold. However, in the case of Y_2O_3 mold, there is not enough strength to handling due to the silica-free binder and there occur metal-mold reactions to some extent, too.

2.2 Evaluation of alpha-case reaction

The wax patterns for the examination of α-case reactions were made by pouring molten wax into a simple cylindrical silicon rubber mold (Ø15 × 70 mm). Subsequently, the patterns were inspected and dressed to eliminate any imperfection or contamination, and coated with Al_2O_3 slurry. The Al_2O_3 shell molds were dried at a controlled temperature (298 K ± 1 K) and a relative humidity (40% ±1%) for 4 hrs. Dipping, stuccoing and drying procedures were repeated three times. After the primary layer coating, the patterns were coated with the back-up layers by the chamotte. To prevent the shell cracks, the dewaxing process of the shell molds were carried out at around 423 K and 0.5 MPa in a steam autoclave. Finally, the shell molds were fired at 1,223 K for 2 hrs.

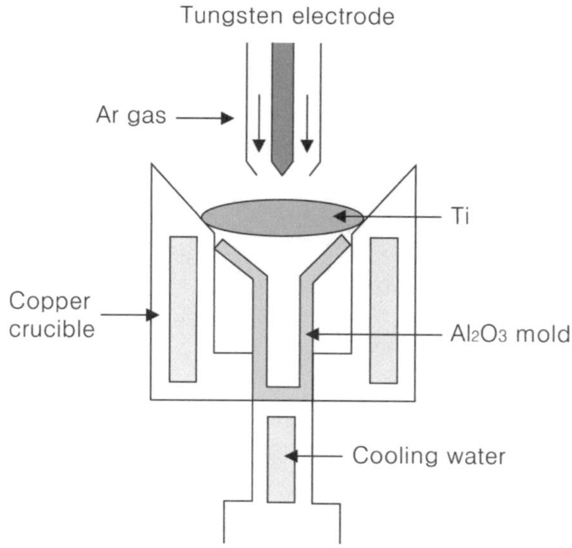

Fig. 2. Schematic diagram of drop casting procedure of titanium with a plasma arc melting furnace.

In order to prevent any contamination from refractory crucibles, the specimens for the of α-case formation examination were prepared in a plasma arc melting (PAM) furnace with drop casting procedure. The pressure in the PAM was controlled at about 1.33×10^{-1}Pa by a rotary pump and then in order to minimize the effect of oxygen contamination, high purity argon was backfilled to the pressure of 4.9×10^{3} Pa.

After the drop casting, Ti castings were taken out of the molds, sectioned, polished and etched using a Keller solution [5]. The microstructure in the reaction region of the castings was observed using Olympus PME3 microscopy, and its hardness was measured using a Mitutoyo MVK-H2 microvickers hardness tester with the condition of 100 g load and 50 μm intervals. For a closer examination of the alpha-case formation, the distribution state of the elements analysis at the alpha-case region was performed by SHIMADZU EPMA 1600 and the phase structures of reaction products were identified by JEOL JEM-3011 TEM.

3. Thermodynamic calculation of alpha-case formation mechanism

3.1 Evaluation of the alpha-case reaction

In this research, Al_2O_3 was selected for the mold material because the standard free energy changes of the formation of its oxides are more negative than that of TiO_2. In addition, Al_2O_3

features suitable strength, permeability and collapse-ability, which can ensure dimensional accuracy of castings. Fig. 3 shows the microstructure and hardness profile on the surface of pure titanium investment castings poured into Al_2O_3 mold.

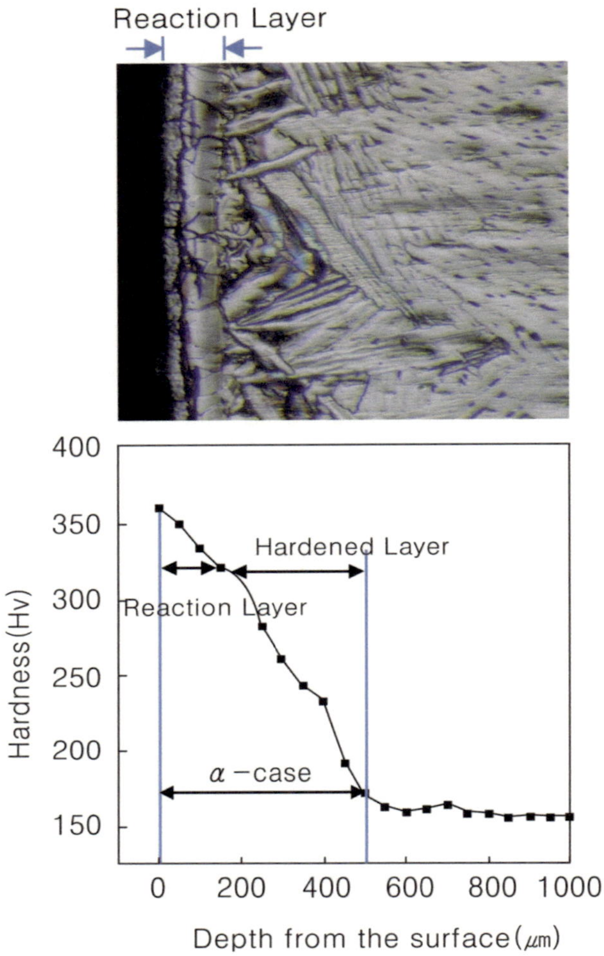

Fig. 3. Microstructure of the interface between Ti and Al_2O_3 mold, and hardness profile.

The microstructure of a distinct reaction layer is about 200 μm thick on the interface between Ti and Al_2O_3 mold. In addition to the reaction layer, there shows a hardened layer about 300 μm thick. The reaction layer and the hardened layer together are called the α-case of titanium castings. However, when the molten titanium (around 2,000 K) is poured into the Al_2O_3 investment mold, and if the α-case results between Ti and the interstitial oxygen, the reaction could be described as follows:

$$Ti(l)+O_2(g)=TiO_2(s) \quad \Delta G^{o}TiO_2= -585.830 \text{ kJ/mol} \tag{1}$$

$$4/3Al(l)+O_2(g)=2/3Al_2O_3(s) \quad \Delta G^{o}_{Al2O3}= 1,028.367 \text{ kJ/mol} \tag{2}$$

$$Ti(l)+2/3Al_2O_3(s)=TiO_2(s)+4/3Al(l) \quad \Delta G^o_F= +99.748 \text{ kJ/mol} \tag{3}$$

The above calculations utilized the joint of army-navy-air force (JANAF) thermochemical tables [6]. According to the equation (3), the α-case formation by interstitial oxygen cannot occur spontaneously. Therefore, the reason why the α-case reaction is generated cannot be explained by the conventional α-case formation mechanism.

3.2 Alpha-case formation mechanism

For the clear examination of the α-case formation mechanism, the distribution of the elements of the α-case and its chemical composition were investigated by EPMA elemental mapping as shown in Fig. 4. On the α-case region, the oxygen element was uniformly distributed. Also, the Si element originating from the colloidal silica binder was scarcely detected on the surface.

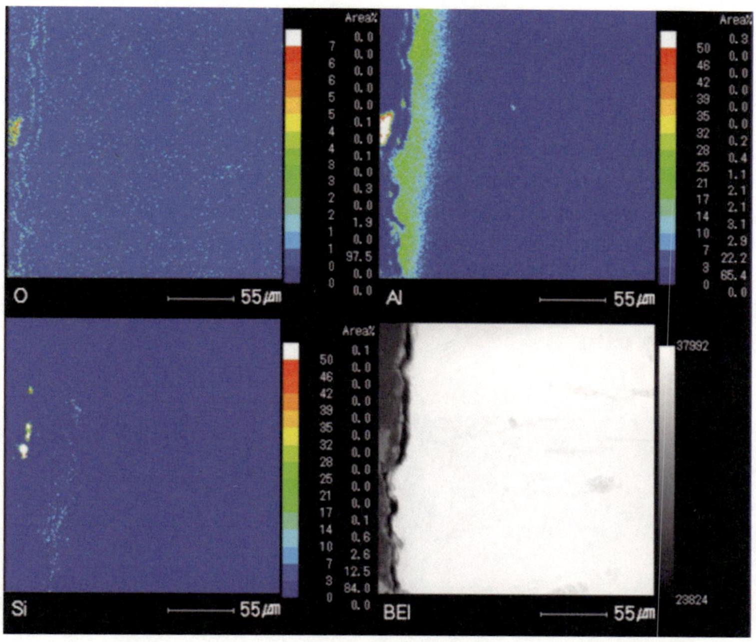

Fig. 4. Comparison of elemental mapping images of O, Al and Si in Ti castings into Al_2O_3 mold and BEI image by EPMA.

However, the concentrated Al contamination layer about 30 μm thick was detected on the interface. The EPMA mapping result shows that not only the interstitial oxygen elements but also the substitutional Al elements dissolved from the mold material affect the metal-mold reactions. Until recently, the effect of substitutional metallic element dissolved from mold materials has been ignored as negligibly small [7]. However, Fig. 3 and 4 indicate that the effect of metallic elements cannot be overlooked any more.

The phase identification of the detected Al element will be the very core of α-case formation mechanism. The phases of the detected Al on the interface were examined by transmission electron microscopy (TEM). In order to observe α-case region precisely, the titanium castings specimen was grinded from inside to the surface until about 30 μm and final thinning was carried out using ion milling. Fig. 5 is the cross-sectional bright field TEM image of the α-case region.

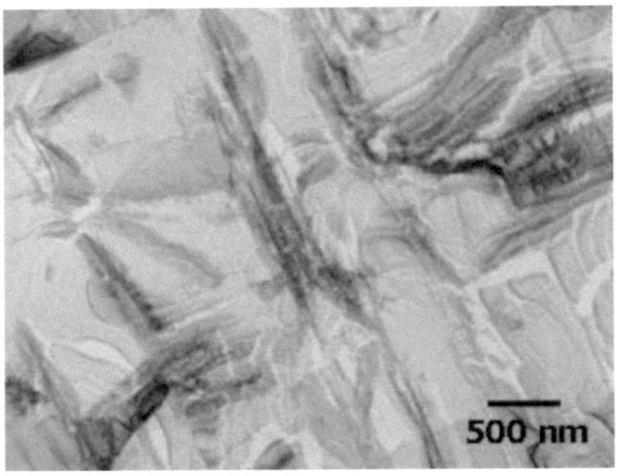

Fig. 5. Bright field TEM image of the α-case region.

To examine what kinds of phases are presented in the α-case region, C2 aperture was temporarily removed. Fig. 6 (a) shows ring and spot patterns on the TEM image without C2 aperture. The definite contrast of continuous ring pattern could not be found on the TEM image since the ring pattern was an extremely small size polycrystalline TiO_2 phase. In the case of spot patterns on Fig. 6 (b), the contrast could be distinguished on the TEM image. And the indexed pattern was a hexagonal close-packed (HCP) phase in the $[2\bar{1}\bar{1}0]$ beam direction as shown Fig. 6 (b).

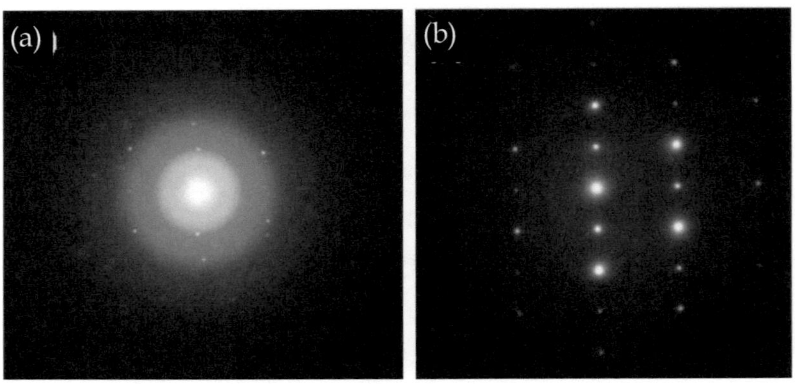

Fig. 6. Results of analytical transmission electron microscopy of (a) ring and spot patterns on the TEM image without C2 aperture on the bright field image, and (b) spot pattern on the TEM image was HCP phase in the $[2\bar{1}\bar{1}0]$ beam direction.

The convergent beam electron diffraction (CBED) analysis was carried out for the verification of the phase of spot patterns with the primitive cell volume method [7]. The primitive cell volume is characteristic material constant. So, the phase of interest can be easily identified by comparing the measured primitive cell volume of an unknown phase using CBED patterns with the known values of the possible phases. This method is accurate (error range of <10%) for the phase identification. In the CBED pattern where zero order Laue zone (ZOLZ) disk and high order Laue zone (HOLZ) ring are present, by measuring the distance of ZOLZ disk from the transmitted beam to the diffracted beam (D_1, D_2), an internal angle (ANG) between D1 and D2, and HOLZ ring's radius (CRAD), the primitive cell volume of unknown phase can be easily determined using the equation (4). Camera Length (CL) was calibrated with Au standard sample at 200 keV.

$$Cell\ volume = \frac{CL^2 \cdot \lambda^3}{D_1 \cdot D_2 \cdot \sin(ANG)\left[1 - \cos\left\{\tan^{-1}\left(\frac{CRAD}{CL}\right)\right\}\right]} \quad (4)$$

Fig. 7 (a) shows the CBED pattern of the HCP phase. There are two possible HCP phases of Ti and Ti_3Al. The measured primitive cell volume (61.08 ang^3, CL=521.4 mm, D_1=D_2=6.5 mm, ANG=60°, CRAD=23 mm) from CBED pattern corresponded to the theoretic value of Ti_3Al (66.60 ang^3). This phase identification was in accordance with EDS spot analysis as shown in Fig. 6 (b). The phase of the detected Al from EPMA mapping was Ti_3Al phase. Thus, considering the microstructure, hardness profile, EPMA mapping and TEM, the α-case is not wholly TiO_2, but TiO_2 and Ti_3Al between Ti and Al_2O_3 mold.

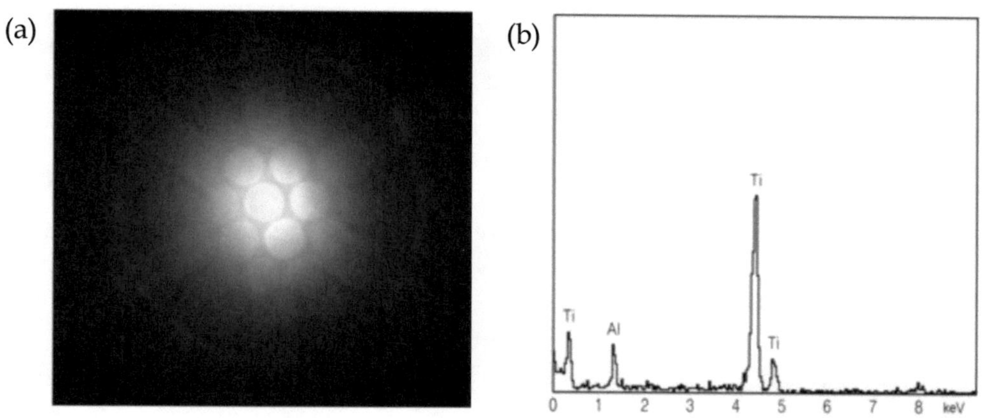

Fig. 7. Convergent beam electron diffraction analysis with the primitive unit cell volume method (a) The measured primitive unit cell volume, 61.08 ang^3 (CL=521.4 mm, D_1=D_2=6.5 mm, ANG=60°, CRAD=23 mm, λ=0.0251 Å) and (b) EDS spot analysis result.

In this study, from the synthesis of the microstructure, hardness profile, EPMA mapping and TEM, it could be confirmed that the α-case is formed by not only interstitial oxygen element but also substitutional metallic elements dissolved from mold materials as shown in Fig. 8.

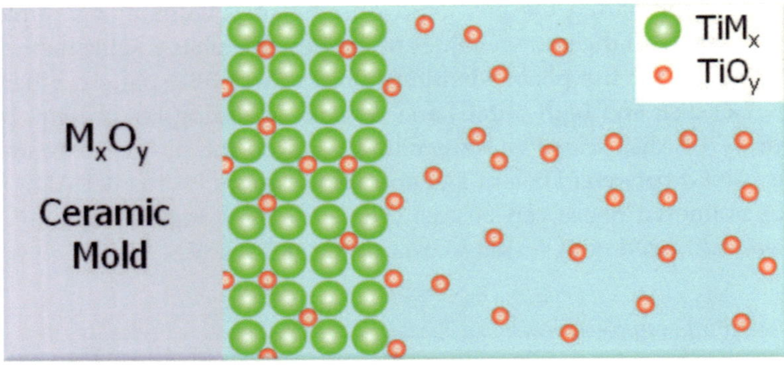

Fig. 8. Schematic diagram of the interstitial and substitutional α-case formation mechanism of titanium castings.

4. New alpha-case formation mechanism with alpha-case controlled mold

4.1 Alpha-case controlled stable mold fabrication

The α-case formation mechanism was applied to the development of α-case controlled stable mold and verified by titanium alloy castings with the α-case controlled mold materials. To synthesize the α-case formation products, 10 wt% to 50 wt% of about 45 μm titanium powders were added into Al_2O_3 for the purpose of eliminating the α-case. After blending, the slurry was prepared with the Al_2O_3 based powder. The α-case controlled stable mold manufacturing and titanium casting procedures were carried out at the same condition as mentioned above. The synthesized phases between titanium and Al_2O_3 were identified by RIGAKU X-ray diffractometer.

4.2 New alpha-case formation mechanism

Although the above mentioned Al_2O_3 has the proper strength, permeability and collapse-ability enough to ensure dimensional accuracy of castings, it could not be applied for mold due to its severe α-case formation. However, in this study, for the maximization of performance and cost-effectiveness of the investment mold, the α-case controlled mold was designed on Al_2O_3 base.

In consideration of the interstitial and substitutional elements in α-case formation mechanism, 10 wt% to 50 wt% of titanium powders were blended with Al_2O_3 for the previous synthesis of α-case reaction products into the mold. Then the blended powders were mixed with the colloidal silica and agitated.

Fig. 9 shows the X-ray diffraction result of 50 wt% titanium mixed Al_2O_3 powders before curing. The composition of the starting titanium and Al_2O_3 remained unchanged after being mixed with colloidal silica. After the mixing and agitation, the mold materials were cured at 1,223 K for 2 hrs. According to XRD analysis, TiO_2 and Ti_3Al phases were in-situ synthesized on the Al_2O_3 base between titanium and Al_2O_3 powders as shown in Fig. 10.

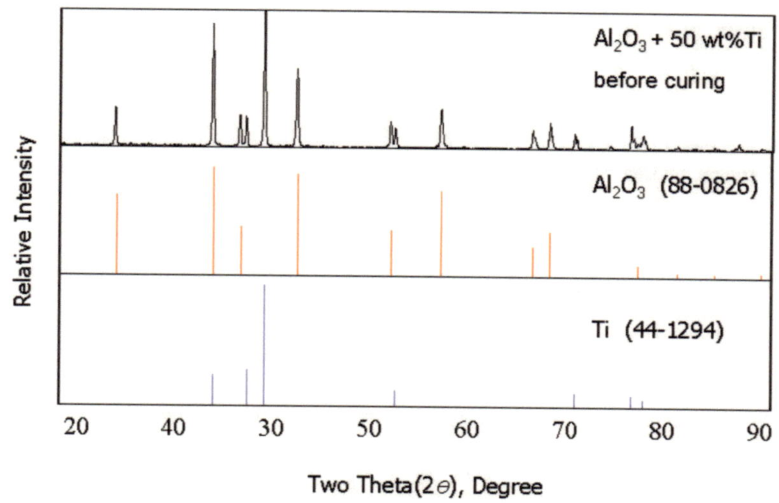

Fig. 9. Illustration of the diffraction patterns and replots for identifying phases in 50 wt% titanium and Al_2O_3 mixed with colloidal silica before curing.

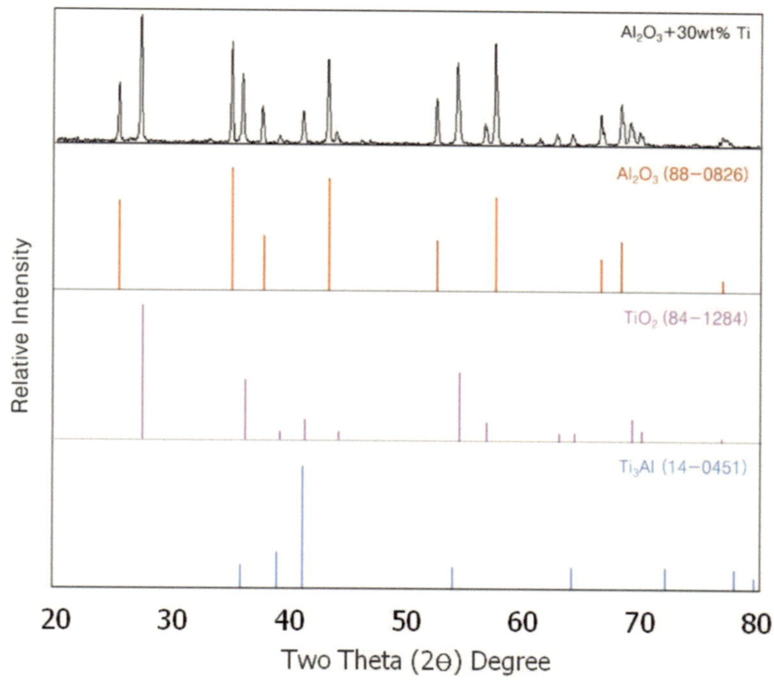

Fig. 10. X-ray diffraction pattern of synthesized α-case controlled stable mold materials.

The scan angles were from 20° to 80° and the pattern was verified with JCPDS card numbers 88-0826 (Al_2O_3), 84-1284 (TiO_2) and 14-0451 (Ti_3Al) [8]. No trace of the starting titanium reagent was detected by XRD analysis. Fig. 11 shows the variation of XRD patterns with the amount of titanium powders. The component ratio of TiO_2 and Ti_3Al phases increased gradually with the addition of titanium. Consequently, the synthesis of the α-case formation products TiO_2 and Ti_3Al phases into Al_2O_3 based mold could be possible by the addition of titanium powders.

The effect of α-case controlled stable mold which contained the interstitial metal-mold reaction product TiO_2 and the substitutional metal-mold reaction product Ti_3Al was verified with titanium casting.

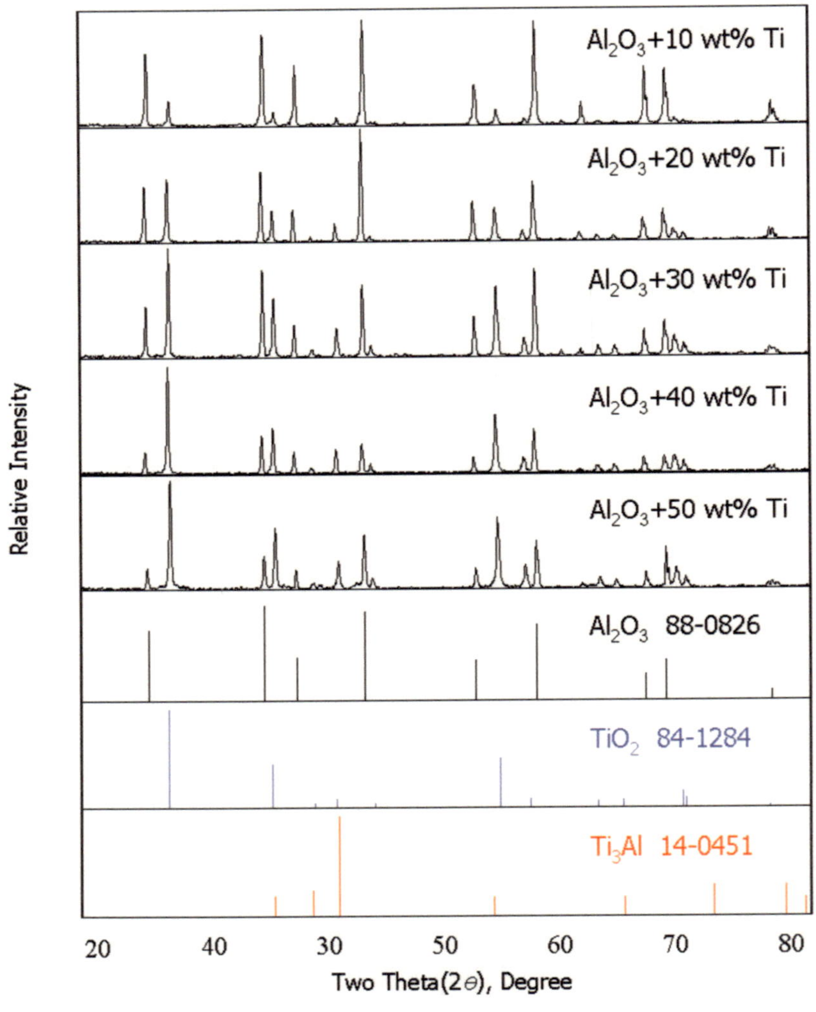

Fig. 11. Illustration of the diffraction patterns and replots for identifying phases from 10 wt% to 50 wt% titanium and Al_2O_3 after curing at 1,223 K for 2 hrs.

Fig. 12 (a) indicates that in the case of CaO stabilized ZrO_2, the expensive and thermally stable mold, the externals of titanium castings lost metallic luster as a result of the α-case formation reactions. However, in the α-case controlled stable mold, its characteristic luster of titanium was well preserved as shown in Fig. 12 (b).

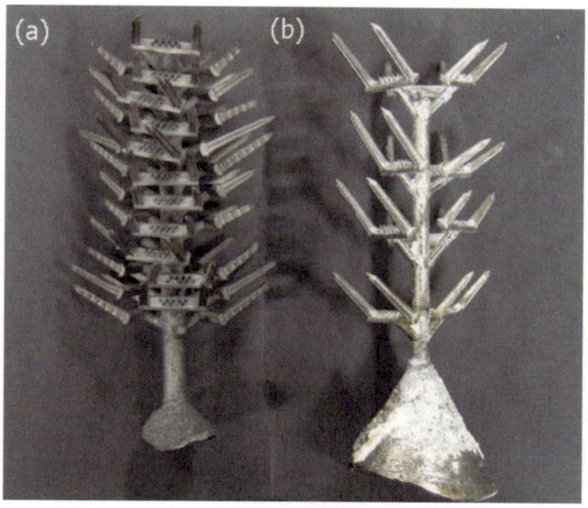

Fig. 12. Comparison of externals of titanium castings (a) CaO stabilized ZrO_2 mold and (b) the α-case controlled stable mold.

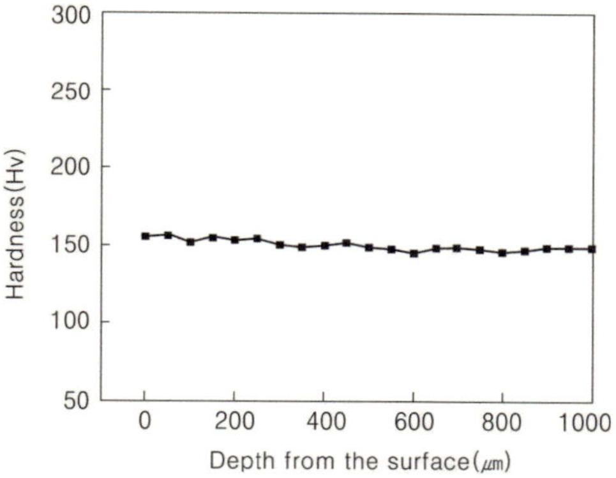

Fig. 13. Microstructure and hardness profile between pure titanium and α-case controlled stable mold.

This external examination result of the α-case controlled stable mold is in good accordance with the microstructure and hardness profile as shown in Fig. 13. The effect of prevention of α-case formation can be obtained after addition of more than 2 wt% titanium.

4.3 Noble synthesis of the alpha-case controlled stable mold

The α-case controlled stable mold is much less expensive than the conventional mold materials for titanium such as ZrO_2, CaO stabilized ZrO_2 and Y_2O_3, and the complete control of α-case is possible. However, in this study, the noble synthesis route was conceived, which is more cost-effective than the above route by the inexpensive TiO_2 and aluminum raw materials.

The homogeneous powders containing 30 mol% TiO_2 powders (anatase, 45 μm, 99% pure) and 70 mol% aluminum powders were prepared by blending. After blending, the powders were mixed with colloidal silica and agitated. The α-case controlled stable mold manufacturing and titanium casting procedures were carried out at the same condition as mentioned above. The 3:7 molar ratio of TiO_2 and aluminum was chosen to yield the Al_2O_3 and TiAl final product after in-situ synthesis according the following reaction.

$$3TiO_2 + 7Al \rightarrow 2Al_2O_3 + 3TiAl \quad (5)$$

In order to synthesize the Al_2O_3 and TiAl phases, the mixed TiO_2 and aluminum powders were cured at the 1,323 K for 2 hrs. However, the equation (5) is an ideal reaction, and thus it is very difficult to synthesize the ideal product by the simple blending and after agitation.

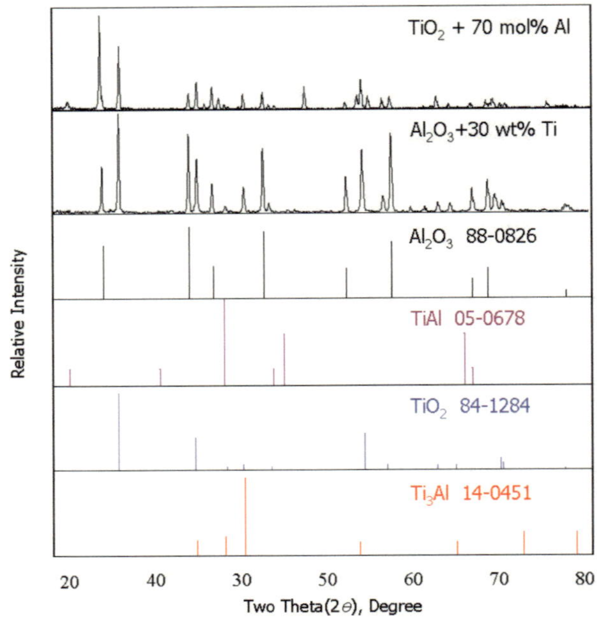

Fig. 14. Illustration of the diffraction patterns and replots for identifying the synthesized phases between TiO_2 and aluminum powders.

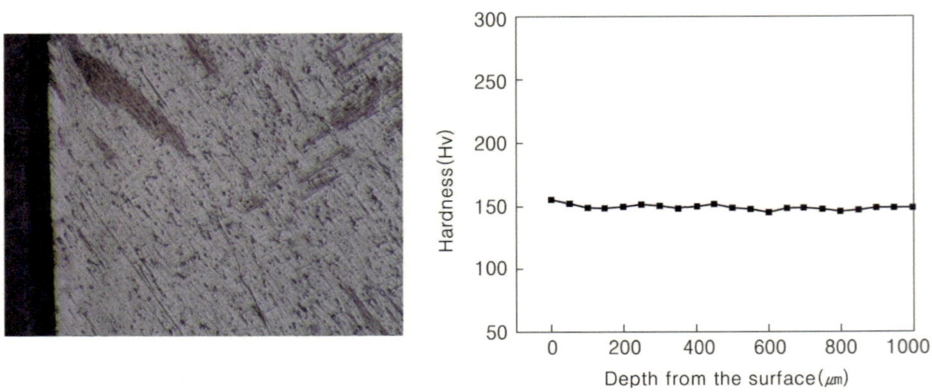

Fig. 15. Microstructure and hardness profile between pure titanium and noble route α-case controlled stable mold.

According to the XRD analysis as shown in Fig. 14, the major phases are anatase TiO_2 and rutile TiO_2 as well as the in-situ synthesized Al_2O_3, Ti_3Al and TiAl phases since the partially non-reacted anatase TiO_2 remained and the TiAl, Ti_3Al intermetallic compounds and rutile TiO_2 were synthesized between anatase TiO_2 and aluminum powders. As a result, the synthesis which is similar to the titanium powders added α-case controlled stable mold can be obtained by the economical route. The α-case free titanium casting can be possible by the noble route synthesized α-case controlled stable mold which contained the α-case formation products such as Al_2O_3, TiO_2, TiAl and Ti_3Al phases as shown in Fig. 15.

5. Conclusion

By the conventional α-case formation mechanism regardless of thermodynamic approaches, the alpha-case generation cannot be explained. In order to ascertain the reason, α-case formation mechanism was closely examined and from the mechanism, two kinds of α-case controlled stable molds were developed.

1. Regardless of thermodynamic approaches, about 500 μm thick α-case was generated between titanium and Al_2O_3 mold. The reason why the α-case generated cannot be explained by the conventional α-case formation mechanism, which is known to be formed by the interstitials, especially oxygen dissolved from mold materials.
2. In spite of having used pure titanium, the concentrated aluminum contamination layer about 30 μm thick was detected on the interface by the EPMA elemental mapping. From the results of the TEM phase identification, the phase of the detected aluminum from EPMA mapping was identified as Ti_3Al phase.
3. Considering the microstructure, hardness profile, EPMA mapping and TEM analysis, it could be confirmed that the α-case is formed not only by interstitial oxygen element but also by substitutional metallic elements dissolved from mold materials.
4. The synthesis of the TiO_2 and Ti_3Al phases which are the α-case reaction products between titanium and Al_2O_3 mold can be obtained by the simple curing of Al_2O_3 mold added titanium powders. The complete control of α-case formation can be possible with the α-case controlled stable mold.
5. The α-case free titanium casting can be possible by the noble route synthesized α-case controlled stable mold which is in-situ synthesized between TiO_2 and aluminum powders to obtain the α-case formation product such as Al_2O_3, TiO_2, TiAl and Ti_3Al phases.
6. Consequently, the economical net-shape forming of titanium alloys without α-case formation and the verification of a newly established α-case formation mechanism, can be possible using the α-case controlled stable molds.

6. References

[1] Matthew JD.; *Titanium a Technical Guide*, ASM International, ISBN 978-0871706867, OHIO USA

[2] Christoph L. & Manfred P.; *Ttitanium and Titanium Alloys* : Wiley-VCH, ISBN 3-527-30534-3, Weinheim, Germany

[3] Joseph RD. (editor) ; *Metals Handbook Vol. 2*, ASM International, ISBN 0871703785, OHIO USA

[4] Doru. S. (editor); *Metals Handbook Vol. 15,* ASM International, ISBN 0871700212, OHIO USA
[5] Petzow G.; *Metallographic Etching*, ASM International, ISBN 0871706334, OHIO USA
[6] Chase M.W., Davis C.A., Downey J.R., Frurip D.J., McDonald R.A. & Syverud A.N.; *JANAF Thermochemical Tables*, American Chemical Society and American Institute of Physics, ISBN 1-56396-831-2, New York, USA
[7] Williams D.B. & Carter C.B., *Transmission Electron Microscopy*, Plenum Press, ISBN 978-0306453243, New York, USA
[8] *PCPDFWIN Version 2.1*, JCPDS-International Centre for Diffraction Data, 2000.

3

Genesis of Gas Containing Defects in Cast Titanium Parts

Vladimir Vykhodets[1], Tatiana Kurennykh[2] and Nataliya Tarenkova[2]
[1]*Institute of Metal Physics, Ural Division, Russian Academy of Sciences,*
[2]*VSMPO-AVISMA Corporation,*
Russia

1. Introduction

At present, gas-containing inclusions of metallurgical origin like Ti-O, Ti-N, and Ti-O-N are one of the main problems of titanium production. These defects negatively influence the mechanical properties of titanium alloys, as the mechanical properties (hardness and plasticity) of gas-containing inclusions differ appreciably from the corresponding characteristics of alloy matrices. Therefore, cracks nucleate in defect-matrix contact zone upon loading, which results in catastrophic fracture of heavily loaded parts. For example, the tragic consequences of failures of aircraft engine parts owing to the presence of similar effects are well known. The technology of vacuum arc remelting (VAR) of titanium alloys is most vulnerable from the view point of the formation of gas-containing defects, and, at the same time, it dominates in the overall production of titanium alloys. Efforts over many years have not resulted in the development of a technology for the defect-free production of titanium alloys by the VAR method and the titanium defect problem known also as "hard alpha" remains a troublesome aspect of the use of titanium alloys. Attempts to remove or modify the hard-alpha inclusions by diffusion homogenization of ingots had failed due to the long time required for the appropriate heat treatment at high temperature. Therefore, the genesis of defects under the conditions of industrial processes, their identification in each specific case, classification, and the elaboration of measures for reducing the formation of defects are topical questions.

A generally accepted concept has been developed (Bellot & Mitchel, 1994), according to which the sources of defects are particles of titanium and its alloys with a high concentration of oxygen and nitrogen atoms (the predominant element is nitrogen) and liquidus temperatures exceeding the smelting process temperature. Such particles can be formed at different stages of the preparation of batch materials for smelting. It may be supposed that all the components of a batch, such as titanium sponge, titanium return production wastes, and titanium containing alloy additions, are potential sources of defects. Smelting products with rather high concentrations of oxygen and nitrogen atoms may also be sources of defects. In melting oxygen and nitrogen atoms pass into a melt due to diffusion, the melting temperature of particles enriched in light elements decreases, the inclusion size decreases too, and conditions for the their dissolution are created. This scenario takes place for most particles enriched in oxygen and nitrogen atoms. Nevertheless, some of them remain solid by the end of melting although titanium sponge (plus titanium return production wastes, and alloy additions) are melted twice for many applications and even three times for critical aeroengine parts.

Studies that have been conducted on the defects in titanium alloys may be conditionally classified into two groups: works with model defect sources and model smelting processes and works based on the analysis of the characteristics of defects revealed for industrial processes. At present, the model approach is predominant. However, the patterns established in model experiments may be not adequate for the processes that occur during smelting under industrial conditions, owing to the difference between the characteristics of real and model defect sources, industrial and experimental smelting modes, and also of the corresponding equipment.

In our opinion, the possibilities of the model approaches have been almost exhausted. This is shown by the fact that the model approaches has not led to substantial progress in the understanding of the genesis of defect formation under industrial conditions and the development of technologies for the defect-free production of titanium alloys. In this connection, we place our emphasis on gaining information on the processes of defect formation exclusively on the basis of the data characterizing industrial processes in the present work. In this case, it is impossible to study a statistically representative sample of defect sources, as, on average, a single defect is formed in several millions of particles of batch materials, so the study of a statistically significant ensemble of defects of industrial origin was used as the basis of this approach. This was achieved by the long term monitoring of a process in which the VAR technology was used. We know only one work of a similar type in the literature. This is a report that was published by the Material Development Department of the TRW Company (Grala, 1968). Metallographic and micro X-ray spectral studies of seventeen metallurgical defects in a Ti–6Al–4V alloy were performed in this work. According to the results of this work, the formation of defects may be caused by an increased concentration of atoms of both light elements (oxygen and nitrogen) and aluminum.

2. Materials and methods

The detection of inclusions in metal and their localization in semi products (rods, billets, plates, and sheets) made from titanium alloys was conducted by the method of ultrasonic testing. After the device received an echo signal indicating the presence of an internal defect we marked and cut out a specimen containing the zone of the signal with increased amplitude, whereupon specimens were subjected to mechanical processing (layer by layer metal stripping) accompanied by metallographic studies. As a result, the zone of the maximum echo signal was brought into the visible plane of a metallographic polished specimen. These polished specimens were the objects of metallographic studies; measurements of the microhardness and the concentration of oxygen, nitrogen, and alloying metal atoms in the matrix and the material of defects were also made.

The measurements of the microhardness in the zone of defects were performed with the use of a device with a Vickers diamond tip. The load of the indentor was 200 g. The concentrations of the atoms of alloying metals (mainly, aluminum and vanadium) in revealed defects were determined by X-ray spectral analysis using the energy dispersion method on an EDAX scanning electron microscope. The statistical error in the measurement of the aluminum concentration was 10% of the measured value for aluminum concentrations of less than 5.0% (hereinafter, all concentrations are given in weight percents). For higher aluminum concentrations, the error was 0.5% and the error in the measurement of the vanadium concentration was approximately 0.5%.

The concentrations of oxygen and nitrogen atoms in specimens were determined by nuclear microanalysis technique (NRA) on a 2 MV Van de Graaf accelerator with the use of the $^{16}O(d,p_1)^{17}O$ and $^{14}N(d,\alpha_1)^{12}C$ reactions at an incident beam particle energy of 0.9 MeV. The method in its traditional variant (Vykhodets et al., 1987), in which the diameter of the incident beam of the accelerator was 1.0 or 2.0 mm, was used for the measurement of the average concentrations of light element atoms in a matrix of the titanium alloys. Defects usually had smaller sizes. In this connection, we developed a variant of the NRA method with enhanced locality (Vykhodets et al., 2006) to study them. Using this method, we measured the average concentrations of oxygen and nitrogen in defects 0.1–2.0 mm in size. In this case, each specimen that was measured was equipped with an individual collimator. Its dimension was chosen with consideration for the size of an analyzed zone, i.e., a defect, in a certain specimen. A scheme of NRA experiment is shown in Fig. 1.

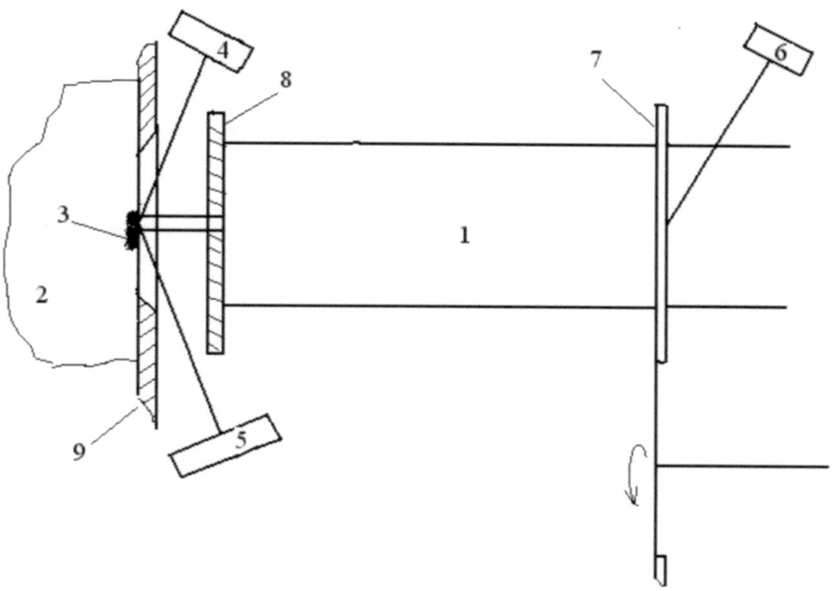

Fig. 1. Scheme of NRA experiment: 1 – incident beam; 2 – specimen; 3 – defect to be investigated; 4,6 – silicon surface-barrier detectors of the system of monitoring incident beam; 5 - silicon surface-barrier detector, which register spectrum of the products of nuclear reactions for the studied specimen; 7 – propeller of the system of monitoring incident beam; 8 - individual collimator; 9 – plate with specimens.

Using the NRA method, we measured the concentrations without destruction of a specimen down to a depth of nearly 2.0 μm. This allowed us to distinguish the effects of the volume and surface (uncontrollable) alloying of alloys with the atoms of light elements in the nondestructive profile analysis. These appeared, in particular, at the stage of the preparing polished specimens and could propagate in a specimen down to a depth of nearly 0.5 μm. The effects of the surface alloying of specimens with atoms of oxygen and nitrogen are excluded from consideration in all the results below. In this work, the sensitivity in the measurements of the concentrations of oxygen and nitrogen was approximately 0.01%, the statistical error in the measurements of concentrations was at a level of several percent of a measured value. The mounting of the individual collimators 8 and the specimens 2 with the defects onto the plate 9 was performed under an optical microscope before the plate was placed into the vacuum chamber of the accelerator unit. As a result, the precision with which the beam hit the defect was 0.01 mm. The typical spectrum of nuclear reactions is shown in Fig. 2.

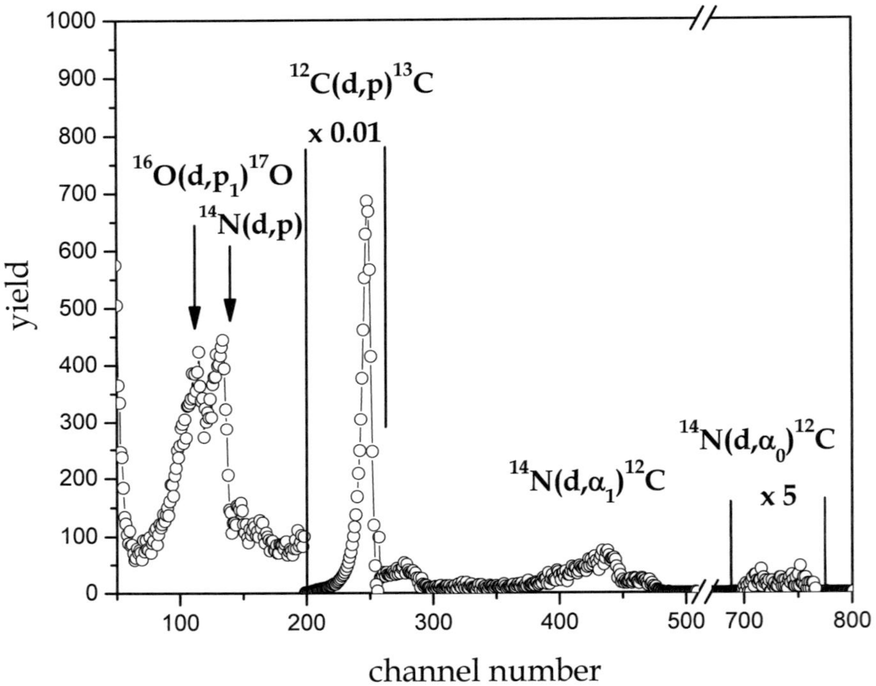

Fig. 2. The typical spectrum of nuclear reactions.

In total, more than 100 metallurgical defects revealed in the Ti–6Al–4V alloy were studied. This grade dominates in the overall production of titanium alloys. Defects revealed in other grades of titanium alloys were also investigated. In all the cases, the alloys were obtained by the VAR. Inclusions varied in their shapes and sizes and, to a first approximation, it was possible to distinguish spherical and extended defects. The causes of such variations have not been established as yet and this question is not discussed in this work. The minimum size of the studied effect was predominantly within the interval from 0.1 to 2.0 mm. The length of extended defects could amount to several centimeters. The full range of studies was conducted not only for defects, but also for the matrix surrounding a revealed defect. These investigations were performed for verification, and their results did not reveal any particular features. The microstructure, microhardness, and the concentrations of gaseous impurities and metal atoms near defects were identical to those for the zones remote from defects. Some of the results represented in this work were previously published in (Tarenkova et al., 2006; Vykhodets et al., 2007; Vykhodets et al., 2011), but the corresponding data were obtained on an appreciably smaller sample of defects.

The above mentioned characteristics of defects (concentrations of gaseous impurities and aluminum, microhardness etc.) are traditional in the description of defects in titanium alloys. In this work, we included also the data on defect coordinates in ingots into the defect database. They are also important for the understanding of the genesis of gas-containing defects since in the industrial VAR technology the rate of dissolution of defect sources in different ingot parts may differ essentially. This is connected with different temperature of the liquid pool in different parts of the ingot and with the use of magnetic stirring in the VAR technology. This operation results in different liquid flow velocity around solid defect sources in the melt. Earlier, in model experiments (Bellot et al., 1997; Reddy, 1990; Schwartz, 1993; Mitchell, 1984) it was found that the dissolution rate of defect sources strongly depends both on the temperature of the liquid pool and the degree of stirring. For example, the measured dissolution rate of nitrogen-containing inclusions for intense stirring was 10 times faster than those without stirring and the dissolution rate doubled for the temperature rise of about 100^{0}C. In our study, the data on the distribution of defects in ingots were obtained by means of the following procedure. Under production conditions, for each metal volume we made a flow chart of technological operations. After detecting the defect in semiproduct or finished product, the flow chart allowed finding the position of the metal volume in the ingot from which the semiproduct or finished product was produced. Further calculations allowed determining two coordinates of a defect in the ingot: distances from a defect to the ingot base and the ingot axis. Accuracy of coordinate determination in the ingots was about several millimeters.

3. Experimental results and discussion

3.1 Oxygen and nitrogen concentration in defects

Histograms of the distributions of defects over the concentrations of nitrogen and oxygen are shown in Figs. 3 and 4. Almost all the revealed defects contained excess concentrations of oxygen and nitrogen in comparison with the volume of titanium alloys; the excess ranged from several times to several tens of times larger. In almost of cases, the concentration of oxygen was slightly higher than that of nitrogen. At the registered concentration level of gaseous impurities, the alloying of inclusions by nitrogen is the predominant factor in the formation of defects. This is associated with the fact that the alloying of titanium by nitrogen leads to an appreciably higher increase in the liquidus temperature of a material in comparison with the alloying of titanium by oxygen. This result, which indicates the predominant role of nitrogen in the formation of gas-containing defects, agrees with the data from study (Wood, 1969). The data obtained on the concentration of gaseous impurities in defects turned out to be insufficient for the verification of the concept, according to which all sources of defects did not dissolve in the liquid pool because of a high concentration of nitrogen and oxygen in them. There are two main reasons for this. Firstly, the range of possible values of smelting temperatures for an industrial process is not known with certainty. Secondly, the concentration of gaseous impurities could not be measured strictly at the end of smelting since during ingot cooling or other high-temperature treatment processes the concentration of light element atoms decreased due to their diffusion from a

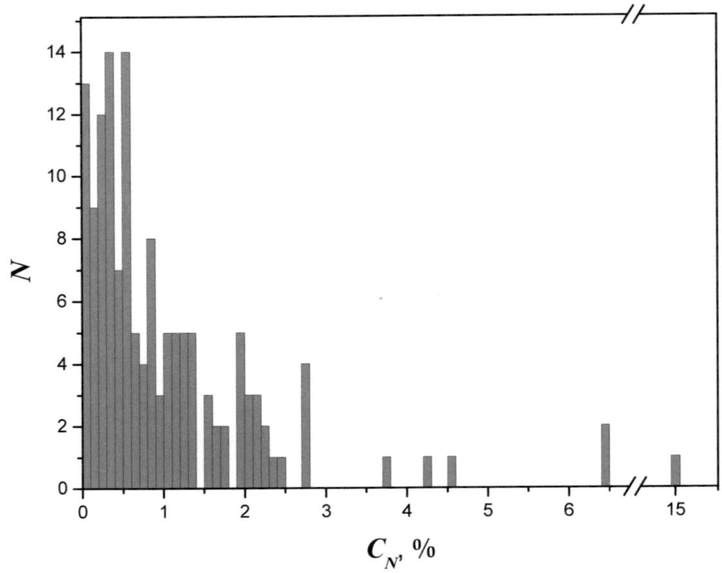

Fig. 3. Distribution of defects over the concentration of nitrogen in them: N is the number of defects; C_N is the average concentration of nitrogen in a defect.

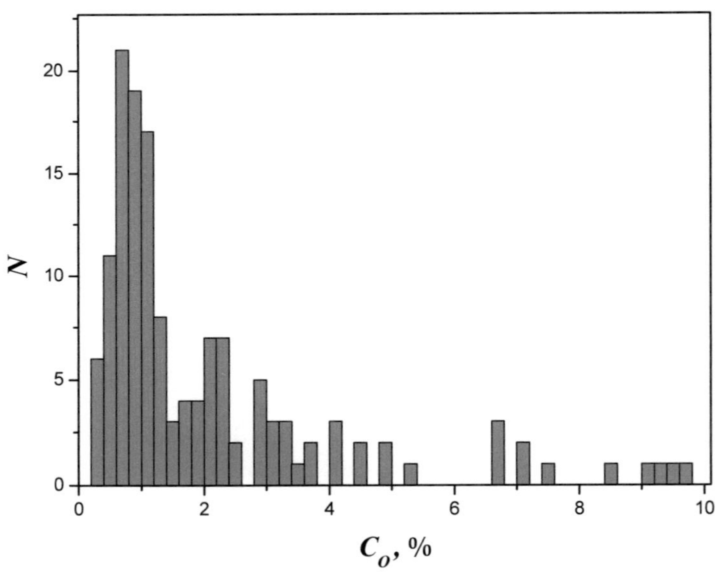

Fig. 4. Distribution of defects over the concentration of oxygen in them: N is the number of defects; C_O is the average concentration of oxygen in a defect.

defect to a solid titanium alloy. Fig. 5 presents a calculated histogram of defect distribution on the liquidus temperature of the defect material. The liquidus temperatures T_L were found using the data of binary diagrams of states for the systems Ti-O and Ti-N (Bellot & mitchell, (1994); Fromm & Gebhardt, 1976). Here, an additive approximation was used, namely, it was assumed that in Ti-N and Ti-O-N systems nitrogen atoms raise the liquidus temperature of the defect material by the same value if nitrogen concentrations are equal in these systems. The same additive approximation was used for Ti-O and Ti-O-N systems. The effect of vanadium and aluminum on the liquidus temperature of the defect material was not considered since it is much weaker than that of nitrogen and oxygen.

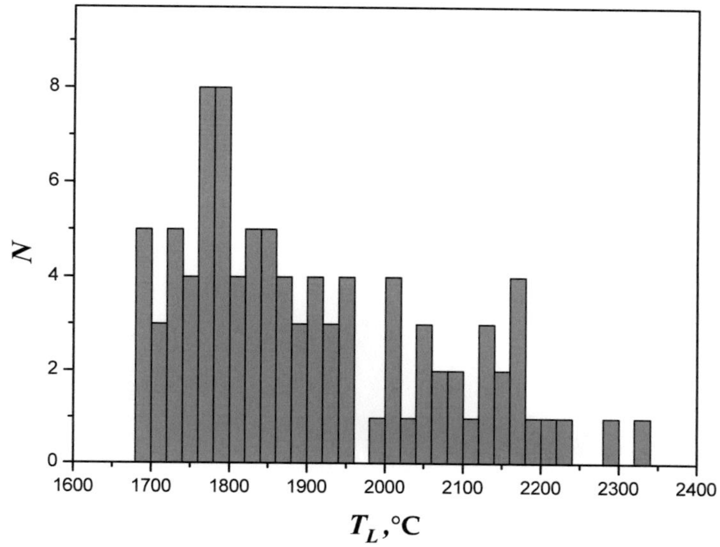

Fig. 5. Distribution of defects over the liquidus temperature of the defect material: *N* is the number of defects; T_L is the calculated liquidus temperature of the defect material.

The temperature of smelting is usually assumed to be close to 1900°C. In study (Bellot et al., 1997), a lower value (1780°C) is reported. Anyway, it can be stated that defects were revealed, the material of which has the liquidus temperature both higher and lower than that of the smelting process. This conclusion can hardly change the rough approximations used for the calculation of the liquidus temperature of the defect material. The existence of defects with liquidus temperatures below the temperature of smelting in principle can be due to diffusion processes occurring in the system upon completion of smelting.

On the whole, in most cases the concentration of nitrogen in defects was higher than in the alloy matrix. For such defects, the data on nitrogen and oxygen concentrations are not in conflict with the existing viewpoint that the sources of defects are titanium particles and its alloys with high concentrations of oxygen and nitrogen. At the same time, prolonged monitoring of the industrial technology revealed about ten defects, in which the concentration of nitrogen did not exceed that in the alloy matrix. Naturally, such defects cannot be considered as gas-containing defects; they turned out to be abnormal also in some other respects. We shall dwell on this question in greater detail in section 3.3.

3.2 Microhardness

Under the conditions of industrial production, the microhardness is traditionally measured in each case that a defect is revealed. These data are usually used as an indicator of the presence of light element atoms in defects. In this study, we did not observe any univocal correlation between the microhardness and the concentration of gaseous impurities in defects. This can be seen from Fig. 6, which illustrates the dependence of the microhardness on the total concentration of oxygen and nitrogen in defects. For definiteness, in Fig. 6, we give only the minimum values of the microhardness in the body of a defect because of a slight scatter observed in the measurements of the microhardness.

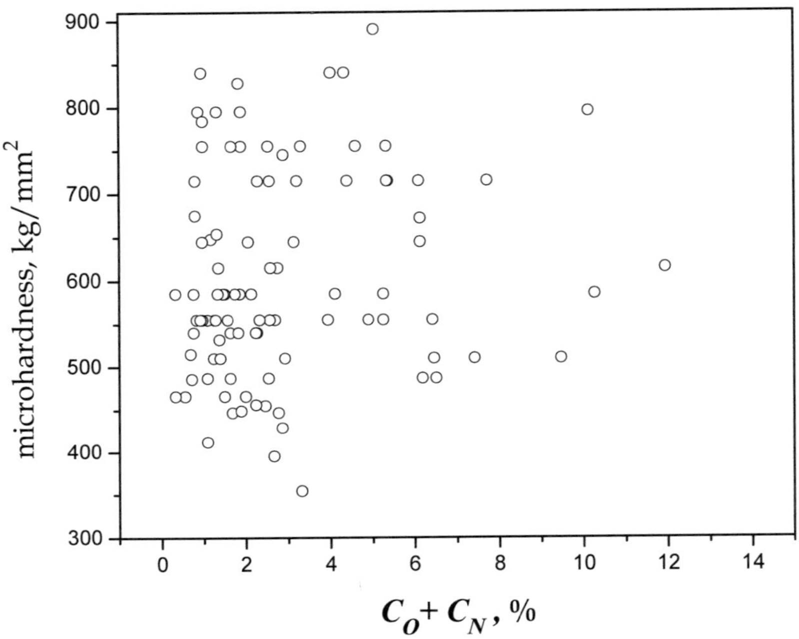

Fig. 6. Microhardness of defects versus total concentration of nitrogen and oxygen ($C_O + C_N$) in them.

We also detected several defects with a very high microhardness from ~ 900 to 2600 kg/mm², the corresponding data are not shown in Fig. 6. These defects proved to be abnormal in several other respects. As in the case of the results on the concentration of nitrogen in defects, this circumstance will be used in this work for the classification of revealed defects. The result shown in Fig. 6 is not trivial. A linear or close to linear dependence of the microhardness on the concentration of gaseous impurities is usually observed (David et al., 1979). From Fig. 6, it follows that in the case of industrial defects, the value of microhardness cannot serve as a criterion of the presence of gaseous impurities in them. On the whole, the only pattern that can be noted with respect to the microhardness is that its value in defects is usually higher that that in the matrix (nearly 350 kg/mm²).

3.3 Aluminum concentration at the centre of defects

As different batch materials before smelting and smelted alloys contain various quantities of alloying elements, the patterns of the concentrations of alloying element atoms in defects may be used for the identification of defect sources. A similar idea was suggested in study (Bewlay & Gigliotti, 1997); this approach is widely used in industry. Let us estimate the prospects of this approach for the Ti-6Al-4V alloy. This alloy mainly contains atoms of aluminum (from 5.50 to 6.75%) and vanadium (from 3.50 to 4.50%).

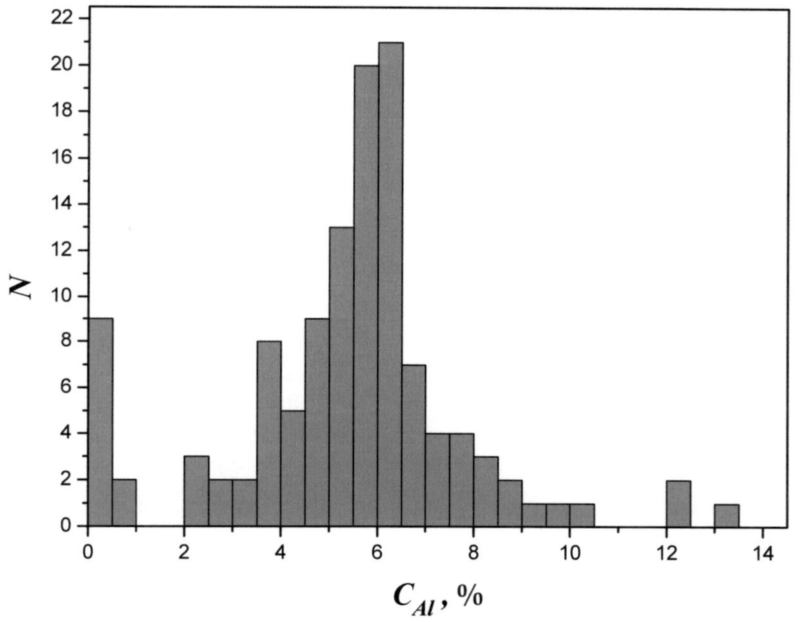

Fig. 7. Distribution of defects revealed in the Ti-6Al-4V alloy over the concentration of aluminum: *N* is the number of defects; C_{Al} is the aluminum concentration at the center of a defect.

Aluminum and vanadium are also present in the same quantities in titanium production wastes. Titanium sponge does not contain appreciable amounts of alloying elements; only atoms of aluminum (20.0%), oxygen, and titanium are present in the oxygen containing addition alloy before smelting. A much more difficult situation with the concentration of aluminum and vanadium takes place in the smelting products, whose particles can pass into a melt in the process of smelting. According to the data of several reports of the VSMPO-AVISMA Corporation, the concentration of vanadium and aluminum in smelting products obtained in the smelting of the Ti-6Al-4V alloy may be 2.0-4.2 and 4.7-20.0%, respectively. On the whole, it is possible to talk of higher aluminum concentrations in smelting products in comparison with those in a melted alloy.

The histogram of the distribution of defects over the aluminum concentration at their centers is shown in Fig. 7. The presence of a pronounced peak in the distribution may be associated with the fact that all the probable defect sources fall within the corresponding concentration interval. It can be seen that the genesis of only 26 defects can reliably be established on the basis of the measurement results on aluminum concentration. These were defects with an aluminum concentration of less than 4.0%; their origination from titanium sponge does not raise any peculiar doubts. The remaining revealed defects cannot be identified with the use of the data shown in Fig. 7. The vanadium concentration at the center of defects was also measured. As shown by analysis, the data of vanadium do not clarify the situation at all, so they are not given here. Hence, we can state that the approach that is used in practice for the identification of defect sources and based on the measurement of the concentration of aluminum and vanadium in defects has no any substantial grounds. It is not inconceivable that progress in this matter may be made with the use of additional criteria. First of all, we may use the analysis of the dependence of the aluminum concentration at the center of a revealed defect on its linear size.

To analyze the diffusion processes that occur during the smelting of titanium alloys in the subsystem of aluminum atoms we shall use the solution of the second Fick equation for the diffusion from a finite sized body with conjunctive boundaries in the form (Fromm & Gebhardt, 1976)

$$\frac{C - C^0}{C^S - C^0} = A\left(Dt/l^2\right), \tag{1}$$

where C is the concentration of an impurity (aluminum in our case) at the center of an inclusion at the end of smelting; C^0 is the initial concentration of an impurity in an inclusion; C^S is the concentration of an impurity at the interface between an inclusion and melt; A is the function of the parameter Dt/l^2 ; D, t and l are the diffusion coefficient of an impurity, the annealing time and the linear size of a specimen, respectively.

The precise dependence $A(Dt/l^2)$ for a cylindrical specimen (defect in our case) is shown in Fig. 8 (curve). This dependence also has a similar shape for other regular bodies. Certainly, in the case of defect sources that appear in the smelting of titanium alloys, it is necessary to take into account the fact that their sizes are reduced in the process of smelting. Because of the absence of a precise solution of the second Fick equation in the case of variable dimensions of an inclusion, we shall obtain an approximate solution of the problem for the dependence of the aluminum concentration at the center of a defect, whose dimensions are varied in the process of smelting. The calculated dependences $A(l)$ obtained at several certain values of the parameter Dt are illustrated in Fig. 8 (straight lines). It can be seen that the dependences $A(l)$ are close to linear ones within the interval of the most appreciable changes, i.e.,

$$A(l) = A(0) - \eta l\,. \tag{2}$$

The points in one of the straight lines indicate the precise values of the function $A(l)$. It can be seen that the deviation of the points from a linear dependence is slight. In this case, expression (1) may be applied also for the variable size of a defect with the substitution of A for its average value obtained for the entire interval of changes during smelting.

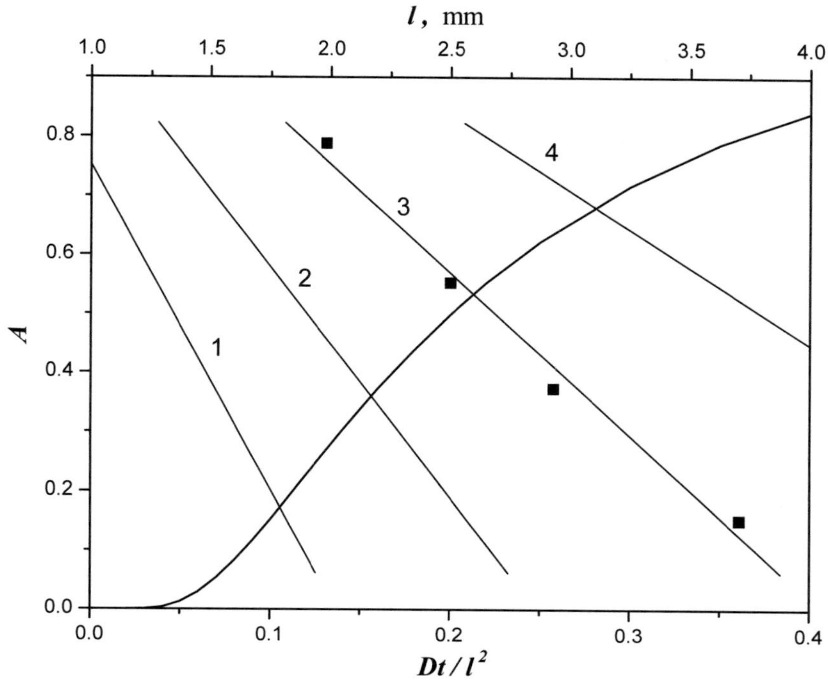

Fig. 8. Calculated dependences $A(Dt/l^2)$ and $A(l)$ for cylindrical specimens. The dependences $A(l)$ were calculated at Dt of $3.44 \cdot 10^{-5}$, $6.88 \cdot 10^{-5}$, $13.8 \cdot 10^{-5}$, and $27.5 \cdot 10^{-5}$ mm^2 (respectively: 1, 2, 3, 4).

Below, we will use the following notation: l is the radius of a cylindrical defect source before smelting, d is the diameter of a revealed defect, and Δ is the thickness of the defect source layer dissolved in the process of smelting. Taking into consideration that

$$l = \frac{d}{2} + \Delta , \tag{3}$$

we obtain

$$C_{Al} = C^0 - \left(C^0 - C^S\right)\left[A(0) - \frac{1}{2}\eta\Delta\right] + \frac{1}{2}\left(C^0 - C^S\right)\eta d \tag{4}$$

Hence, according to expression (4), the dependence of the aluminum concentration C_{Al} at the center of a revealed defect on its linear size d must be linear. It is self evident that this issue concerns the situation where the concentrations C^0 and C^S are equal for all defect sources. Otherwise, a band of the values of the concentration C_{Al} must be observed in experiments. Equation (4) may be used for calculating the thickness Δ of the defect source layer dissolved in the process of smelting. No fitted parameters are required to perform the corresponding calculations. As shown by the calculations, the value of $A(0)$ does not depend on the parameter Dt and is determined only by the shape of a defect source. For example, $A(0) \approx 1.49$ for a cylindrical specimen and $A(0) \approx 1.31$ for a plate (l was accepted as the radius of a cylinder or the half thickness of a plate). In this connection, the results of the calculations for the dissolved layer thickness Δ depend on the shape of a defect. This creates certain problems, as defect sources can actually have any shape.

Note that the approach for obtaining the experimental data on the thickness Δ of the defect source layer dissolved in the process of smelting and on the size l of a defect source before smelting for the industrial technology was first found in this work. Information of this kind is important for the development of methods that reveal the probable defect sources in batch materials. Moreover, the dissolution rate V of defect sources in the process of industrial smelting can be estimated in a linear approximation with the use of the value Δ as

$$V = \frac{\Delta}{t}. \quad (5)$$

Previously, the defect dissolution rate, as mentioned in section 2, was determined only in model experiment (Bewlay & Gigliotti, 1997; Bellot et al., 1997; Reddy, 1990; Schwartz, 1993; Mitchel, 1984), and its strong dependence on the temperature of the liquid pool and the liquid flow velocity around solid particle was established.

Since reliable identification of the sources of 26 defects proved to be feasible, the additional identification method based on the analysis of the linear dependence of the aluminum concentration C_{Al} at the center of a defect on its size d (expression (4)) might be approved on the same defects. Actually, linear dependences (4) are expected to be fulfilled only qualitatively. This is connected with the following circumstances. To obtain such dependences, it was necessary to clarify the notion "linear size of a defect," which is especially topical for extended defects. In specimens with complex shapes, the rate of diffusion processes is generally determined by the minimum linear size of a specimen, so it is chosen as the parameter d. Besides, the specificity of experiments performed in this work consisted in the fact that the concentrations of light elements and aluminum could not be measured at the end of smelting. During the cooling of ingots and other high temperature treatment processes, the atoms of impurities diffuse from defects into the solid titanium alloy and on the contrary, thus leading to a decrease or increase in the concentration of impurities in defects. In section 3.1 it was pointed out that the concentrations of oxygen and nitrogen in the revealed defects were apparently much smaller than those at the end of smelting. However, for aluminum this effect is expected to be appreciably weaker or negligible since the diffusion coefficients of aluminum in solid titanium are almost five orders of magnitude smaller than those of oxygen and nitrogen (Le Claire & Neumann, 1990; Le Claire, 1990). Besides, there are other reasons for the scatter of experimental data on $C_{Al}(d)$. In particular, parameters $A(0)$, C^S, and C^0 in expression (4) will vary within certain intervals for different defects and different smelting processes. For example, as already noted above, the concentration of aluminum in the melt varied in different smelting processes from 5.50 to 6.75%.

The experimental data on $C_{Al}(d)$ for these 26 defects are illustrated in Fig. 9. It can be seen that there is no pattern in the corresponding data. The situation changes radically if defects with ordinary (below 800 kg/mm^2) and high (above 900 kg/mm^2) microhardness values are considered separately. As seen from Fig. 9, a linear dependence $C_{Al}(d)$, which results from the diffusion model, is observed for each individual group of defects in this case.

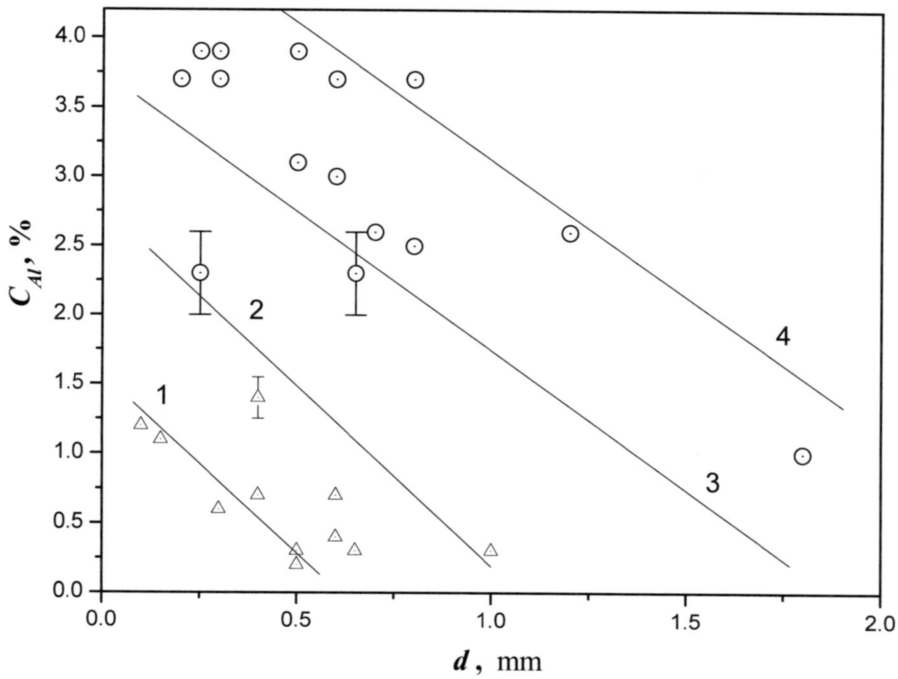

Fig. 9. The aluminum concentration C_{Al} at the centre of defects that were revealed in the Ti-6Al-4V alloy and originated from titanium sponge as a function of the minimum linear size of the defect: triangles are data for defects with microhardness above 900 kg/mm²; circles are data for defects with microhardness below 800 kg/mm².

The concentration intervals for defects with ordinary and high microhardness are denoted in Fig. 9 as 1-2 and 3-4, respectively. It may be supposed that these two groups of defect sources differ from each other in the diffusion coefficients *D* of aluminum atoms in them (these coefficients are lower for defects with high microhardness). In turn, the difference between the values of *D* may be caused by different phase states of inclusions with high and ordinary microhardness values.

The results displayed in Fig. 9 are of interest mainly for two reasons. Firstly, they showed that the value of microhardness should be considered in the classification of defect sources. Secondly, the accuracy with which the theoretical linear dependence $C_{Al}(d)$ was fulfilled proved to be high. This occurred in spite of a large number of factors determining the scatter of $C_{Al}(d)$ data. The width of the concentration intervals denoted in Fig. 9 as 1-2 and 3-4 was 1.25%. This value agrees in particular with the technical specifications for the concentration range of aluminum in the Ti-6Al-4V alloy (from 5.50 to 6.75%). Note however that a single point in Fig. 8 falls outside of the established dependence $C_{Al}(d)$. The microhardness of this defect was about 800 kg/mm², C_{Al} = 2.3%, and d = 0.25 mm. This shows that small and rare deviations from the theoretical dependence $C_{Al}(d)$ can nevertheless take place.

In Fig. 10, the data on $C_{Al}(d)$ are shown for almost all the defects revealed in the Ti–6Al–4V alloy. Only a single defect is not represented in this figure. Its minimum size (3.0 mm) appreciably exceeded the sizes of the other defects, so it was unreasonable to provide the data on this defect. In connection with this exclusion, it should be noted that this defect was not inconsistent with the common patterns, which will be considered below. From Fig. 10, it can be seen that almost all the defects fit into the determined intervals of aluminum concentrations. In Fig. 10, they are denoted as 1–2, 3–4, 5–6, and 7–8. The first two intervals have already been discussed above. Interval 7–8, as well as intervals 1–2 and 3–4, correspond to the slanting linear dependence $C_{Al}(d)$. It makes sense to relate this to the sources of defects, in which the concentration of aluminum before smelting was higher in comparison with that in a melt. They might be the particles of smelting products and addition alloy. Interval 5–6 is horizontal. According to the diffusion model, it corresponds to the sources of defects in which the concentration of aluminum before smelting was close to that in the liquid pool. These may be particles of smelting products and return production wastes. The width of the concentration intervals was equal for all the corridors (1–2, 3–4, 5–6, and 7–8).

Let us discuss the sense of these results in connection with the problem of the identification of defect sources. First, we consider the case of small defects. From Fig. 10, it can be seen that such defects can be equally related to the intervals denoted as 3–4, 5–6, and 7–8. This means that in the case of small defects, their identification with the use of the discussed scheme is impossible, i.e., these defects may be formed from the particles of titanium sponge, return production wastes, smelting products, and addition alloy.

Further, let us consider medium and large sized defects. Here, the situation looks to be much more unambiguous. First, it is possible to note that only the defect sources that are characterized by a nearly equal and high aluminum concentration before smelting fitted in the interval 7–8. Such a condition is satisfied by the particles of addition alloy with an aluminum concentration of 20.0%. Hence, we can state that the defects that originate from the addition alloy are presented within interval 7–8. The absence of defects that formed from smelting products within the interval 7–8 and their presence within interval 5–6 seems to be unnatural. This allows us to think that defects that formed from smelting products are not represented within the interval 5–6, i.e., in the case of large defects, the 5–6 interval is caused by the formation of defects from return production wastes.

Let us consider the defects that did not fit in any of the intervals distinguished in Fig. 10. Only four similar defects were found. One of them had microhardness higher than 900 kg/mm^2. The abnormal behavior of such defects has already been mentioned. The other three defects, whose data are denoted by asterisks in Fig. 10, have nitrogen concentrations that are lower than that in the matrix. It was previously mentioned that such inclusions cannot be reliably classified as gas-containing defects. We suppose that these defects were formed from the particles of smelting products and that their residence in a liquid pool might be very short, for which reason they fall outside of the diffusion patterns. Another defect, which is within the interval 5–6 in Fig. 10, should be classified with inclusions of the same type. The results for this defect are denoted by an asterisk. Consequently, it can be seen that all the cases in which the data on defects lie outside the distinguished intervals 1–2, 3–4, 5–6, and 7–8, are accompanied by abnormal properties of the defects (very high microhardness or very low nitrogen concentration).

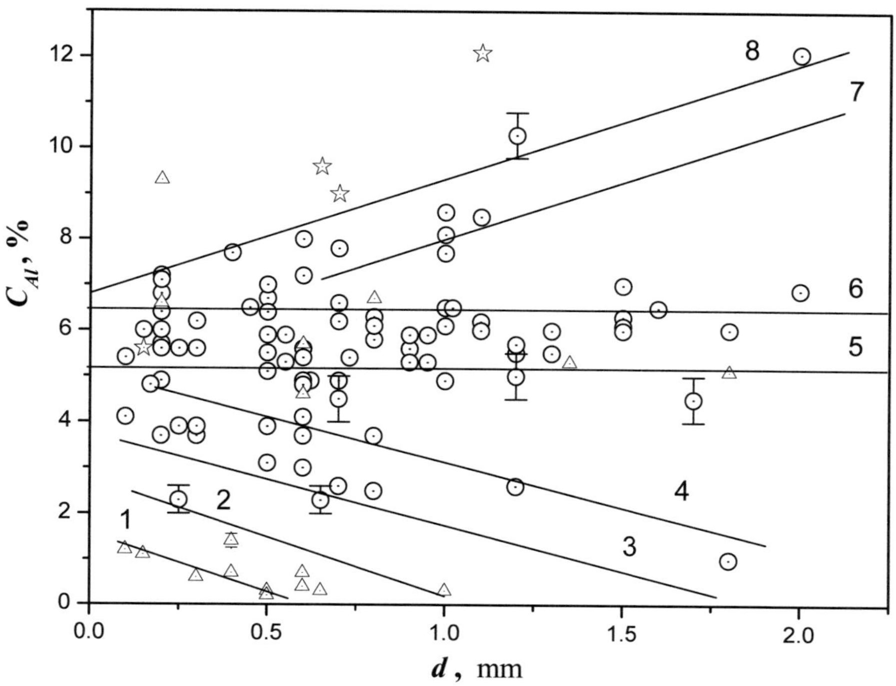

Fig. 10. The aluminum concentration C_{Al} at the centre of defects that were revealed in the Ti-6Al-4V alloy as a function of the minimum linear size of the defect: triangles are data for defects with microhardness above 900 kg/mm²; circles are data for defects with microhardness below 800 kg/mm²; asterisks are data for defects in which the nitrogen concentration does not exceed that in the matrix.

So, a method for the identification of the sources of defects in the industrial VAR technology was developed in this work. Diffusion laws form the basis of this method. This scheme has proven to be very simple for practical application. The identification of the source of a defect revealed in the smelting of the Ti–6Al–4V alloy requires a very limited data set containing the aluminum concentration at the center of the defect, its minimum size, the microhardness of the material of the defect, and the average nitrogen concentration in the body of the defect. Under industrial conditions, this information is usually available, so this method can be recommended for practical application. Note also that during the prolonged monitoring of an industrial process we did not observe defects whose data contradict this method.

Overall, the identification proved to be unfeasible for approximately 20% of the defects. According to the results of the identification, return production wastes were the source of approximately half of all the revealed defects, and nearly a quarter of them originated from titanium sponge. The remaining defects were formed from the particles of addition alloy and smelting products.

Further, using expressions (3)–(5) and the data shown in Fig. 10, the thickness of the defect source layer dissolved in the process of smelting and the dissolution rate V of defect sources were estimated. The average value of the thickness Δ was about 2 – 3 mm. At the same time, the data obtained do not exclude a broad interval of Δ values during the industrial process: from 0 to 8.5 mm. With the corresponding estimates, the shape of the defects and the concentration of aluminum in the liquid pool varied from 5.50 to 6.75%. Besides, it was taken into consideration that the concentration of aluminum C^S at the boundary between the defect source and the melt can be smaller than the corresponding equilibrium concentration. With this approach, the interval of Δ values was obviously overestimated. In reality, it will be smaller.

The average value of the dissolution rate V was approximately 1.0 μm/s. The estimation was performed under the assumption that the residence time of any melt volume in the liquid phase was 2500 s. This value of the average dissolution rate is very low. Model experiments (Bellot et al., 1997) showed that the dissolution rate of the nitrogen-containing source of defects is 2.2, 4.0, and 6.7 μm/s at the temperature of the liquid pool of 1700, 1800, and 1900°C, respectively, when the liquid flow velocity around solid particle in the liquid pool was 1 cm/s. Even when the liquid flow velocity around solid particle was 0.1 cm/s and the temperature was 1800°C, it was 2.0 μm/s. One of the possible directions for defect formation reduction is enhancement of the dissolution rate of defect sources. The prospects of this trend will be discussed in section 3.4.

3.4 Location of defects in ingots

The experimental data on the location of defects in ingots were obtained for the ingots from 650 to 1000 mm in diameter and from 1200 to 2750 mm in height. Hereafter we use the following non-dimensional coordinates of defects: r/R (r is the distance from a defect to the ingot axis, R is the ingot radius) and h/H (h is the distance from a defect to the ingot base, H is the ingot height). The subjects of investigation were the distributions of defects over r/R and h/H coordinates, as well as the probability of formation of one, two, and more defects in an ingot. The experimental data on the location of defects in the ingots are presented in Figs. 11, 12 and 13. In Fig. 14, the number of defects revealed in the unit of the ingot volume is shown as a function of the non-dimensional radius r/R. The data for Fig. 14 were obtained by calculation from primarily data displayed in Fig. 13.

In a real process, the defect formation probability is $p = N/N_h \ll 1$, where N and N_h are the numbers of defects and heats, respectively. If the formation of a defect is an independent event the probability of formation of several defects in one ingot $p(n)$ is

$$p(n) = p^n, \tag{6}$$

where n is the number of defects in the ingot. Since the formation of a defect is a very rare event ($p \ll 1$), the formation of two or more defects in one ingot in this mechanism is almost impossible. However, experimental results indicated that one, two and even more defects can exist in ingots at the close probability.

Altogether we have discovered 205 defects in 86 ingots. 42 ingots contained one defect, 27 ingots contained two defects, 7 ingots had three defects, and in 10 ingots there were more than three defects. These results show that generally the formation of each defect cannot be considered as an independent event .

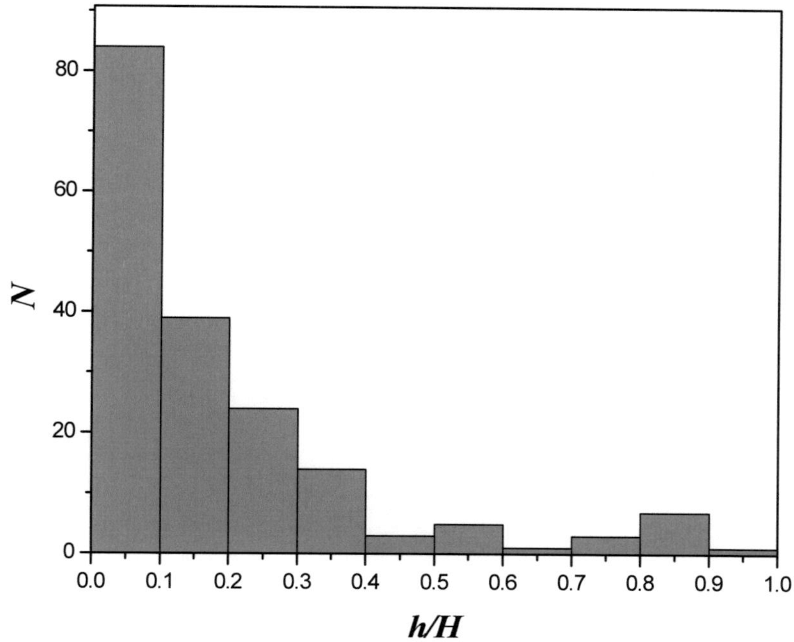

Fig. 11. Distribution of defects over their non-dimensional coordinate *h/H*: *N* is the number of defects; *h* is the distance from a defect to the ingot base; *H* is the ingot height.

For those cases when two defects were revealed in an ingot simultaneously, we analyzed the differences in *h* and *r* coordinates of these defects. In the majority of cases, such defects were very close to each other (the difference in *h* and *r* coordinates usually did not exceed several millimeters). Only in one case the difference in the *h* coordinates was very large (more than 600 mm). Based on this result we can assert that the sources of defects usually do not move over large distances during smelting.

From Figs. 11 and 14 it is seen that the distribution of defects over the ingot height and radius is strongly inhomogeneous. The defects are located chiefly near the ingot base and axis. Since the sources of defects do not move over large distances during smelting as mentioned above, it may be stated that the dissolution rate of defect sources near the ingot base and the axis is much smaller than the average dissolution rate in the ingot. This may be due to a decreased temperature of the liquid pool and (or) a decreased liquid flow velocity around solid particle in the melt. The ingot volumes with *h/H* values from 0 to 0.2 and *r/R* from 0 to 0.1 turned out to be most critical for defect formation. In this range of *h/H* values (20% of the metal volume) 68% of defects were revealed, while in the above range of *r/R* values (1% of the metal volume) there were revealed 17% of defects. These findings are indicative of strong inhomogeneity of defect distribution in the ingots. The number of defects in the unit volume of the metal for the volume bounded by *h/H* values from 0 to 0.2 and *r/R* values from 0 to 0.1 (0.2% of the ingot volume) turned out to be almost 60 times higher than the average value in the ingots.

Relying on these findings we can propose two mechanisms of defect formation. One of them suggests fluctuations in the characteristics of defect sources, i.e. the sources of defects

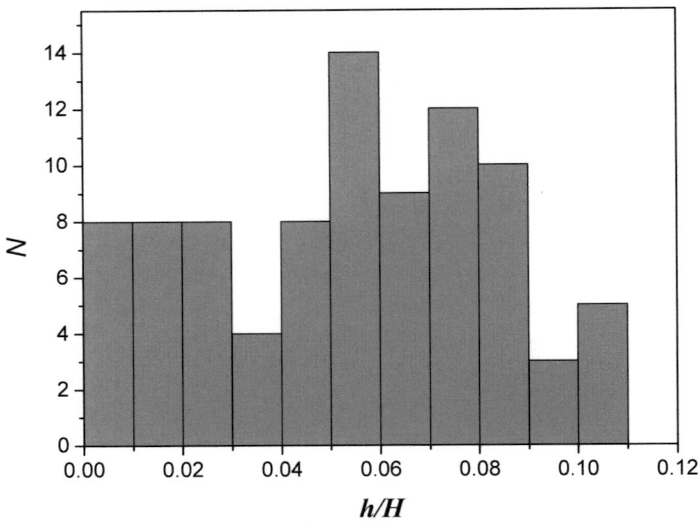

Fig. 12. Distribution of defects over their coordinate h/H near the ingot base: N is the number of defects; h is the distance from a defect to the ingot base; H is the ingot height.

contained in the charge materials have unfavorable characteristics for complete dissolution of defect sources during smelting. Such characteristics can include large sizes of defect sources and/or high concentrations of gaseous impurities in them. According to this mechanism two or more defects could form in one ingot and in one place of the ingot because some defect sources fell into two, three or more fragments during smelting. For this mechanism it is difficult to estimate the probabilities $p(n)$ theoretically.

The second mechanism suggests fluctuations in the dissolution rate of defect sources during smelting. Such fluctuations can be associated with the existence of microvolumes with a decreased temperature of the liquid pool and/or microvolumes with a decreased liquid flow velocity around solid particles in the melt during smelting. According to this mechanism the formation of several defects in one ingot and in one place of the ingot is due to the value of fluctuation for the dissolution rate of defect sources. For this mechanism it is also difficult to determine theoretically the probabilities $p(n)$. Thus, there are no available data at present to give preference to one of the considered fluctuation mechanisms. At the same time, this question is a topical problem. Its solution determines the direction of the main efforts for reducing defect formation in the VAR technology. The first mechanism suggests scientific and technological measures for enhancing the quality of charge materials. With the second mechanism, efforts should be focused on the stabilization of the smelting regime. The results obtained on the distribution of defects over the ingot height (Fig. 11) agree with the reasoning (Bellot & Mitchel, 1994) that the temperature of the liquid pool is minimal near the ingot base. Probably, this cause for enhanced defect formation in the VAR technology cannot be eliminated. The results obtained on the distribution of defects over the ingot radius agree with the model experiments (Bellot et al., 1997) on the examination of the dissolution rate of gas-containing inclusions as a function of liquid flow velocity around solid particles in the melt. In the existing technology, the rotation axis of the magnetic field coincides with the axis of the ingot during the whole period of smelting. This is likely to be one of the reasons why more defects are formed near the ingot axis than at the ingot periphery. Therefore, a possible measure for defect formation reduction is variation of the position of the magnetic field rotation axis during smelting. The estimates showed that with the use of this technique the defect formation can be reduced not more than in 2.5 times. In the estimation we used the data of Fig. 13 and Fig. 14. Besides, the following concepts were accepted.

First, different liquid flow velocities around solid particles in the liquid pool was considered to be the only reason for inhomogeneous distribution of defects over the ingot radius. Second, it was postulated that variation of the position of the magnetic field rotation axis can provide a high liquid flow velocity around solid particles in all parts of the ingot.

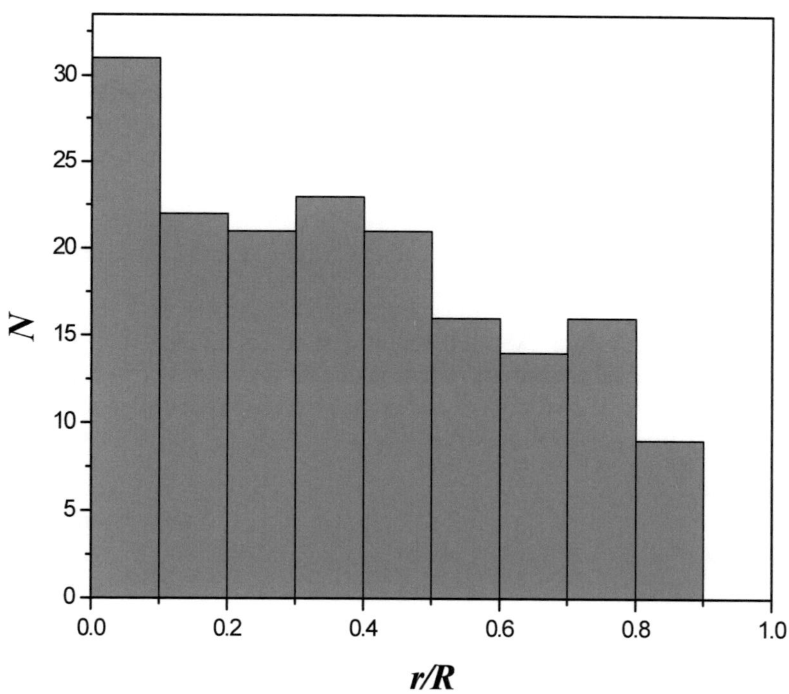

Fig. 13. Distribution of defects over their non-dimensional coordinate r/R: N is the number of defects; r is the distance from a defect to the ingot axis; R is the ingot radius.

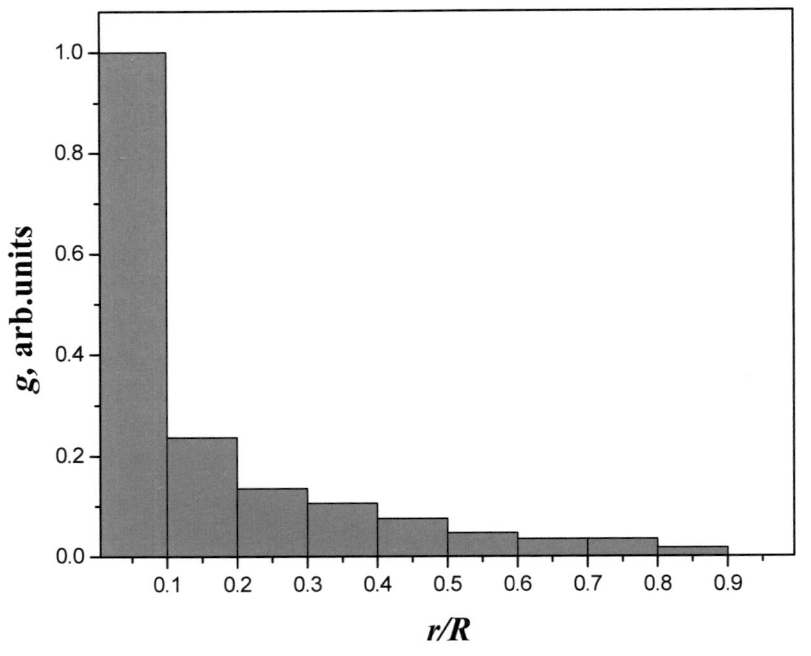

Fig. 14. Distribution of defects over their non-dimensional coordinate *r/R*: g is the number of defects in the unit of the ingot volume; *N* is the number of defects; *r* is the distance from a defect to the ingot axis; *R* is the ingot radius.

4. Conclusion

A database on the characteristics of metallurgical gas-containing defects in titanium alloys was obtained. This was achieved by the long term monitoring of a industrial process in which the vacuum arc remelting (VAR) technology was used. It is the most complete one that is available at present. It contains data on location of defects in ingots, the sizes of defects, their microhardness, and the concentrations of nitrogen, oxygen, and aluminum in them. Almost all the data are related to the Ti–6Al–4V alloy.

Almost all the revealed defects contained excess concentrations of oxygen and nitrogen in comparison with the matrix of titanium alloys; the excess ranged from several times to several tens of times larger. In almost of cases, the concentration of oxygen was slightly higher than that of nitrogen. At the registered concentration level of gaseous impurities, the alloying of inclusions by nitrogen is the predominant factor in the formation of defects. On the whole, the data on the concentrations of nitrogen and oxygen in the defects agree with a generally accepted concept, according to which the sources of defects are particles of titanium and its alloys with a high concentration of oxygen and nitrogen atoms.

It was shown that the approach that was used in practice for the identification of defect sources based on the measurement of the concentration of aluminum and vanadium in defects has no any substantial grounds.

The distribution of defects over the height and radius of ingots was studied; it turned out to be strongly inhomogeneous. The defects were located predominately near the ingot base and the ingot axis. The number of defects in the unit volume of the metal in the lower part and near the axis of the ingots was almost 60 times greater than the average value in the ingots. The inhomogeneous distribution of defects in the ingots is due to different dissolution rate of defect sources during smelting in different parts of the ingot. In turn, this results from a low temperature of the liquid pool in the lower part of the ingots and a low liquid flow velocity around solid particles near the ingot axis.

A large number of cases have been registered when not one but two, three or even more defects are formed simultaneously in the ingot. The probabilities of formation of one and several defects were comparable. Usually, several defects in the ingot were located very close to each other. These findings indicate that generally the formation of each defect cannot be considered as an independent event.

A method for identification of the sources of metallurgical defects in the smelting of titanium alloys has been developed. It is based on the diffusion mechanism for the modification of defect sources in smelting and the established dependences of the aluminum concentration at the center of a defect on its size. The identification scheme included data on microhardness and the average concentration of gaseous impurities in the zone of a defect. The efficiency of identification using this method was 80%.

5. Acknowledgment

This work was supported by the project No 11-2-06-AVI of the Basic Research Program of the Ural Division of Russian Academy of Sciences "Identification of stages which are responsible for formation of gas-containing metallurgical defects for the industrial VAR technology of titanium alloys".

6. References

Bellot, J.P. & Mitchell, A. (1994). Hard-alpha particle behaviour in a titanium alloy liquid pool. In: *Light Metals*, U. Mannweiler (Ed.), 1.187-1.193, The Minerals, Metals & Materials Society, ISBN 0-87339-264-7

Bellot, J.P. et al. (1997). Dissolution of Hard-Alpha Inclusions in Liquid Titanium Alloys, Metallurgical and Materials Transactions B, Vol.28b, pp. 1001 -1010, ISSN 1073-5615 (print version)

Bewlay, B.P. & Gigliotti, M.F.X. (1997). Dissolution rate measurements of TiN in Ti6242, Acta Mater., Vol.45, No. 1, pp. 357-370, ISSN 1359-6454

David, D. et al (1979). Etude de la diffusion de'oxygene dans le titane α oxyde enter 700°C et 950°C, J.Less-Comm. Met., Vol.65, No. 1, pp. 51-69, ISSN 0022-5088

Fromm, E. and Gebhardt, E. (1976). *Gase und kohlenstoff in metallen*, Springer-Verlag, ISBN 3540072551, Berlin, Heidelberg, New York

Grala, E. M. (1968). *Characterization of Alpha Segregation Defects in Ti-6Al-4V Alloy*, Technical Report AFML-TR-68-304, AD0845805, TRW, Inc. Cleveland, OH

Le Claire, A.D. & Neumann, G. (1990). In: *Diffusion in Solid Metals and Alloys*, Mehrer H. (Ed.), Vol.III-26, pp. 85- 212, Landolt-Börnstein, Springer-Verlag, ISBN3-540-50886-4, Berlin

Le Claire, A.D. (1990). In: *Diffusion in Solid Metals and Alloys*, Mehrer H. (Ed.), Vol.III-26, p. 471, Landolt-Börnstein, Springer-Verlag, ISBN3-540-50886-4, Berlin

Mitchell, A. (1984). Final Report to General Electric Corp., pp. 10-29, University of British Columbia, Vancouver, Canada

Reddy, R.G. (1990). Kinetics of TiN Dissolution in Ti Alloys. In: *Electron Beam Melting and Refining State of the Art,* R. Bakish (ed.), 119-127, NV: Bakish Materials, ISBN-10 9992384719, ISBN-13 978-9992384718, Reno

Schwartz, F. (1993). Technical Report YKOG, No 3028/93, pp. 2-7, SNECMA, Paris, France

Tarenkova, N.Yu. et al. (2011), Formation of Gas-Saturated Defects in Titanium Alloys during Vacuum-Arc Remelting, Russian Metallurgy (Metally), Vol.2011, No. 2, pp. 127–132, ISSN 0036-0295

Vykhodets, V.B. et al. (1987). Oxygen diffusion in α-Ti. I. Anisotropy of oxygen diffusion in α-Ti, The Phys. Metal & Metallogr., Vol.64, pp. 127-133, ISSN 0031-918X

Vykhodets, V.B. et al. (2006). Studying distribution of gaseous impurities and carbon in titanium alloys using nuclear microanalysis, The Physics of Metals and Metallography, 2006, V.101, No.3, pp. 267-275, ISSN 0031-918X

Vykhodets, V.B. et al. (2007). Genesis of Metallurgical Defects in Titanium Alloys, Doklady Physical Chemistry, Vol.416, part 2, pp. 285-288, ISSN PRINT 0012-5016, ISSN ONLINE 1608-3121

Vykhodets, V.B. et al (2011). Identification of the sources of gas containing metallurgical defects during the smelting of titanium alloys, Russian Journal of Nondestructive Testing, Vol.47, No. 3, pp. 176–188, ISSN 1061 8309

Wood, F. W. (1969). *Elimination of Low-Density Inclusions in Titanium Alloy Ingots,* Final technical report 1 Sep 67-30 Aug 68, AD0852028

Part 2

Properties of Titanium Alloys Under High Temperature and Ultra High Pressure Conditions

4

Titanium Alloys at Extreme Pressure Conditions

Nenad Velisavljevic[1], Simon MacLeod[2] and Hyunchae Cynn[3]
[1]Los Alamos National Laboratory
[2]Atomic Weapons Establishment
[3]Lawrence Livermore National Laboratory
[1,3]USA
[2]UK

1. Introduction

The electronic structures of the early transition metals are characterised by the relationship that exists between the occupied narrow *d* bands and the broad *sp* bands. Under pressure, the *sp* bands rise faster in energy, causing electrons to be transferred to the *d* bands (Gupta et al., 2008). This process is known as the *s-d* transition and it governs the structural properties of the transition metals.

At ambient conditions, pure Ti crystallizes in the 2-atom hcp, or α phase crystal structure (space group $P6_3/mmc$) and has an axial ratio (c/a) ~ 1.58. Under pressure, the α phase undergoes a martensitic transformation at room temperature (RT) into the 3-atom hexagonal, or ω phase structure (space group P6/mmm). The appearance of the ω phase at high pressure raises a number of scientific and engineering issues mainly because the ω phase appears to be fairly brittle compared with the α phase, and this may significantly limit the use of Ti in high pressure applications. Furthermore, after pressure treatment the ω phase appears to be fully, or at least, partially recoverable at ambient conditions, thus raising questions as to which is the lowest thermodynamically stable crystallographic phase of Ti at RT and pressure.

This chapter deals with the behavior of Ti alloys under extreme pressure and temperature conditions. Our recent results are presented and comparison is made to data available in the open literature. We volume compressed Ti-6Al-4V (an α alloy) and Ti-Beta-21S (a β alloy) in a series of RT diamond anvil cell (DAC) angle-dispersive X-ray diffraction (ADXD) experiments, to investigate the effects of alloying and the pressure environment on phase stability and the transformation pathway. However, before describing the results of these experiments in detail, we present a brief review of the current state of knowledge of Ti at high pressure.

2. Ti at high pressure

The phase relations of Ti have been studied extensively at high pressure. As indicated in the introduction, there is great interest in the properties of Ti at high pressure and temperature.

The RT $\alpha \rightarrow \omega$ phase transition has been observed to occur between 2 GPa and 12 GPa, depending on the experimental technique, the pressure environment, and the sample purity. Although most of the recent work has involved the use of diamond anvil cells (DACs) to volume compress Ti in the static regime (for example, Ming et al., 1981; Akahama et al., 2001; Vohra et al., 2001; Errandonea et al., 2005), many studies have also utilised large volume presses to statically compress Ti (Jamieson 1963; Jayaraman et al., 1963; Bundy 1963; Bundy 1965; Zhang et al., 2008), and to a lesser extent, shock techniques have been employed to compress Ti in the dynamic regime (see Gray et al., 1993; Trunin et al., 1999; Cerreta et al., 2006).

2.1 Dynamic compression of Ti

The effects oxygen and other interstitial impurities on ω phase formation in Ti has been studied using shock compression (Gray et al., 1993 & Cerreta et al., 2006). In these studies, samples with oxygen content 360 ppm (high purity Ti) and 3700 ppm (A-70) were shocked whilst simultaneously probed using a real-time velocity interferometer system for any reflector (VISAR) diagnostic and then analysed post-shock. A phase transition was reported at 10.4 GPa for the high purity sample, whereas the A-70 sample did not show any evidence of a phase change up to 35 GPa (Gray et al., 1993 & Cerreta et al., 2006). A post shock (11 GPa) recovered high purity sample retained 28% of the ω phase (Gray et al., 1993). The suppression of the $\alpha \rightarrow \omega$ stress transformation in Ti is likely caused by the presence of interstitial oxygen (Cerreta et al., 2006). Greef et al. performed an analysis of early Ti shock measurements, but were unable to allow for sample purity in their study, since the oxygen content was not measured in these early experiments, and as a consequence placed the $\alpha \rightarrow \omega$ phase transition at ~ 12 GPa (Greef et al., 2001).

2.2 Static compression of Ti

2.2.1 Room temperature compression

Using DACs and X-ray diffraction, the experimental technique RT $\alpha \rightarrow \omega$ phase transition in commercially available purity Ti has been observed to occur between 2 GPa and 12 GPa - in cases where data was collected during pressure release large transformation pressure hysteresis is observed and in some cases the ω phase can be recovered at ambient conditions (Vohra et al., 1977; Ming et al., 1981; Vohra et al., 2001; Akahama et al., 2001; Errandonea et al., 2005). The effect of the pressure environment (and therefore uniaxial stress) on the $\alpha \rightarrow \omega$ transformation in Ti was studied in a series of DAC and angle dispersive X-ray diffraction (ADXD) experiments (Errandonea et al., 2005). By embedding Ti samples in a variety of pressure-transmitting-media (PTM), Errandonea et al were able to demonstrate that the pressure at which the Ti transformed into the ω phase was found to increase with an increase in the hydrostaticity of the PTM. For an argon PTM (the most hydrostatic PTM used in the experiment), the $\alpha \rightarrow \omega$ transition occurred in the pressure range between 10.5 GPa and 14.9 GPa, and for the least hydrostatic environment (that is, no PTM), the $\alpha \rightarrow \omega$ transition occurred under pressure between 4.9 GPa and 12.4 GPa (Errandonea et al., 2005). A coexistence of the α and ω phases over a largish pressure range was also observed, in agreement with earlier findings (Ming et al., 1981).

Vohra et al. investigated the effects of the impurity levels of oxygen on the $\alpha \rightarrow \omega$ transition (Vohra et al., 1977). In this non-hydrostatic DAC study (no PTM), the oxygen content of the Ti samples was varied between 785 ppm and 3800 ppm (by weight) and the corresponding transition pressure was measured between 2.9 GPa and 6.0 GPa.

Ti has been volume compressed at RT (no PTM) in a DAC to a pressure of 220 GPa and found to follow the transformation pathway $\alpha \rightarrow \omega \rightarrow \gamma \rightarrow \delta$ (Vohra & Spencer, 2001; Akahama et al., 2001). The ω phase structure was observed to be stable to pressures greater than 100 GPa. The reported intermediate orthorhombic γ phase (transforming between ~ 116 GPa - 128 GPa) possessed a distorted hcp crystal structure (space group Cmcm), and the orthorhombic δ phase (transforming at ~ 145 GPa) was of a distorted bcc type structure (space group Cmcm) (Akahama et al., 2001). These observations for Ti are at variance with the other Titanium Group transition metals Zr and Hf, which have been observed to follow the $\alpha \rightarrow \omega \rightarrow \beta$ pathway at RT (Jayaraman et al., 1963; Xia et al., 1990a; Xia et al., 1990b). However, Ahuja et al. have observed the coexistence of a bcc-like structure (referred to by the authors as β') with the ω phase, during the compression of Ti at RT from 40 GPa to 80 GPa (Ahuja et al., 2004). In this study, Ti was embedded in a NaCl PTM. The subsequent ADXD patterns could only be fully analysed on the assumption that the additional reflections present in the patterns belonged to the β phase of Ti. Laser heating Ti to between 1200 K and 1300 K at 78-80 GPa resulted in the formation of a new orthorhombic η-phase (space group Fmmm), which on decompression, at RT, transformed back into the β' phase below 40 GPa (Ahuja et al., 2004). On further decompression below 30 GPa, the β' phase reverted back into the ω phase, and this phase could be quenched to ambient conditions (Ahuja et al., 2004).

Most recently, nano-Ti sample (with grain sizes ~ 100 nm) was compressed at RT in a DAC to 161 GPa and the transformation pathway $\alpha \rightarrow \omega \rightarrow \gamma \rightarrow \delta$ was observed (Velisavljevic et al., n.d.). The slightly high $\alpha \rightarrow \omega$ transition pressure of 10 GPa observed (no PTM), may have been caused by the increase in interface to volume ratio, resulting from the reduction in grain size, and leading to increased resistance to shear deformation. Compared to coarse grained Ti, in nano-Ti there may also have been a larger concentration of interstitial impurities near the grain boundaries, which have been shown to help suppress the $\alpha \rightarrow \omega$ structural phase transition (see Hennig et al., 2005). The ω phase was observed to be stable up to ~ 120 GPa, and under compression to 127 GPa resulted in a phase transformation to the orthorhombic γ phase. A further compression resulted in a transition from γ to the δ phase at 140 GPa, in good agreement with a previous study (Akahama et al., 2001). The δ phase was stable up to 161 GPa. Figure 1 shows a stacked plot of nano-Ti ADXD patterns in the ω, γ and δ phases (Velisavljevic et al., n.d.). The metastable ω phase was recovered after pressure treatment, which is consistent with reports from other experiments on Ti, indicating that after pressure release, samples were either recovered in the ω phase or as a mixture of α+ω (see Errandonea et al., 2005 & Vohra et al., 2001). In the case of these nano-Ti experiments, the recovered sample was observed to consist of only the ω phase (Velisavljevic, n.d.). The high pressure behavior of nano-Ti, including the $\alpha \rightarrow \omega \rightarrow \gamma \rightarrow \delta$ structural phase sequence (without the appearance of the β phase), the change in axial c/a ratio with pressure, recovery of ω phase, and change in volume with pressure and the EOS values (Velisavljevic, n.d.), is very similar and consistent with previous experimental results on Ti (Akahama et al., 2001; Vohra et al., 2001; Errandonea et al., 2005;).

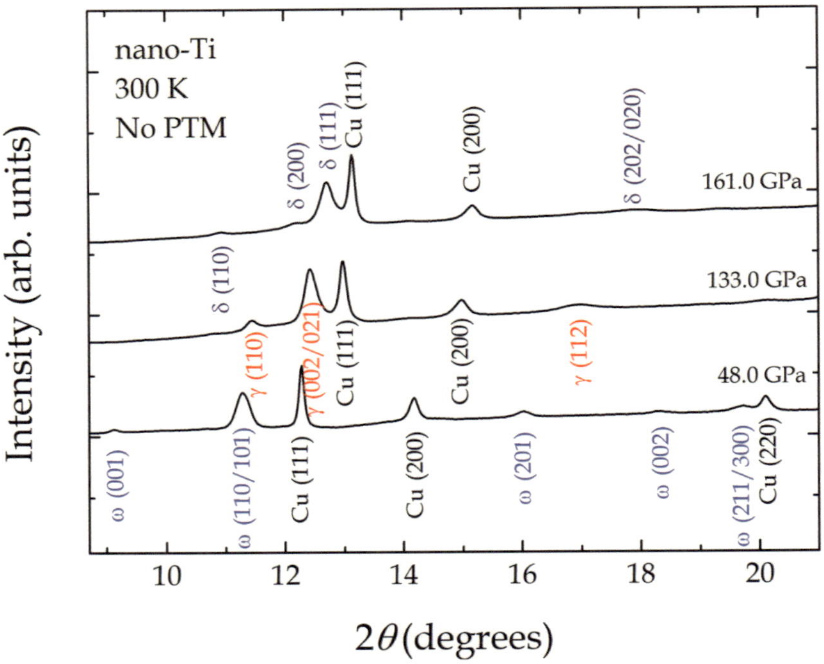

Fig. 1. A stack of ADXD patterns showing pressure induced structural phase transformation $\omega \rightarrow \gamma \rightarrow \delta$ in nano-Ti (Velisavljevic et al., n.d.).

2.2.2 High temperature compression

Very few combined static high-pressure and high-temperature studies have been reported for Ti. Thermally treating Ti to temperatures exceeding 1155 K at room pressure (RP) transforms the α phase into the more densely packed bcc or β phase (space group Im3m), without the intermediate ω or other crystallographic phase occurring.

Jayaraman et al. studied the α → β boundary up to ~ 6.5 GPa and 1100 K, and found that increasing the pressure lowered the α → β transition temperature (Jayaraman et al., 1963). A more extensive study conducted by Bundy, in which Ti was compressed to 16 GPa and heated to 1200 K revealed the location of the boundaries of the three solid phases, α, β and ω (Bundy, 1963 & Bundy, 1965). More recently, Ti was compressed in a cubic anvil apparatus to 8.7 GPa and heated to 973 K (Zhang et al., 2008). The ADXD patterns collected confirmed the high-temperature phase diagram reported by Bundy. The α-ω-β triple point was estimated at 7.5 GPa and 913 K (Zhang et al., 2008), in agreement with the previous estimate of 8.0 ± 0.7 GPa and 913 ± 50 K (Bundy, 1965). Zhang et al. observed the α-phase to undergo an isotropic compression between RP and 7.8 GPa, resulting in the axial ratio (c/a) being constant (1.587) over this pressure range (Zhang et al., 2008).

Errandonea et al. melted Ti in a DAC up to 80 GPa using single-sided laser-heating and the "speckle technique" to determine the onset of melting (Errandonea et al., 2001). No X-ray diffraction patterns were collected in this lab-based study, and so the authors were unable to discriminate between melting from either the β phase or the ω phase. Short term laser-heating appeared to lower the $\alpha \rightarrow \omega$ transition pressure in pure T, as was determined in a follow up DAC study conducted by Errandonea at a synchrotron (Errandonea et al., 2005). At 5 GPa, Ti was transformed into the β phase by laser-heating to 1750 K and 2150 K (above melting), at which point the samples were quenched. A mixture of α and ω phases were obtained at a pressure for which only the α phase existed previously at RT, suggesting that thermal fluctuations may have a similar effect on the $\alpha \rightarrow \omega$ transformation as uniaxial stress.

2.3 Theoretical treatments of Ti at high pressure

Although Ti has received much theoretical attention over the years, we will mention here only studies that are of direct relevance to this chapter. First-principles calculations have been used to generate the phase diagram of Ti (Trinkle et al., 2003; Pecker et al., 2005; Hennig et al., 2005; Trinkle et al., 2005; Verma et al., 2007; Hennig et al., 2008; Mei et al., 2009; Hu et al., 2010).

A multiphase equation of state (EOS) of the three solid phases (α, β and ω), the liquid phase and gas phase, was calculated up to 100 GPa and predicted (based on experiment) the ω-β-liquid triple point at around 45 GPa and 2200 K (Pecker et al., 2005). Verma et al. predicted the $\omega \rightarrow \gamma \rightarrow \beta$ pathway for Ti, and found the δ phase to be energetically unstable under hydrostatic conditions (Verma et al., 2007). The γ phase was found to be elastically stable between 102 GPa and 112 GPa. However, under non-hydrostatic conditions, the authors predicted the γ phase to exist over a larger pressure range (Verma et al., 2007). The influence of anisotropic stresses under very non-hydrostatic conditions may support the existence of the δ phase reported in DAC experiments (Akahama et al., 2001; Vohra et al., 2001; Velisavljevic, n.d.).

The actual mechanism behind the martensitic transformation between the α, β and ω phases in Ti has been explored in a series of molecular dynamics simulations (Trinkle et al., 2003; Trinkle et al., 2005; Hennig et al., 2005; Hennig et al., 2008). The authors propose the lowest energy pathway for the $\alpha \rightarrow \omega$ transition to be the TAO-1, ("titanium alpha to omega"), in which atoms in the α phase transform through small shuffles and strains into the ω phase, without going through a metastable intermediate phase (Trinkle et al, 2003 & Trinkle et al., 2005). The presence of impurities in Ti such as O, N and C can affect the energy barrier to the ω phase and suppress the $\alpha \rightarrow \omega$ transformation (Hennig et al., 2005). Hennig et al. predicted the phase boundaries for Ti up to 15 GPa, and the α-β-ω triple point at 8 GPa and 1200 K. (Hennig et al., 2008).

Mei et al. studied the thermodynamic properties and phase diagram of Ti and predicted the RP $\alpha \rightarrow \beta$ transition at 1114 K and the α–ω–β triple point at 11.1 GPa and 821 K (Mei et al., 2009). The $\alpha \rightarrow \omega$ transition was predicted at 1.8 GPa, slightly lower than experimental measurements. Hu et al. performed a detailed calculation to predict the phase diagram, thermal EOS and thermodynamic properties of Ti (Hu et al., 2010). The axial ratio of the α phase was predicted to be almost invariant with pressure, in agreement with the anvil study of Zhang et al., and with calculation (Zhang et al., 2008 & Mei et al., 2009), but not with a DAC study in which the compression was found to be anisotropic (Errandonea et al., 2005). Hu et al. calculated the RT $\alpha \rightarrow \omega$ transition at 2.02 GPa and the triple point at 9.78 GPa and

931 K, which is in close agreement with experiment (Bundy 1965 & Zhang et al., 2008). The slope of the α-ω boundary (dT/dP = 81 K/GPa) differs significantly to that measured by Zhang et al. in their cubic anvil study (dT/dP = 345 K/GPa) (Zhang et al., 2008). The slope of the ω-β boundary was calculated to be 2.4 K/GPa (Hu et al., 2010), in good agreement with earlier measurements (Bundy, 1965). The predicted RT ω → γ transformation at 110 GPa concurred with the calculation of Verma et al. (Verma et al., 2007).

For reference, we show in figure 2 a representation of the Ti phase diagram as a function of pressure and temperature, based loosely on the experimental and theoretical studies discussed in this review.

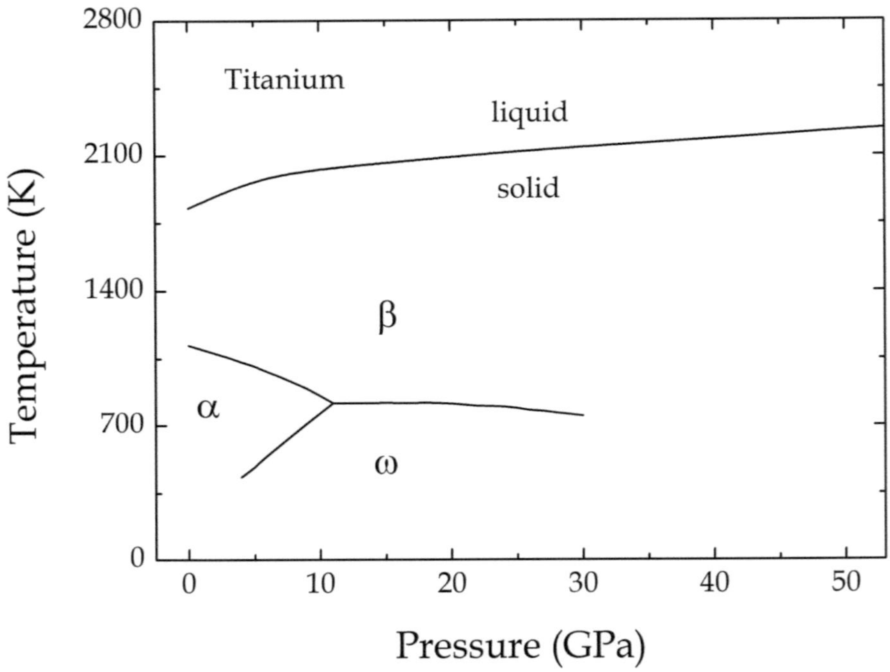

Fig. 2. The *P-T* phase diagram of Ti (this is a representation based on published work).

3. Ti alloys at high pressure

Ti alloys are usually classified according to their phase stability. An α alloy consists mainly of an α-phase stabilizing element such as Al, O, N or C, which has the effect of extending the range of the more ductile α-phase field to higher temperatures and higher pressures. Similarly, a β alloy contains a β-phase stabilizing element such as Mo, V or Ta, the presence of which will shift the β phase field to lower temperatures. Ti alloys containing a combination of both α-phase and β-phase stabilizing elements are by far the most widely used alloys commercially. At ambient conditions, the α+β alloys possess a β-phase fraction by volume that lies somewhere between 5 and 50% and so crystallizes predominantly in the α phase. Of particular importance is the α+β alloy Ti-6Al-4V (wt%), which has found many commercial applications as a result of its superior material properties. The phase stability in

Ti alloys is governed by the presence of impurities, and for a ternary alloy such as Ti-6Al-4V, the substitutional impurities, Al (which is an α-phase stabilizer) and V (a β-phase stabilizer), influence the onset of the α → ω phase transformation by changing the *d* electron concentration in the alloy (Vohra, 1979). The addition of Al reduces the *d* band concentration, whilst the addition of V increases it by one, thus resulting in an overall reduction in the *d* band concentration. Interstitial impurities such as O, N and C can retard the α → ω transformation. *Ab initio* calculations have shown that the presence of these impurities can affect the relative phase stability and the energy barrier of the phase transformation (Hennig et al., 2005). The presence of impurities in the commercial Ti alloys A-70 and Ti-6Al-4V, particularly O and Al, suppresses the onset of α → ω phase transformation by increasing the energy and energy barrier of ω relative to α (Hennig et al., 2005). Thus, the stability range of the α phase at RT is increased. Hennig et al. predicted the RT α → ω phase transformations in A-70 and Ti-6Al-4V to occur at 31 GPa and 63 GPa respectively (Hennig et al., 2005).

3.1 Ti-6Al-4V at high pressure

As the most prevalent Ti alloy currently in commercial and industrial usage, it is perhaps not surprising that Ti-6Al-4V has received the most attention of all the Ti alloys at high pressure (Rosenberg & Meybar, 1981; Gray et al., 1993; Cerreta et al., 2006; Chesnut et al., 2008; Halevy et al., 2010; Tegner et al., 2011; Tegner et al., n.d). The alloying of a metal can have substantial effects on its properties, which of course is the desired effect, and so it is important that we understand how the alloy responds to extremes of pressure.

3.1.1 Dynamic compression of Ti-6Al-4V

In the dynamic regime, the Hugoniot curve of Ti-6Al-4V was generated up to 14 GPa using powder-gun driven shock waves and manganin stress gauges (Rosenberg & Meybar, 1981). A break in the stress-particle velocity curve near 10 GPa was indicative, the authors proposed, of a possible phase transformation, though they were unable to state unequivocally that the transformation was α → ω (Rosenberg & Meybar, 1981). As part of a study to examine the effects of alloy chemistry on ω phase formation in Ti alloys, Gray et al., shocked Ti-6Al-4V (oxygen content 0.18 by weight %) up to 25 GPa and used VISAR to measure wave profiles (Gray et al., 1993). No evidence of a α → ω phase transformation was detected using VISAR, and using neutron diffraction to analyse the recovered specimen did not reveal the presence of ω-phase structure.

3.1.2 Static compression of Ti-6Al-4V

The first study of Ti-6Al-4V in a DAC was reported by Chesnut et al., in which a sample was loaded into a 4:1 methanol: ethanol PTM and compressed to 37 GPa (Chesnut et al., 2008). The ambient conditions volume of Ti-6Al-4V (predominantly in the α phase) was measured using ADXD and found to be V_0 = 17.208 Å^3/atom. Chesnut et al. observed Ti-6Al-4V to undergo the α → ω phase transition at ~ 27.3 GPa (Chesnut et al., 2008). The ω phase was observed to be stable to 37 GPa (the pressure limit of the experiment). The volume change across the phase boundary, at around 1%, was considered too small to detect in shock experiments and may explain why this transformation has yet to be observed in dynamically driven Ti-6Al-4V (Rosenberg & Meybar, 1981 & Gray et al., 1993). Fitting a 3rd order Birch-Murnaghan EOS (Birch 1952) to the data generated an isothermal bulk modulus (which is a measure of the incompressibility of a material) for the α phase of $K = 125.24$ GPa and the pressure derivative of the isothermal bulk modulus, $K' = 2.409$.

Halevy et al. compressed a sample of Ti-6Al-4V to 32.4 GPa in a DAC, and using energy dispersive X-ray diffraction (EDXD), did not observe a transformation to the ω phase (Halevy et al., 2010). A Vinet EOS (Vinet et al., 1987) fit to the experimental data returned values for the isothermal bulk modulus and the pressure derivative of $K = 154 \pm 11$ GPa and $K' = 5.4 \pm 1.4$. No mention was made of a PTM being used in this study.

3.1.2.1 Effects of pressure media on the phase relations of Ti-6Al-4V

In the most recent DAC study of Ti-6Al-4V, conducted by two of us (MacLeod and Cynn), powdered polycrystalline samples were embedded in a variety of PTMs to investigate the effects of the pressure environment on the RT α → ω phase transformation (Tegner et al., 2011; Tegner et al., n.d.). ADXD data were collected at the High Pressure Collaborative Access Team (HP-CAT) beamline 16-IDB at the Advanced Photon Source (APS) in Chicago, for Ti-6Al-4V samples embedded in neon, 4:1 methanol: ethanol and mineral oil. The oxygen content of the Ti-6Al-4V was 0.123 by weight %. The ambient conditions volume in the α phase was measured to be V_0 = 17.252 Å^3/atom. We observed the α → ω phase transformation to occur at 32.7 GPa for Ti-6Al-4V in the neon PTM, 31.2 GPa for the methanol: ethanol PTM and 26.2 GPa for the mineral oil PTM (in order of decreasing hydrostaticity in the PTM). At elevated pressures, ultimately all PTMs will become non-hydrostatic in nature (see Klotz et al., 2009, for a general discussion on the hydrostatic limits of various pressure media) and so it becomes more difficult to quantify the dependence of the α → ω phase transformation pressure in Ti-6Al-4V based on the hydrostaticity of the pressure environment, unlike in Ti where the transition is observed at a much lower pressure (see Errandonea et al., 2005).

A coexistence of the α and ω phases was observed over a largish pressure range (of the order ~ 10 GPa or greater) for the different PTMs, similar to what was observed for Ti. A stacked plot of ADXD patterns, showing the structural response of Ti-6Al-4V to applied pressure, in a neon PTM, is shown in figure 3. Reflections from the Ti-6Al-4V sample in both α and ω phases are present, together with reflections from neon (PTM) and copper (the pressure marker).

In figure 3, at 30.7 GPa, we observe the (100), (002), (101), (102) and (110) peaks that are characteristic of the α phase. The dominant (110/101) reflections corresponding to the ω phase appear at ~ 32.7 GPa (between the α phase (002) and (101) peaks) and then gradually grow in magnitude with pressure, whilst simultaneously the α phase peaks diminish in intensity, until the pressure reaches ~ 44 GPa. By 44 GPa, the α → ω transformation is virtually completed. The ω phase (001), (201) and (210) reflections appear at a slightly higher pressure than the (110/101) peaks, at around 36 GPa to 39 GPa. In all, up to 10 ω phase peaks were indexed in this study. It is clear from figure 3 that both the α and ω phases coexist over a large pressure range, of the order of 10 GPa (between ~ 32.7 GPa and ~ 44 GPa). We observed similar behaviour for Ti-6Al-4V embedded in methanol: ethanol and mineral oil, and also for an experiment with no PTM (Tegner et al., 2011). For Ti-6Al-4V embedded in methanol: ethanol, we decompressed our DAC from 75 GPa to ambient and observed the ω phase to gradually revert back to the α phase.

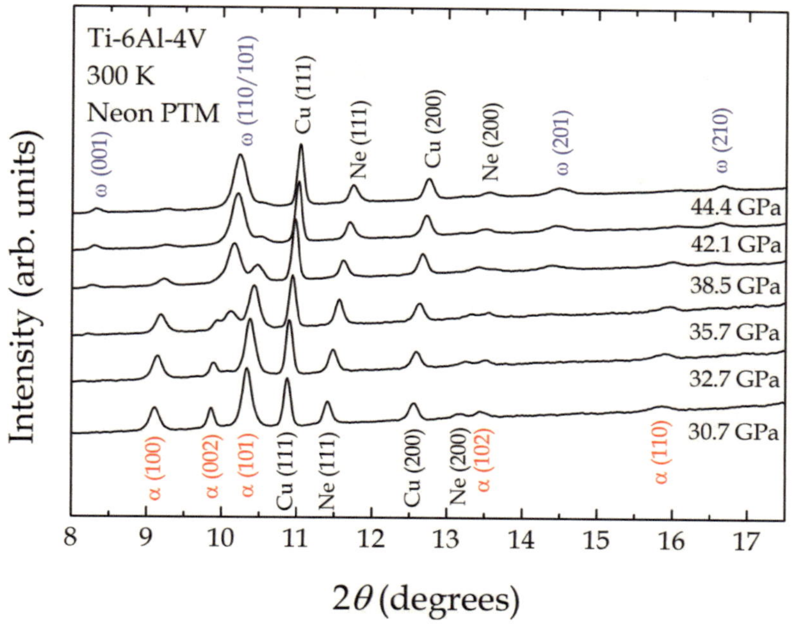

Fig. 3. A stack of ADXD patterns showing structural change in Ti-6Al-4V with increasing pressure (Tegnet et al., 2011).

We determined the pressure in our experiments by analysing the reflections from the pressure markers (either Ta or Cu) and using a known EOS from previous shock measurements. In the case of the Cu marker used in the compression of Ti-6Al-4V in a neon PTM (figure 3), we used a well known Cu shock study (Carter et al., 1971).

We now show in the *P-V* plot in figure 4 our measurements for Ti-6Al-4V embedded in a methanol: ethanol PTM, alongside previous DAC measurements (Chesnut et al., 2008 & Halevy et al., 2010). There is good agreement between the Chesnut et al. and our (Tegner et al., n.d.) measurements (both using a methanol: ethanol PTM), but not with the Halevy et al measurements (there was no mention of a PTM being used in their study). We measured the $P_{\alpha\to\omega}$ = 31.2 GPa, and an isothermal bulk modulus of K = 115 ± 3 GPa and pressure derivative of K' = 3.22 ± 0.22 after fitting a Vinet EOS (Vinet et al., 1987) to the data (Tegner et al., 2011). The volume change across the α–ω phase boundary was measured to be less than 1%, in agreement with the previous methanol: ethanol study (Chesnut et al., 2008).

The axial ratio (c/a) for the α phase of Ti-6Al-4V (at ambient conditions) was measured to be 1.602, which is slightly higher than that reported for pure Ti (1.587), but is an expected result due to the presence of the α stabiliser Al (possessing a smaller atomic radius than Ti) in the alloy.

For the α phase of Ti-6Al-4V, in a methanol: ethanol PTM, we found the axial ratio to be almost constant between 1.600 and 1.602 up to 42 GPa, see figure 5. The measured c/a ratios for the ω phase are also effectively constant between 34 GPa and 74 GPa, varying between 0.616 and 0.617. We find similar results for a loading of Ti-6Al-4V in a neon PTM. These measurements are in broad agreement with those reported by Errandonea et al. for pure Ti

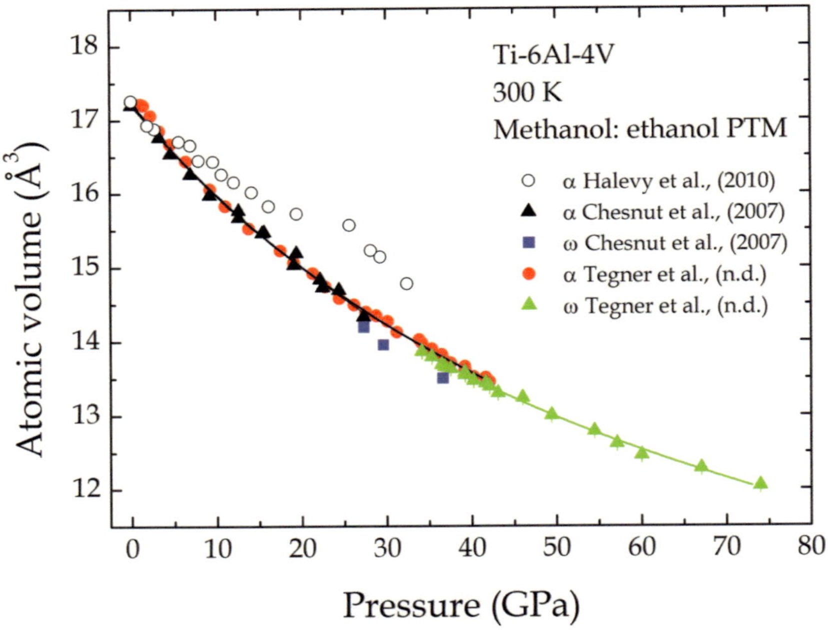

Fig. 4. The *P-V* plot for Ti-6Al-4V in a methanol: ethanol PTM (Chesnut et al., 2008 & Tegner et al., 2011) and also for an unspecified PTM (Halevy et al., 2010).

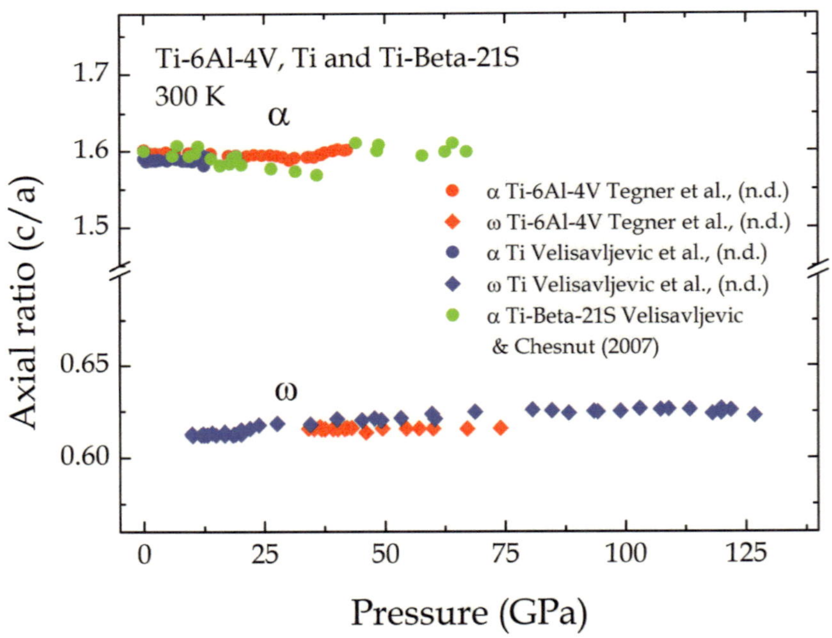

Fig. 5. The axial ratios (*c/a*) for the α and ω phases of Ti-6Al-4V in a methanol: ethanol PTM (Tegner et al., n.d.), Ti with no PTM (Velisavljevic et al., n.d.) and Ti-Beta-21S in a methanol: ethanol PTM (Velisavljevic & Chesnut, 2007).

(Errandonea et al., 2005). We also include for reference in figure 5 our axial ratio results for nano-Ti with no PTM (Velisavljevic et al., n.d.) and Ti-Beta-21S in a methanol: ethanol PTM (Velisavljevic & Chesnut, 2007) (see section 3.2). The *c/a* ratio for nano-Ti in the ω phase had a steady value of 0.612 initially, and as the pressure was increased above 20 GPa, this value increased slightly to 0.626 and then levelled off above 80 GPa. Under compression, the ω phase of Ti-6Al-4V in a methanol: ethanol PTM was observed to be stable to ~ 115 GPa (the pressure limit in this experiment).

We observed, in further RT volume compression experiments of Ti-6Al-4V embedded in neon and mineral oil PTMs, a gradual transformation from the ω phase to the body-centred-cubic β phase (space group Im3m) (Tegner et al., 2010 & Tegner et al., n.d.). For both neon and mineral oil PTMs, the transformation is completed between 115 GPa and 125 GPa. The β phase is formed by the splitting of the alternating (001) plane along the *c* axis of the ω phase into two (111) planes of the β phase (Xia et al., 1990a). All the β phase peaks are therefore contained in the ω diffraction pattern (that is, the peaks are coincident). With no detectable volume change from ω → β, it was not possible to ascertain over what pressure range both phases coexisted. The disappearance of the ω phase (001), (210) and (002) peaks was the only indication that a solid-solid phase transformation had occurred. We observed the β phase of Ti-6Al-4V to be stable to at least 221 GPa (the pressure limit of the experiment) in a mineral oil PTM (Tegner et al., 2011) and stable to 130 GPa (the pressure limit of the experiment) in the neon PTM (Tegner et al., n.d.).

The *P-V* plot for Ti-6Al-4V in a mineral oil PTM is shown in figure 6. No intermediate phases were observed in this experiment. For comparison, we include the *P-V* data for nano-Ti (Velisavljevic, n.d.), Ti-6Al-4V (Chesnut & Velisavljevic, 2008) and (see section 3) Ti-Beta-21S (Velisavljevic & Chesnut, 2007). Verma et al. proposed that the observations of the δ phase in Ti DAC experiments were likely caused by the presence of non-hydrostatic stresses and that the transition sequence ω → γ → β is thermodynamically preferable, with the intermediate γ phase existing over a range of at least 10 GPa (Verma et al., 2007). In our Ti-6Al-4V study, we collected ADXD patterns with pressure steps ~ 5 GPa and found no evidence for an intermediate γ phase in Ti-6Al-4V. Nor was there any evidence for an orthorhombic δ phase. The Ti-6Al-4V sample was embedded in a PTM (albeit the very non-hydrostatic mineral oil and neon PTMs above 100 GPa). Our own calculations predict the transformation pathway for Ti-6Al-4V to be α → ω → β, with α → ω occurring at 24 GPa and ω → β at ~ 105 GPa, in agreement with experiment (Tegner et al., 2010). Above 110 GPa we find that the orthorhombic γ and δ phases relax to the cubic β phase under hydrostatic conditions. As far as we know, all Ti DAC experiments compressed above 100 GPa have had samples compressed in the absence of a PTM (Vohra et al., 2001; Akahama et al., 2001; Velisavljevic et al., n.d.).

In figure 7, an integrated ADXD pattern is shown of Ti-6Al-4V in a neon PTM, collected at 129 GPa (Tegner et al., n.d.). The peaks are indexed as the β-phase reflections (110), (200), (211) and (220) and face-centred cubic (fcc) neon (111), (002) and (220). The pressures were calculated using a previous static high pressure EOS study of neon (Dewaele et al., 2008). We performed a Rietveld analysis of our diffraction patterns to confirm the crystal structure of Ti-6Al-4V to be the β phase.

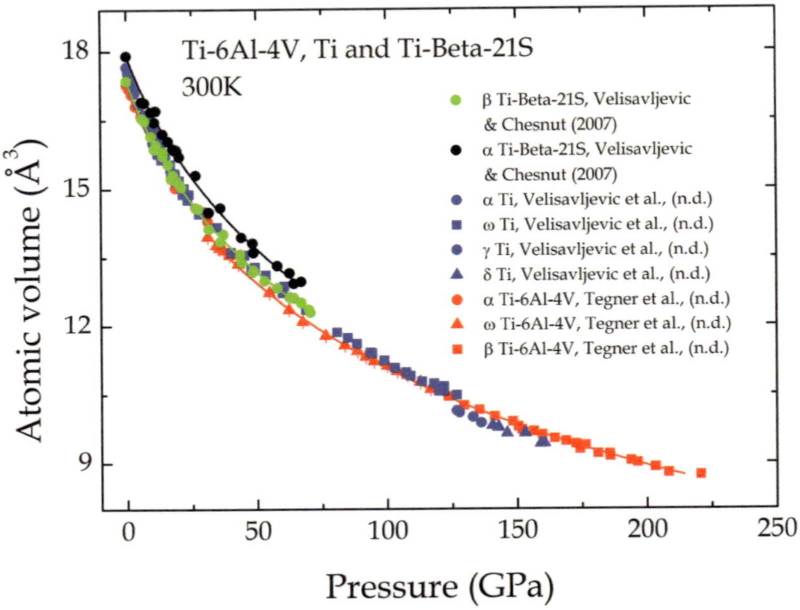

Fig. 6. The *P-V* plot for Ti-6Al-4V in a mineral oil PTM (Tegner et al., n.d.), Ti with no PTM (Velisavljevic et al., n.d.) and Ti-Beta-21S in a methanol: ethanol PTM (Velisavljevic & Chesnut, 2007).

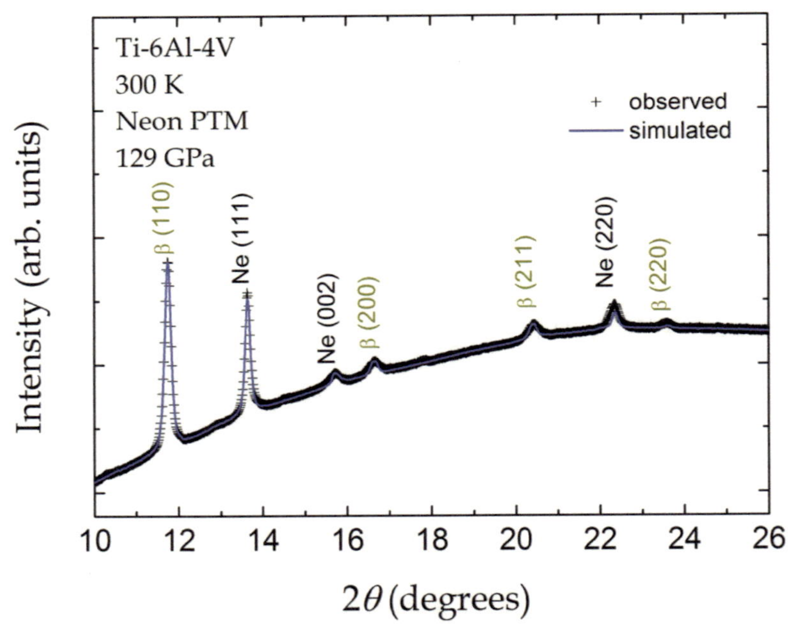

Fig. 7. An integrated ADXD pattern and Rietveld analysis for β (bcc) phase Ti-6Al-4V at 129 GPa and embedded in a neon PTM (Tegner et al., n.d.). The neon was used as both a PTM and pressure marker.

The observation of the α → ω → β transformation pathway at RT for Ti-6Al-4V suggests that the slope of the ω-β phase boundary (see boundary in figure 2) is negative at high pressures, or that there are two separate areas of β phase separated by the ω phase (see for example, Xia, 1990a).

3.2 Absence of the ω phase in the Ti-beta-21S alloy at high pressure

In cases where a sufficient amount of alloying elements are introduced, known as Mo equivalent (Mo_{eq}), alloys with up to 50% β phase can be recovered after temperature treatment (Bania, 1993). The large β phase concentration can have a significant effect on the pseudo-elastic response (Zhou et al., 2004), while also effecting structural phase stability at high pressures. One example is the Mo rich Ti-Beta-21S (also known as TIMETAL21S®) alloy, which is a β-stabilized alloy with 15 wt.% Mo, 3 wt.% Al, 2.7 wt.% Nb, 0.2 wt% Si, and the remainder made up by Ti. The standard "solution-treated-and-aged" (STA) heat treatment, in which the samples were heated to 1098 K, held for 30 minutes, cooled to RT (air-cooled equivalent rate), subsequently heated to 828 K, held for 8 hours, and again cooled to RT, produced a sample resulting in a mixture of α+β phases, with 29% - 42% in the α phase (Velisavljevic & Chesnut, 2007; Honnell et al., 2007).

An initial ADXD pattern collected at ambient conditions clearly shows the sample is composed of both the α and β phases (Velisavljevic & Chesnut, 2007), as shown in figure 8. A close examination of the diffraction pattern does not indicate the existence of any peaks that could not be attributed to either the α or β phase, and in particular, there is no evidence of ω phase in the sample. Cold compression of the sample in a methanol: ethanol PTM, in a DAC, shows very little change at low pressure. Up to ~ 11 GPa both α and β phases appear to be stable, including the axial ratio of the α phase, which remains fairly constant near the initial value of $c_0/a_0 = 1.601$. Comparison of the axial ratio of the α phase of Ti-Beta-21S and the c_0/a_0 value of 1.602 reported for the α phase of Ti-6Al-4V indicates that the inclusion of Mo, which has a larger atomic radius than Ti, does not have a significant effect on the initial axial ratio. However, the measured volume $V_0 = 17.912$ Å^3/atom ($a_0 = 2.956 \pm 0.002$ Å and $c_0 = 4.734 \pm 0.025$ Å) for the α phase of Ti-Beta-21S is larger than both the $V_0 = 17.355$ Å^3/atom ($a_0 = 3.262 \pm 0.004$ Å) measured for the β phase of Ti-Beta-21S and values reported for Ti-6Al-4V, which would suggest that a significant amount of alloying elements are still present in the α phase portions of the sample. Above 11 GPa, anisotropic compression of the α phase is observed, which leads to a steady decrease in the c/a ratio down to 1.568 at 36 GPa, as shown in figure 5. A sudden increase in the axial ratio up to 1.613 at 44 GPa, followed by a steady value up to 67 GPa is then observed. Over this same pressure range, besides a steady decrease in volume, no significant changes are observed for the β phase. The sudden change in the axial ratio in the 36-44 GPa region could denote a potential first order isostructural phase transition. Over the same pressure region of 36-44 GPa, no evidence of appearance of the ω phase or any other new phases could be detected.

With pressure increased to 58 GPa, the sample remains stable, as a mixture of α+β phases. However, above this pressure a relative change in the measured peak intensities between the two phases indicates the onset of a structural phase transition (Velisavljevic & Chesnut, 2007). A comparison of the intensity change of the (102) α-phase peak and the

(200) β-phase peak indicates a steady disappearance of the α phase and a transition of the sample to 100% β phase above 67 GPa. The ADXD patterns collected at 18 GPa and 48 GPa in figure 8 show the mixture of the α+β phases. At 71 GPa in figure 8, only β phase peaks are now present. From the experimental data, a Vinet EOS fit returned values of K = 119 GPa and K' = 3.4 for the α phase and K = 109 GPa and K' = 3.8 for the β phase (Honnell et al., 2007), and for a Birch-Murnaghan EOS fit, values of K = 117 GPa and K' = 3.4 for the α phase and K = 110 GPa and K' = 3.7 for the β phase (Velisavljevic & Chesnut, 2007) were obtained. Although, as previously mentioned, there is a volume difference observed between the α phase of Ti-Beta-21S and Ti-6Al-4V, the overall compressibility and EOS of Ti-Beta-21S are in good agreement with the various EOS values generated for Ti-6Al-4V, as shown in figure 6.

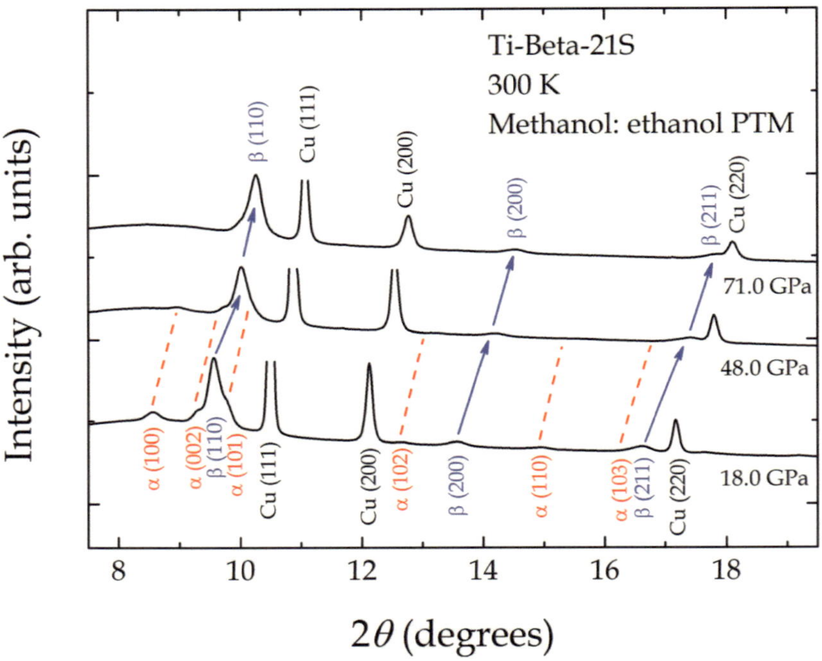

Fig. 8. A stack of ADXD patterns showing structural change in Ti-Beta-21S with increasing pressure. Initially sample is composed of mixture α+β phase and with pressure increase sample transforms completely to β phase – ω or other intermediate phases were not observed at any point. Additional peaks in ADXD spectra belong to Cu, which was used as a pressure marker.

For data collection, and the values reported here for Ti-6Al-4V and Ti-beta-21S, synchrotron source monochromatic X-ray beams were used. The image plate detectors available at the synchrotrons had pixel sizes of 100 μm² and so the diffraction patterns were generated with a resolution ~Δd/d = 10^{-3}. As a consequence, the uncertainties in the measured volume data were of the order of ~0.3%. This is consistent with the different values reported for the volume of Ti-6Al-4V at ambient conditions by the various authors. For example, Tegner et al. reported V_0 = 17.252 Å³/atom (Tegner et al., 2010) whereas

Chesnut & Velisavljevic measured a 0.25% smaller value, $V_0 = 17.208$ Å^3/atom (Chesnut & Velisavljevic, 2007).

4. Discussion

A *P-T* phase diagram for Ti (Errandonea et al., 2001; Errandonea et al., 2005; Pecker et al., 2005; Zhang et al., 2008; Mei et al, 2009; Hu et al., 2010) and Ti-6Al-4V (Chesnut et al., 2008, Tegner et al., 2010) summarising the current state of knowledge of the phase relations of these systems up to 125 GPa is shown in figure 9. There is good agreement between experiment and theory for the location of the Ti α–ω phase boundary, and also the melt curve, but the location and slope of the ω–β boundary is still in dispute and requires more study for clarification. Phase stability and the effects of anisotropic stresses on the α → ω transition in Ti is an issue that also requires more attention at high temperature. Very little is known about Ti alloys at high pressure, and even less at high pressure and high temperature. In figure 9, we suggest possible phase boundaries for Ti-6Al-4V as a dashed blue line. We are unsure about the exact location, or slope even, of the α–ω and α– β boundary for Ti-6Al-4V at high temperature.

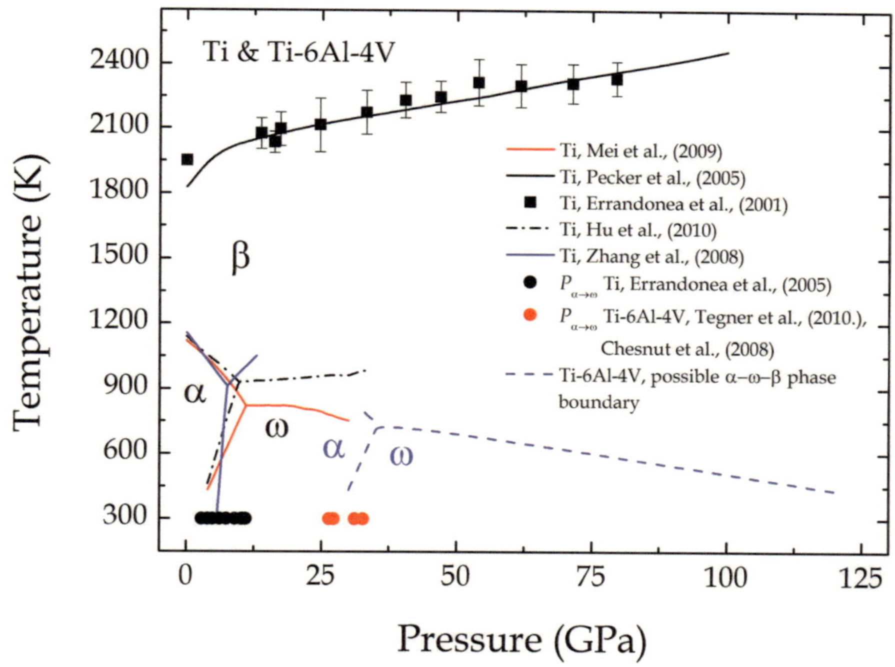

Fig. 9. The combined *P-T* phase diagram of Ti (Errandonea et al., 2001; Errandonea et al., 2005; Pecker et al., 2005; Zhang et al., 2008; Mei et al, 2009; Hu et al., 2010) and Ti-6Al-4V (Chesnut et al., 2008, Tegner et al., 2010). The possible phase boundaries for the α+β alloy Ti-6Al-4V (wt%) are suggested by us as the dashed blue line.

By comparing our high pressure Ti-Beta-21S (Velisavljevic & Chesnut, 2007), Ti-6Al-4V (Chesnut et al., 2008; Tegner et al., 2010; Tegner et al., n.d.) and nano-Ti (Velisavljevic et al., n.d.) data, we observe that the addition of alloying elements can have a significant influence on the structural phase transition sequence in these metals at high pressure and temperature. For example, the presence of alloys and interstitial impurities in Ti-6Al-4V suppresses the onset of the $\alpha \rightarrow \omega$ phase transformation, thus ensuring the predominance of the α-phase alloy over a much larger pressure range than exists for pure Ti, which is desirable in industrial and commercial applications. The main effect observed in Ti-Beta-21S is the complete suppression of the brittle ω phase at high pressures.

However, it is also important to recognize that changes in the electronic configuration cause changes in crystal structures. These changes can be induced either by an increase in pressure or an increase in the occupancy of the *d* bands. The structural trend exhibited by the 3*d*, 4*d*, and 5*d* transition metals is well known. The variation in the electronic configuration affects crystal structures and mechanical properties such as micro-structures and dislocations. For industrial applications, it is the machineability and superior mechanical properties that are of paramount interest. Pressure allows us to measure the differences in crystal structures induced by changes in the electronic configuration.

Among the Group IV elements, similar structural changes occur. Group IV elements and their alloys apparently favor a transformation pathway $\alpha \rightarrow \omega \rightarrow \beta$ at high pressure, based on recent DAC experiments and theoretical calculations. The intermediate structures, δ and γ, appear to be metastable and are shear driven. Based on the experimental results one can conclude that high pressure structural phase transitions in Ti and Ti alloys are highly susceptible to loading conditions and stress distribution, as shown from experiments using various PMTs. However, stability of various phases can be controlled by other variables, such as alloying, which can change electronic structure by increasing/decreasing *d* band occupancy, inclusion of interstitial impurities, which help reduce shear deformation, and in some cases it appears that shear driven structural phase transitions can be controlled by varying sample grain size as well. Although these factors play a significant role in controlling structural phase transitions they appear to have only a slight effect on the initial compressibility (i.e. EOS) - measured α phase EOS parameters for Ti-Beta-21S, Ti-6Al-4V, Ti, and nano-Ti are all relatively close with values of K = 115-125 GPa and K' = 2.4-3.9. Overall it appears that with stress conditions, grain size, and presence of impurities, there is a systematic shift of the transition pressure $P_{\alpha \rightarrow \omega}$, as the transition pressure increases with improved hydrostaticity of the pressure environment and by grain size reduction.

5. Acknowledgements

SM would like to acknowledge the support of Professor Malcolm McMahon and Dr John Proctor of the Centre for Science at Extreme Conditions (CSEC), Edinburgh University, in collecting the Ti-6Al-4V data. This work was performed under the auspices of the U.S. Department of Energy by Lawrence Livermore National Laboratory in part under contract W-7405-Eng-48 and in part under Contract DE-AC52-07NA27344. LANL is operated by

LANS, LLC for the DOE-NNSA - this work was, in part, supported by the US DOE under contract # DE-AC52-06NA25396. HP-CAT is supported by CIW, CDAC, UNLV and LLNL through funding from DOE-NNSA, DOE-BES and NSF. APS is supported by DOE-BES under Contract No. DE-AC02-06CH11357.

6. References

Ahuja, R.; Dubrovinsky, L.; Dubrovinskaia, N.; Osorio Guillen, J.M.; Mattessini, M.; Johansson, B. and Le Bihan T. (2004). Titanium metal at high pressure: Synchrotron experiments and *ab initio* calculations. *Physical Review B*, Vol.69, pp. 184102-1-184102-4

Akahama, Y.; Kawamura, H. & Le Bihan T. (2001). New δ (Distorted-bcc) Titanium to 220 GPa. *Physical Review Letters*, Vol.87, No.27, pp. 2755031-2755034

Bania, B.J.. (1993). Beta Titanium alloys and their role in the Titanium industry. *In: Eylon D, Boyer R, Koss D, Editors, Beta Titanium alloys in the 1990's TMS*, ISBN 0-87339-200-0, Warrendale, USA, February 1993

Birch, F. (1952). Elasticity and constitution of the Earth's interior. *Journal of Geophysical Research*, Vol.57, pp. 227-286

Bundy, F.P. (1963). General Electric Report No. 63-RL-3481C (unpublished)

Bundy, F.P. (1965). Formation of New Materials and Structures by High-Pressure Treatment. *Irreversible Effects of High Pressure and Temperature on Materials*, ASTM Special Technical Publication No. 374, Philadelphia, February, 1964

Carter, W.J.; Marsh, S.P.; Fritz, J.N. & McQueen, R.G. (1971). The Equation of State of Selected Materials for High-Pressure References. *National Bureau of Standards Special Publication 326: Accurate Characterisation of the High-Pressure Environment*, pp. 147-158, Gaithersburg, Md., October 14-18, 1968

Cerreta, E.; Gray III, G.T.; Lawson, A.C.; Mason, T.A. & Morris, C.E. (2006). The influence of oxygen content on the α to ω phase transformation and shock hardening of titanium. *Journal of Applied Physics*, Vol.100, 013530-1-013530-9

Chesnut, G.N.; Velisavljevic, N. & Sanchez, L. (2008). Static High pressure X-ray Diffraction of Ti-6Al-4V. *Proceedings of the American Physical Society Topical Group on Shock Compression of Condensed Matter – 2007*, pp. 27-30, ISBN 978-0735404694, Kohala Coast, Hawaii, USA, June 24-29, 2007

Dewaele, A.; Datchi, F.; Loubeyre, P. & Mezouar, M. (2008). High pressure-high temperature equations of state of neon and diamond. *Physical Review B*, Vol.77, pp. 094106-1-094106-9

Errandonea, D.; Schwager, B.; Ditz, R.; Gessmann, C.; Boehler, R. & Ross, M. (2001). Systematics of transition-metal melting. *Physical Review B*, Vol.63, pp. 132104-1-132104-4

Errandonea, D.; Meng, Y.; Somayazulu, M. & Häusermann, D. (2005). Pressure-induced α → ω transition in titanium metal: a systematic study of the effects of uniaxial stress. *Physica B*, Vol.355, pp. 116-125

Gray, G.T.; Morris, C.E. & Lawson, A.C. (1993). Omega phase formation in Titanium and Titanium alloys. *Proceedings of Titanium '92: Science and Technology*, ISBN 0873392221, San Diego, USA, June 1992

Greeff, C.W.; Trinkle, D.R. & Albers, R.C. (2001). Shock-induced α-ω transition in titanium. *Journal of Applied Physics*, Vol.90, No.5, pp. 2221-2226

Gupta, S.C.; Joshi, K.D. & Banerjee, S. (2008). Experimental and Theoretical Investigations on *d* and *f* electron Systems under High Pressure. *Metallurgical and Materials Transactions A*, Vol.39A, pp. 1593-1601

Halevy, I.; Zamir, G.; Winterrose, M.; Sanjit, G.; Grandini, C.R. & Moreno-Gobbi, A. (2010). Crystallographic structure of Ti-6Al-4V, Ti-HP and Ti-CP under High-Pressure. *Journal of Physics: Conference Series*, Vol.215, pp. 1-9

Hennig, R.G.; Trinkle, D.R.; Bouchet, J.; Srinivasan, S.G.; Albers, R.C. & Wilkins, J.W. (2005). Impurities block the α to ω martensitic transformation in titanium. *Nature*, Vol.4, pp. 129-133

Hennig, R.G.; Lenosky, T.J.; Trinkle, D.R.; Rudin, S.P. & Wilkins, J.W. (2008). Classical potential describes martensitic phase transformations between the α, β, and ω titanium phases. *Physical Review B*, Vol.78, pp. 054121-1-054121-10

Honnell, K.G.; Velisavljevic, N.; Adams, C.D.; Rigg, P.A.; Chesnut, G.N.; Aikin Jr, R.M & Boettger, J.C. (2007). Equation of State for Ti-beta-21S. Compression of Condensed Matter - 2007, edited by M. Elert, M.D. Furnish, R. Chau, N.C. Holmes and J. Nguyen, Conference Proceedings of the APS topical group on SCCM, (AIP, New York), 2007, Pt. 1, p.55.

Hu, C.-E.; Zeng, Z.-Y.; Zhang, L.; Chen, X.-R.; Cai, L.-C. & Alfè, D. (2010). Theoretical investigation of the high pressure structure, lattice dynamics, phase transition, and thermal equation of state of titanium metal. *Journal of Applied Physics*, Vol.107, pp. 093509-1-093509-10

Jamieson, J.C. (1963). Crystal structure of Titanium, Zirconium, and Hafnium at high pressures. *Science* Vol.140, pp. 72-73

Jayaraman, A.; Klement, W. & Kennedy G.C. (1963). Solid-Solid Transitions in Titanium and Zirconium at High Pressures. *Physical Review*, Vol.131, No.2, pp. 644-649

Klotz, S.; Chervin, J.-C.; Munsch, P. & Le Marchand, G. (2009). Hydrostatic limits of 11 pressure transmitting media. *Journal of Physics D: Applied Physics*, Vol.42, pp. 075413-1-075413-7

Mei, Z.-G.; Shang, S.-L.; Wang, Y. & Liu, Z.-K. (2009). Density-functional study of the thermodynamic properties and the pressure-temperature phase diagram of Ti. *Physical Review B*, Vol.80, pp. 104116-1-104116-9

Ming, L.C.; Manghnani, M. & Katahara, M. (1981). Phase-Transformations in the Ti-V System Under High-Pressure up to 25-GPa. *Acta Metallurgica*, Vol.29, No.3, pp. 479-485

Pecker, S.; Eliezer, S.; Fisher, D.; Henis, Z. & Zinamon, Z. (2005). A multiphase equation of state of three solid phases, liquid, and gas for titanium. *Journal of Applied Physics*, Vol.98, pp. 043516-1-043516-12

Rosenberg, Z. & Meybar, Y. (1981). Measurement of the Hugoniot curve of Ti-6Al-4V with commercial manganin gauges. *Journal of Physics D: Applied Physics*, Vol.14, pp. 261-266

Tegner, B.E.; MacLeod, S.G.; Cynn, H.; Proctor, J.; Evans, W.J.; McMahon, M.I. & Ackland, G.J. (2011). An Experimental and Theoretical Multi-Mbar Study of Ti-6Al-4V. *Mater. Res. Soc. Symp. Proc.*, Vol.1369, Materials Research Society

Tegner, B.E.; MacLeod, S.G.; Cynn, H.; Proctor, J.; Evans, W.J. & McMahon, M.I. (n.d.). Manuscript to be submitted.

Trinkle, D.R.; Hennig, R.G.; Srinivasan, S.G.; Hatch, D.M.; Jones, M.D.; Stokes, H.T.; Albers, R.C. & Wilkins, J.W. (2003). New Mechanism for the α to ω Martensitic Transformation in Pure Titanium. *Physical Review Letters*, Vol.91, No.2, pp. 025701-1-025701-4

Trinkle, D.R.; Hatch, D.M.; Stokes, H.T.; Hennig, R.G. & Albers, R.C. (2005). Systematic pathway generation and sorting in martensitic transformations: Titanium α to ω. Physical Review B, Vol.72, pp. 014105-1-014105-11

Trunin, R.F.; Simakov, G.V. & Medvedev A.B. (1999). Compression of Titanium in Shock Waves. *High Temperature*, Vol.37, pp. 851-856

Velisavljevic, N. & Chesnut, G.N. (2007). Direct hcp → bcc structural phase transition observed in titanium alloy at high pressure. *Applied Physics Letters*, Vol.91, pp. 101906-1-101906-3

Velisavljevic, N. (n.d.). Manuscript to be submitted.

Verma, A.K.; Modak, P.; Rao, R.S.; Godwal, B.K. & Jeanloz, R. (2007). High-pressure phases of titanium: First-principles calculations. Physical Review B, Vol.75, pp. 014109-1-014109-5

Vinet, P.; Ferrante, J.; Rose, J.H. & Smith J.R. (1987). Compressibility of solids. *Journal of Geophysical Research – Solid Earth and Planets*, Vol.92, No.B9, pp. 9319-9325

Vohra, Y.K.; Sikka, S.K.; Vaidya, S.N. & Chidambaram, R. (1977). Impurity Effects and Reaction-Kinetics of Pressure-Induced Alpha-Omega Transformation in Ti. *Journal of Physics and Chemistry of Solids*, Vol.38, No.11, pp. 1293-1296

Vohra, Y.K. (1979). Electronic basis for omega phase stability in group IV transition metals and alloys. *Acta Metallurgica*, Vol.27, No.10, pp. 1671-1674

Vohra, Y.K. & Spencer, P.T. (2001). Novel γ-Phase of Titanium Metal at Megabar Pressures. *Physical Review Letters*, Vol.86, No.14, pp. 3068-3071

Xia, H.; Duclos, S.J.; Ruoff, A.L. & Vohra, Y.K. (1990a). New High-Pressure Phase Transition in Zirconium Metal. *Physical Review Letters*, Vol.64, No.2, pp. 204-207

Xia, H.; Parthasarathy, H.L.; Vohra, Y.K. & Ruoff, A.L. (1990b). Crystal structures of group IVa metals at ultrahigh pressures. *Physical Review B*, Vol.42, No.10, pp. 6736-6738

Zhang, J.; Zhao, Y.; Hixson, R.S.; Gray III, G.T.; Wang, L.; Utsumi, W.; Hiroyuki, S. & Takanori, H. (2008). Experimental constraints on the phase diagram of titanium metal. *Journal of Physics and Chemistry of Solids*, Vol.69, pp. 2559-2563

Zhou, T.; Aindow, M; Alpay, S.P.; Blackburn, M.J.; Wu, M.H. (2004). Pseudo-elastic deformation behavior in a Ti/Mo-based alloy. *Scripta Materialia,* Vol.50, pp. 343-348

5

Hot Plasticity of Alpha Beta Alloys

Maciej Motyka, Krzysztof Kubiak,
Jan Sieniawski and Waldemar Ziaja
Department of Materials Science,
Rzeszow University of Technology
Poland

1. Introduction

Two phase titanium alloys most often are hot deformed, mainly by open die or close-die forging. Desired mechanical properties can be achieved in these alloys by development of proper microstructure in plastic working and heat treatment processes. Irreversible microstructural changes caused by deformation at the temperature in α+β↔β phase transformation range quite often cannot be eliminated or reduced by heat treatment and therefore required properties of products cannot be achieved (Bylica & Sieniawski, 1985; Lütjering, 1998; Zwicker, 1974). Some of the properties of titanium alloys, such as: high chemical affinity to oxygen, low thermal conductivity, high heat capacity and significant dependence of plastic flow resistance on strain rate, make it very difficult to obtain finished products having desired microstructure and properties by hot working. Differences in temperature across the material volume, which result from various deformation conditions (local strain and strain rate) lead to formation of zones having various phase composition (equilibrium α and β phases, martensitic phases α'(α")), morphology (equiaxial, lamellar, bi-modal) and dispersion (fine- or coarse-grained) and therefore various mechanical properties (Kubiak & Sieniawski, 1998).

Obtaining desired microstructure of Ti-6Al-4V titanium alloy using plastic deformation in the α+β↔β phase transformation range is related to proper conditions selection taking into account plastic deformation, phase transformation, dynamic recovery and recrystallization effects (Ding et al., 2002; Kubiak, 2004; Kubiak & Sieniawski, 1998). Grain refinement can be achieved by including preliminary heat treatment in thermomechanical process. Final heat treatment operations are usually used for stabilization of microstructure (they restrict grain growth) (Motyka & Sieniawski, 2010).

Titanium alloys together with aluminium alloys belong to the largest group of superplastic materials used in industrial SPF. Their main advantages are good superplasticity combined with relatively high susceptibility to diffusion bonding. Among them two-phase α+β Ti-6Al-4V alloy has been the most popular for many years as it exhibits superplasticity even after application of conventional plastic working methods (Sieniawski & Motyka, 2007).

2. Deformation behaviour of titanium and its alloys

Titanium has two allotropic forms: Ti_{α} with hexagonal close packed (hcp) crystal structure (up to 882.5°C) and Ti_{β} with body centered cubic (bcc) crystal structure (between 882.5 and 1662°C). Each of the allotropes exhibit different plasticity resulting from its crystal structure and different number of the slip systems (A. D. Mc Quillan & M. K. Mc Quillan, 1956; Zwicker, 1974).

Deformation of α titanium, both at room and elevated temperature, occurs by slip and twinning (Tab. 1). Primary slip systems in α titanium are: $\{1\bar{1}00\} < 11\bar{2}0 >$ and $\{1\bar{1}01\} < 11\bar{2}0 >$ (Fig. 1). Critical resolved shear stress value (τ_{crss}) depends on deformation temperature, impurities content and slip system (Fig. 2). If the relative value of τ_{crss} stress for $\{1\bar{1}00\} < 11\bar{2}0 >$ slip system is set to 1, for $\{1\bar{1}01\} < 11\bar{2}0 >$ system it equals to 1.75, and for $\{0001\} < 11\bar{2}0 >$ system - 1.92. Coarse grained and single crystal α titanium deforms in $\{0001\} < 11\bar{2}0 >$ system, because stacking fault energy and atom packing density are larger in $\{1\bar{1}00\}$ planes than in {0001} planes (Kajbyszew & Krajuchijn, 1967; Zwicker, 1974).

Metal	c/a ratio	Slip plane	Critical resolved shear stress τ_{crss}, MPa	Twinning plane
Zinc	1.856	{0001} $\{1\bar{1}00\}$	0.34 10-15	$\{1\bar{1}02\}$
Manganese	1.624	{0001} $\{1\bar{1}00\}$	4.5 5.1	$\{1\bar{1}02\}$ $\{1\bar{1}01\}$
		$\{1\bar{1}00\}$	39.2	$\{1\bar{1}03\}$ $\{11\bar{2}1\}$
Titanium α	1.587	{0001}	62.1	$\{11\bar{2}2\}$ $\{11\bar{2}1\}$
		$\{1\bar{1}00\}$	13.83	$\{11\bar{2}2\}$ $\{1\bar{1}02\}$

Table 1. Slip and twinning planes and critical resolved shear stress in metals with hcp crystal structure (Kajbyszew & Krajuchijn, 1967)

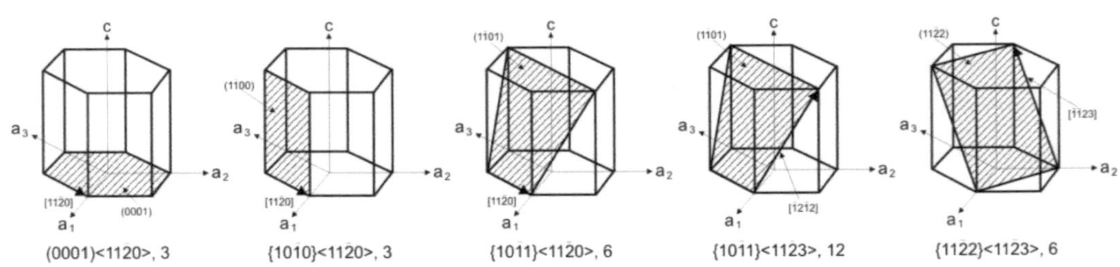

Fig. 1. Slip systems in α titanium (Balasubramanian & Anand, 2002)

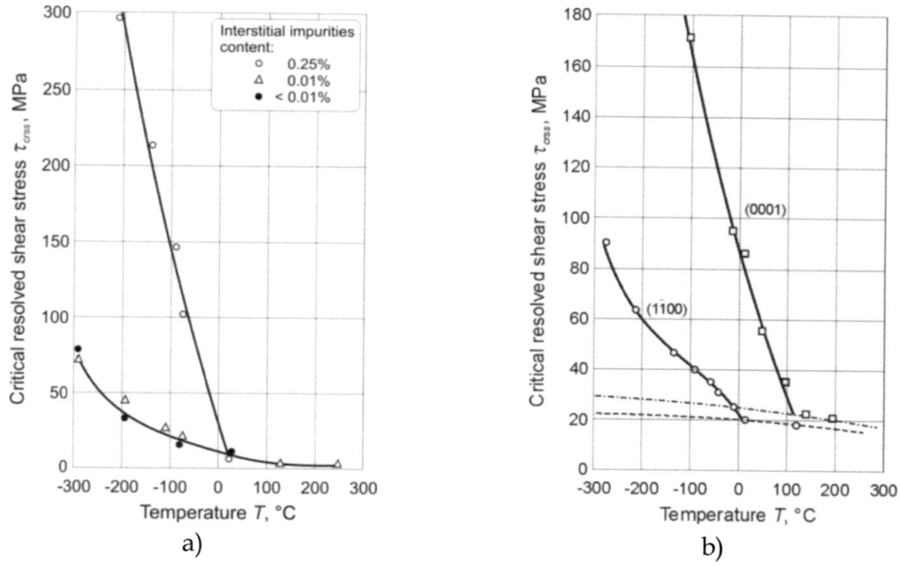

Fig. 2. Critical resolved shear stress as a function of: a) interstitial impurities content and temperature (Zwicker, 1974) b) temperature and slip plane in titanium with 0.1% interstitial impurities content (Levine, 1966)

The number of twinning planes is higher in α titanium than in other metals (except for zirconium) having hcp crystal structure. Twinning in single crystal α titanium occurs in $\{1\bar{1}02\}$, $\{11\bar{2}1\}$, $\{11\bar{2}2\}$, $\{11\bar{2}3\}$, $\{11\bar{2}4\}$ planes, while in polycrystalline α titanium in $\{1\bar{1}02\}$, $\{11\bar{2}1\}$, $\{11\bar{2}2\}$ planes (Fig. 3). The smallest value of critical stress for twinning occurs in $\{1\bar{1}02\}$ plane. It was found that slip and twinning interact with each other. Twinning in $\{11\bar{2}2\}$ plane hinders slip in that plane. Twinning is intensified by increase in deformation, metal purity, grain size and by decrease in temperature (Churchman, 1955).

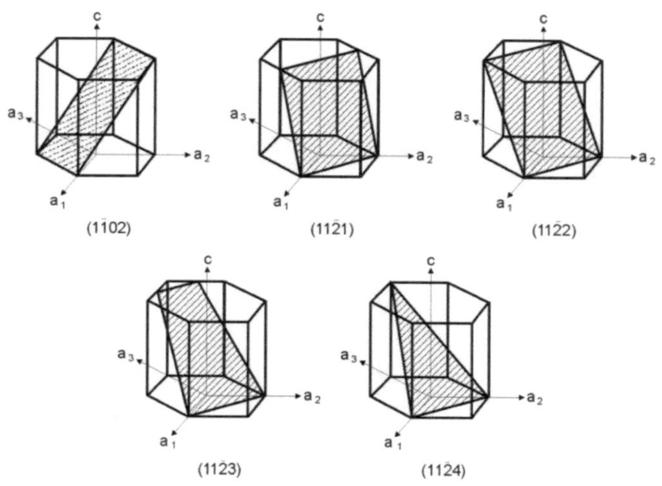

Fig. 3. Twinning planes in α titanium (Balasubramanian & Anand, 2002)

Plastic deformation in β titanium takes place by mechanisms characteristic of metals with bcc structure. In β titanium following slip systems operate: {110} <111>, {112} <111>, {123} <111> along with twinning system {112} <111> (Balasubramanian & Anand, 2002; Wassermann & Grewen, 1962). Alloying elements affect plasticity of β titanium to different degree. Cold strain of β titanium (with high content of Mo, Nb or Ta) can exceed 90%. However addition of ruthenium or rhodium results in very large decrease in plasticity, rendering cold deformation of β titanium practically impossible (Raub & Röschelb, 1963).

Titanium alloys can be classified according to their microstructure in particular state (e.g. after normalizing). It must be emphasized that this classification is questionable, because phase transformations in alloys with transition elements proceed so slowly, that very often the microstructure consistent with phase equilibrium diagram cannot be obtained at room temperature. According to widely accepted classification of the alloys in normalized state following types of titanium alloys can be distinguished (Fig. 4) (Glazunov & Kolachev, 1980):

1. α alloys
   - not heat strengthened
   - heat strengthened due to metastable phases decomposition
2. α+β alloys
   - strengthened by quenching
   - with increased plasticity after quenching
3. β alloys
   - with mechanically unstable β_{MN} phase (decomposing under the stress)
   - with mechanically stable β_{MS} phase (not decomposing under the stress)
   - with thermodynamically stable β_{TS} phase.

Two transition types of alloys can also be distinguished:

1. near α alloys – which contain up to 5% of β stabilizers (microstructure is composed of α phase and small amount, about 3-5%, of β phase),
2. near β alloys – their properties match the properties of α+β alloys with high volume fraction of β phase (the microstructure after solutionizing is composed of metastable β_M phase).

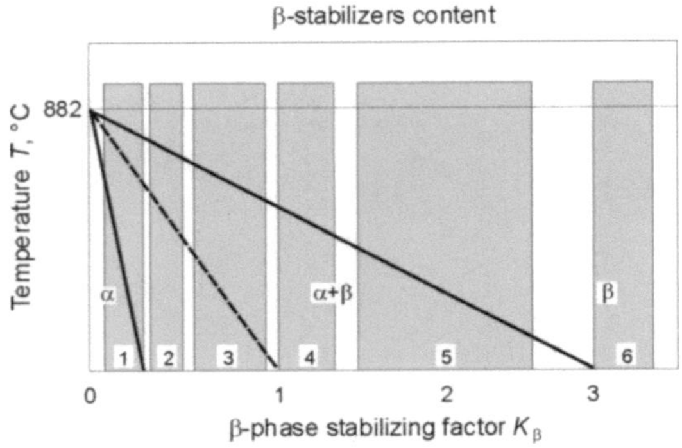

Fig. 4. Classification of titanium alloys on the basis of β stabilizers content: 1 – α alloys, 2 – near α alloys, 3 – martensitic α+β alloys, 4 – transition α+β alloys, 5 – near β alloys, 6 – β alloys (Glazunov & Kolachev, 1980)

Plasticity of commercial pure (CP) titanium and single-phase alloys depends on type and content of impurities, alloying elements, temperature of deformation and strain rate. Increase in impurities and alloying elements content reduces plasticity due to solid solution strengthening (Fig. 5) (Glazunov & Mojsiejew, 1974).

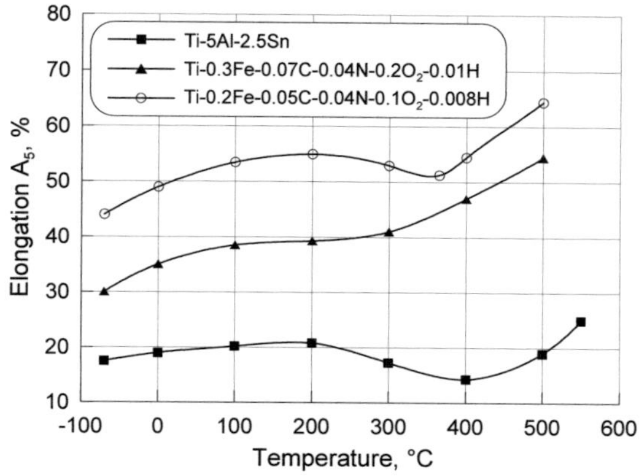

Fig. 5. The influence of temperature on plasticity of CP titanium and single-phase α alloys (Glazunov & Mojsiejew, 1974)

Increase in temperature of deformation reduces plastic flow stress of the CP titanium and single-phase α alloys (Fig. 6).

Critical strain of CP titanium depends on temperature of deformation and strain rate (Fig. 7).

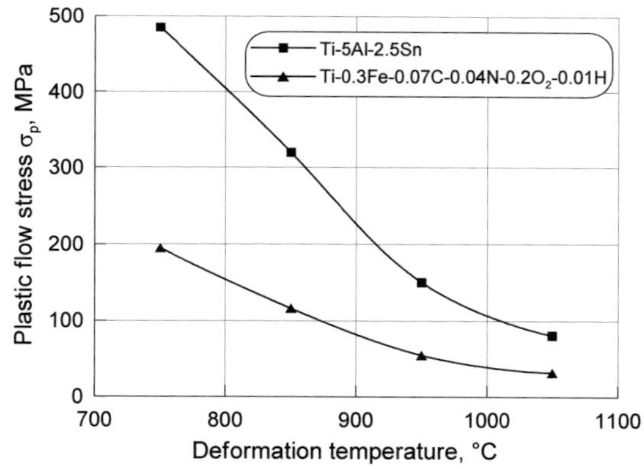

Fig. 6. Dependence of plastic flow stress of CP titanium and α alloys on deformation temperature (Hadasik, 1979)

Roughly 90% of about 70 grades of titanium alloys that are manufactured by conventional methods are two-phase martensitic or transition alloys. They exhibit high relative strength (UTS/ρ), good creep resistance (up to 450°C), corrosion resistance in many environments, good weldability and formability. The most widely used representative of this group of alloys is Ti-6Al-4V showing good balance of mechanical and technological properties.

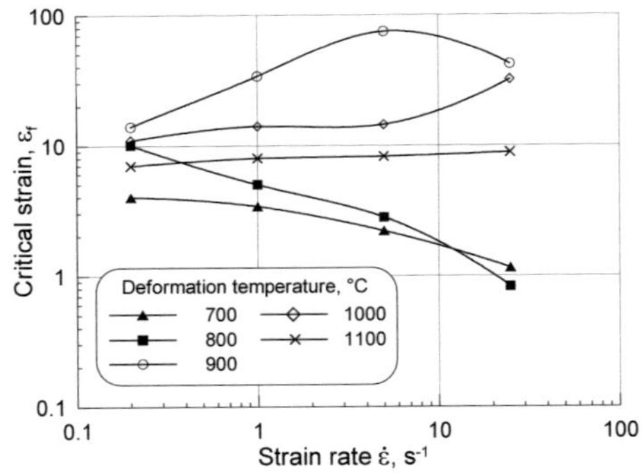

Fig. 7. The effect of temperature of deformation and strain rate on the critical strain of CP titanium (Hadasik, 1979)

The basic technological processes enabling final product manufacturing and development of mechanical properties of titanium alloys are hot working and heat treatment. Application of cold working is limited to the operations of bending (sheets, flat bars, tubes and bars) and shallow drawing of sheets allowing to obtain large elements. Bulk cold forming is not used due to high resistance of titanium and its alloys to plastic flow. (Fig. 8) (Lee & Lin, 1997).

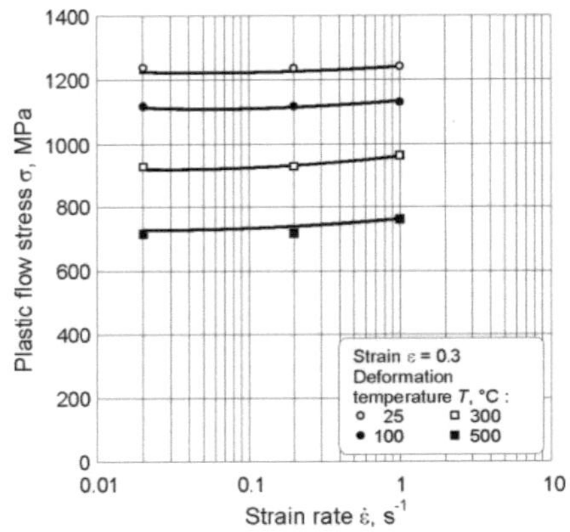

Fig. 8. The influence of strain rate on plastic flow stress of cold worked Ti-6Al-4V alloy (Lee & Lin, 1997)

Plasticity, microstructure and mechanical properties of titanium alloys depend on hot working conditions (Peters et al., 1983; Brooks, 1996):

- heating rate, time and temperature of soaking and furnace atmosphere,
- start and finish temperature of deformation,
- draft in final operations of deformation,
- strain rate,
- cooling rate after deformation.

Application of all or selected operations of hot working and heat treatment (Fig. 9) allows to vary to a large extent the microstructure e.g. morphology and dispersion of phases in α+β titanium alloys (Fig. 10).

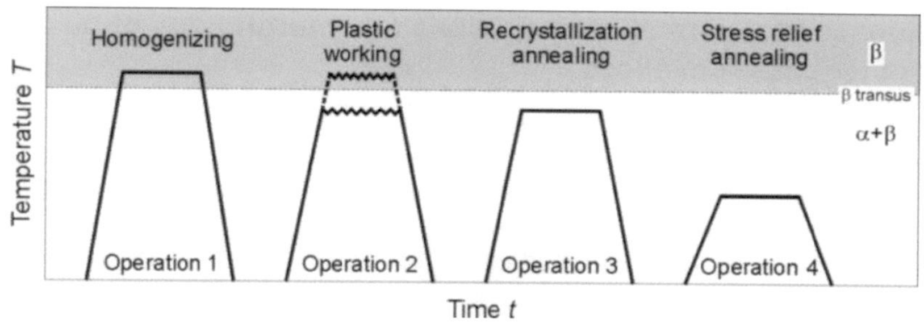

Fig. 9. The schematic diagram of plastic work and heat treatment processes of two phase titanium alloys (Peters et al., 1983)

The processes of hot working and heat treatment of two-phase titanium alloys, e.g. Ti-6Al-4V, allow to obtain various types of microstructure (Fig.10) (Ezugwu & Wang, 1997; Kubiak & Sieniawski, 1998):

- martensitic (or composed of metastable β_M phase),
- globular (fine or coarse-grained),
- necklase (fine or coarse-grained),
- lamellar (fine or coarse-grained),
- bi-modal (with various volume fraction and dispersion of α phase).

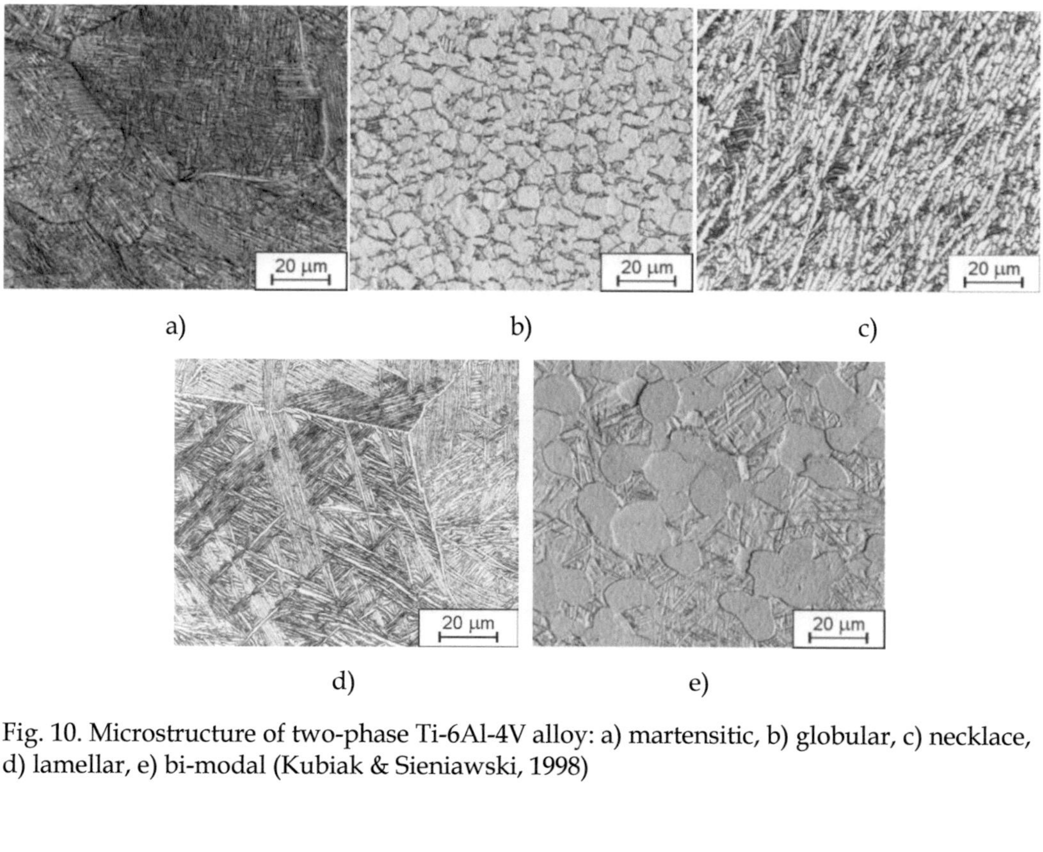

Fig. 10. Microstructure of two-phase Ti-6Al-4V alloy: a) martensitic, b) globular, c) necklace, d) lamellar, e) bi-modal (Kubiak & Sieniawski, 1998)

3. The influence of deformation conditions and morphology of phases on the plasticity of α+β titanium alloys

Hot deformation behaviour of two-phase titanium alloys depends on chemical and phase composition, stereological parameters of microstructure and process conditions (deformation temperature, strain rate, stress and strain distribution). They exhibit analogous dependences of the plastic flow stress on temperature and strain rate like other metals alloys. Increase in strain rate raises plastic flow stress while increase in deformation temperature reduces it (Sakai, 1995).

Increase in temperature of deformation of two-phase titanium alloys reduces plastic flow stress (Fig. 11) more effectively in α+β field than in β phase field, as a result of change in volume fraction of α and β phases (Fig. 12) (Ding et al., 2002; Sheppard & Norley, 1988).

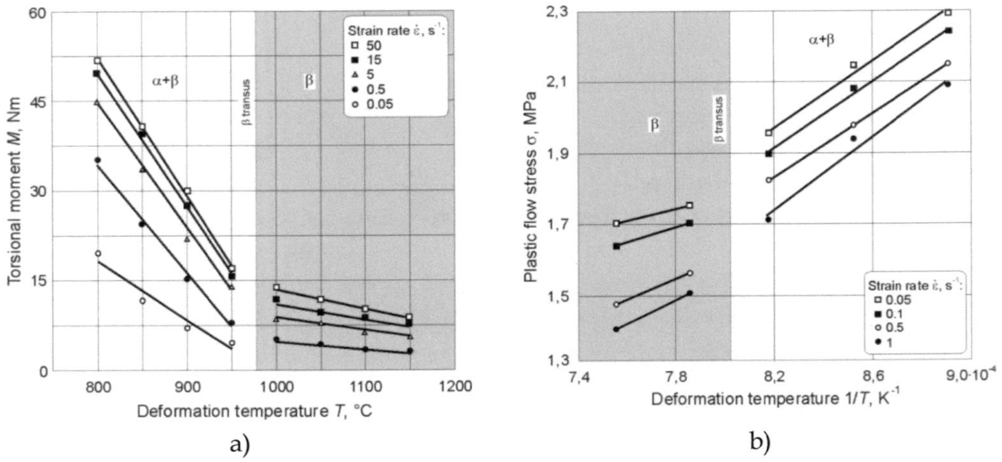

Fig. 11. The effect of temperature of deformation and strain rate on plastic flow stress for Ti-6Al-4V alloy: a) torsion test (Sheppard & Norley, 1988), b) compression test (Ding et. al., 2002)

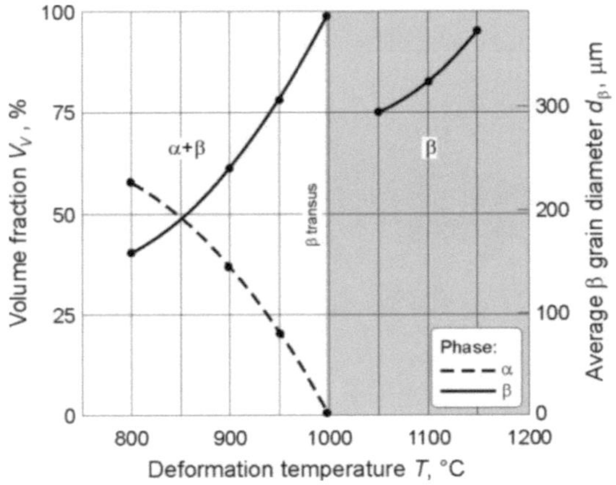

Fig. 12. The effect of temperature on volume fraction of α and β phases in Ti-6Al-4V alloy during torsion test (Sheppard & Norley, 1988)

High sensitivity to strain rate is a characteristic feature of two-phase titanium alloys. The coefficient of strain rate sensitivity depends on temperature of deformation and strain rate (Fig. 13), grain size (Fig. 14), strain magnitude (Fig. 15) and morphology of phase constituents (Fig. 16) (Semiatin et al., 1998).

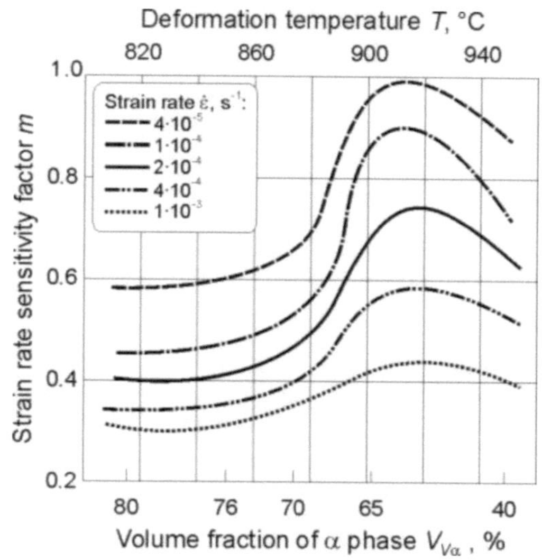

Fig. 13. The dependence of strain rate sensitivity factor m on temperature (volume fraction of α phase) and strain rate for Ti-6Al-4V alloy (Semiatin et al., 1998)

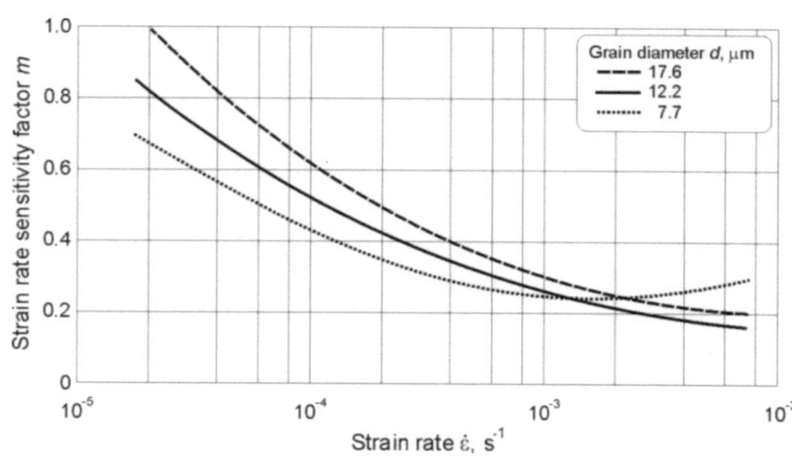

Fig. 14. The dependence of strain rate sensitivity factor m on grain size of α phase for Ti-6Al-4V alloy (Semiatin et al., 1998)

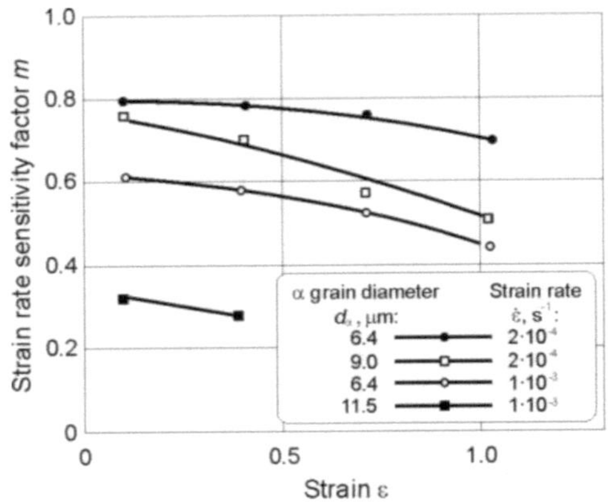

Fig. 15. The dependence of strain rate sensitivity factor m on the strain magnitude for Ti-6Al-4V alloy (Ghosh & Hamilton, 1979)

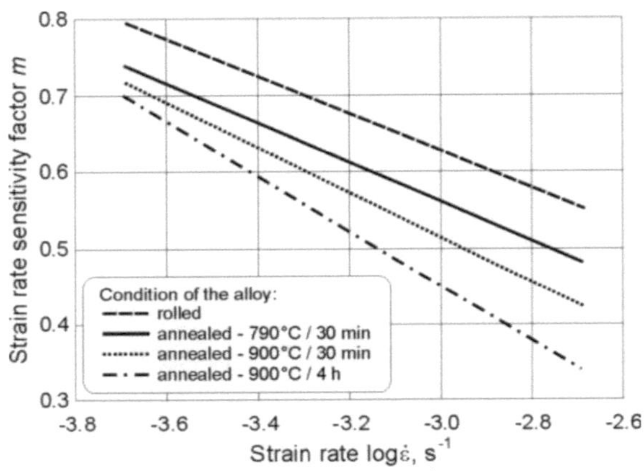

Fig. 16. The dependence of strain rate sensitivity factor m on microstructure morphology (type of heat treatment) for hot-rolled sheets of Ti-6Al-4V alloy (Semiatin et al., 1998)

Volume fraction of α and β phases affects the value of the coefficient of strain rate sensitivity m. For two-phase titanium alloys it reaches maximum value when $V_{v\alpha}=V_{v\beta}\cong 50\%$, at the temperature close to start temperature of $\alpha+\beta\rightarrow\beta$ phase transformation and strain rate in the range of $\dot{\varepsilon} = 4\cdot10^{-5} - 1\cdot10^{-3}\ s^{-1}$. Grain refinement leads to increase in m value and growth of strain magnitude reduces it (Ghosh & Hamilton, 1979).

Deformation of titanium alloys in two-phase range leads to distortion of α and β grains, fragmentation of α grains and their globularization. As a result of these processes (Fig. 17) elongated α phase grains develop with particular orientation which are arranged along direction of maximum deformation (Seshacharyulu et al., 2002).

Fig. 17. The deformation process in α+β field – fragmentation and globularization of α phase lamellae (Seshacharyulu et al., 2002)

These processes occur simultaneously during deformation and have an impact both on texture (Fig. 18) and morphology of phases (Lütjering, 1998).

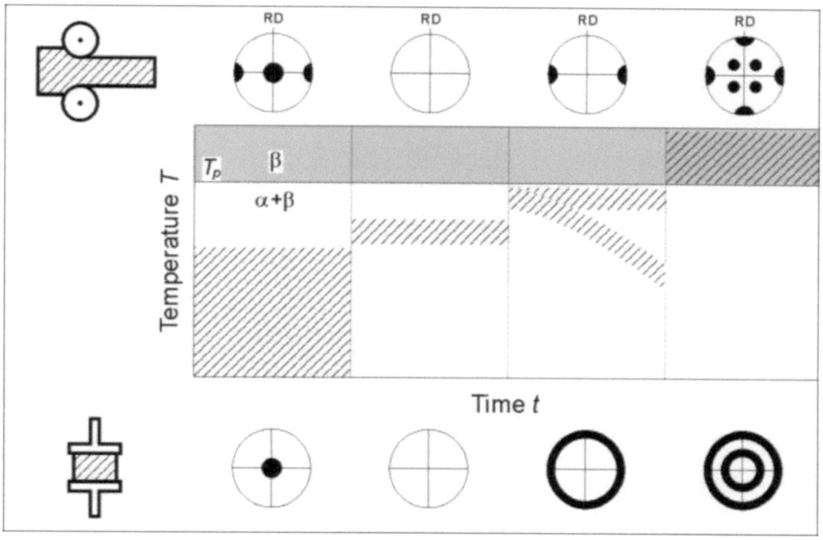

Fig. 18. The texture diagram – pole figure (0002) – of two-phase titanium alloys for various modes and temperature of deformation (Lütjering, 1998)

After deformation in the α+β temperature range and air cooling following microstructure of α phase can be obtained (Fig. 19) (Kubiak & Sieniawski, 1998):

- coarse lamellar – not deformed or slightly deformed large size lamellae of primary α phase are usually present in dead zone of semi-finished products (Figs 19-3 and 19-5),
- fine lamellar – α lamellae formed upon cooling from temperature above the β transus temperature at the cooling rate slightly lower than critical (Fig. 19-1),
- distorted lamellar – primary α lamellae which were fragmented and deformed (Fig. 19-4),
- equiaxed – α grains formed upon slow cooling from the temperature higher than finish temperature of α+β↔β transformation (Fig. 19-2),
- distorted equiaxed – α grains distorted upon deformation below the finish temperature of α+β↔β transformation (Fig. 19-6).

Increase in degree or rate of deformation may lead to local increase in temperature even above finish temperature of α+β↔β transformation and development of martensitic α'(α") phases or colonies of lamellar α and β phases with stereological parameters depending on cooling rate.

Upon deformation of α+β titanium alloys above the finish temperature of α+β↔β transformation dynamic recrystallization occurs. New grains nucleate at primary β grain boundaries and then grow until complete ring arranged along these grain boundaries is formed. New nuclei form also inside formerly recrystallized grains in the ring and the process repeats until new grains completely fill up the 'old' grain (Fig. 20) (Ding et al., 2002).

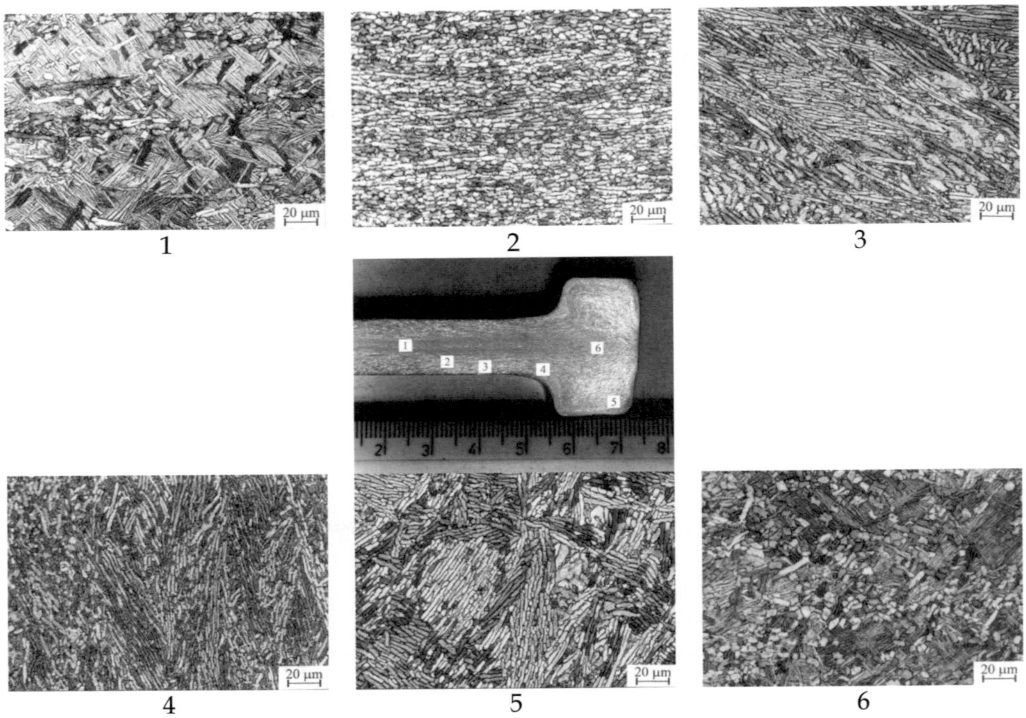

Fig. 19. The microstructure of Ti-6Al-4V alloy after die forging at 950°C (Kubiak & Sieniawski, 1998)

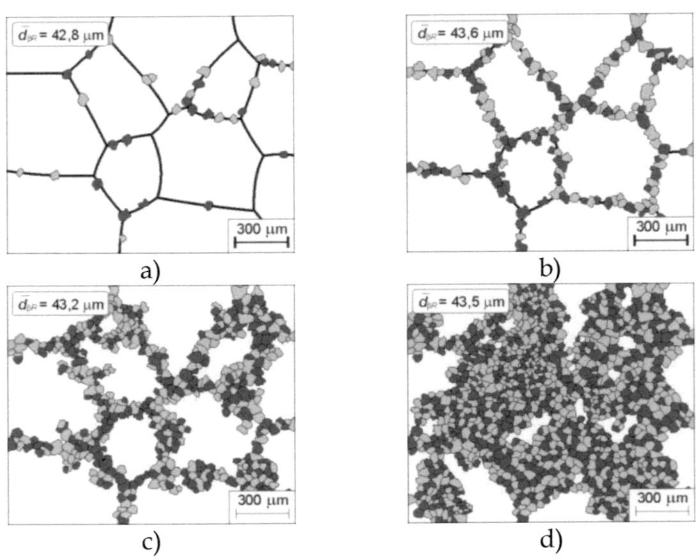

Fig. 20. Simulation of microstructure development for Ti-6Al-4V alloy deformed at 1050°C at the strain rate $\dot{\varepsilon}$ = 1.0 s^{-1} for various strains: a) ε = 0.7; b) ε = 5.0; c) ε = 15, d) ε = 45; $\bar{d}_{\beta R}$ – average diameter of recrystallized β grains (Ding et al., 2002)

Increase in deformation rate decreases the size of recrystallized α grains (Fig. 21). The change in average diameter of primary β grains – $\overline{d}_{p\beta}$ and average thickness of α lamellae – $\overline{g}_{p\alpha}$, induced by deformation, can be described for Ti-6Al-4V alloy by following equations, depending on morphology of primary α and β phases:

- lamellar microstructure $\overline{d}_{p\beta} = 1954.3\ Z^{-0,172}$ (μm) (Seshacharyulu et al., 2002)
- lamellar microstructure $\overline{g}_{p\alpha} = 1406.4\ Z^{-0,139}$ (μm) (Seshacharyulu et al., 2002)
- equiaxed microstructure $\log(\overline{d}_{p\beta}) = 3.22 - 0.16 \log(Z)$ (Seshacharyulu et al., 1999)

where: Z – Zener-Holomon parameter.

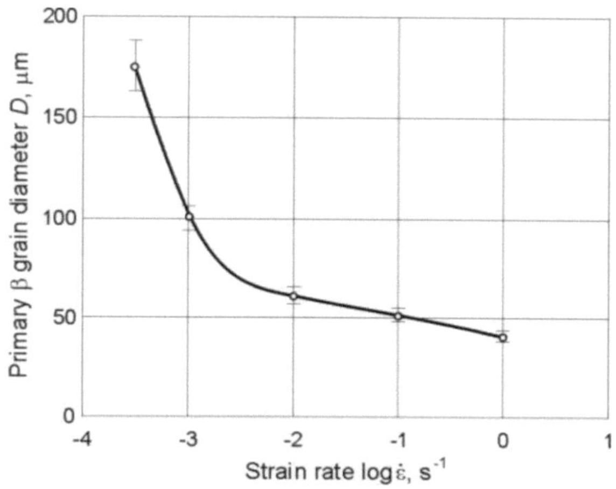

Fig. 21. The effect of strain rate on average diameter of primary β phase grains in Ti 6Al-4V alloy with globular initial microstructure (Seshacharyulu et al., 1999)

However presented model of dynamic recrystallization (Fig. 20) does not take into account the change of shape of primary β grains and deformation of recrystallized β grains. Thus it can be stated that it describes process of metadynamic recrystallization. Upon cooling after dynamic or metadynamic recrystallization following phases can be formed within β grains (depending on cooling rate): colonies of α and β lamellae, equiaxed α and β grains, martensitic α'(α") phases or the mixture of them.

The character of flow curves obtained during sequential deformation of two phase titanium alloys at the temperature in β field confirms occurrence of dynamic processes of microstructure recovery. Increase in time of metadynamic recrystallization reduces strengthening. The character of the curves describing volume fraction of recrystallized β phase confirms the influence of chemical composition on recrystallization kinetics (Fig. 22). Increasing content of alloying elements leads to decrease in rate of recrystallization. Time for recrystallization of 50% of β phase is equal t0,5 ≈ 3.5 s for Ti-6Al-5Mo-5V-2Cr-1Fe alloy and t0,5 ≈ 1.5 s for Ti-6Al-4V alloy.

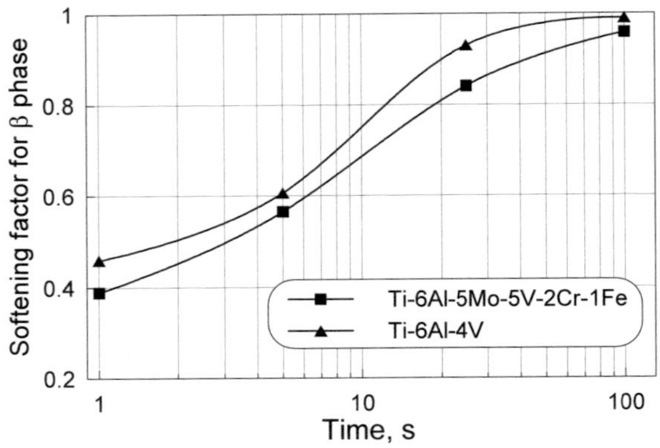

Fig. 22. The dependence of volume fraction of recrystallized β phase on dwell time during sequential deformation of Ti-6Al-4V and Ti-6Al-5V-5Mo-2Cr-1Fe alloys (Kubiak, 2004)

Microstructure of α+β titanium alloys after sequential deformation at the temperature in β field and air cooling is composed of globular and heavily deformed α grains (Fig. 23) in the matrix of the lamellar α and β phases (Fig. 24a, b). Increase in dwell time during sequential deformation leads to reduction of the strengthening effect and moreover to reduction of dispersion of the phases - growth of both globular and deformed α grains. New, recrystallized grains nucleate at the primary β grain boundaries, forming chains (Figs 23b, 24b). The dislocation density in globular and lamellar α phase and β phase is low (Fig. 24c, d).

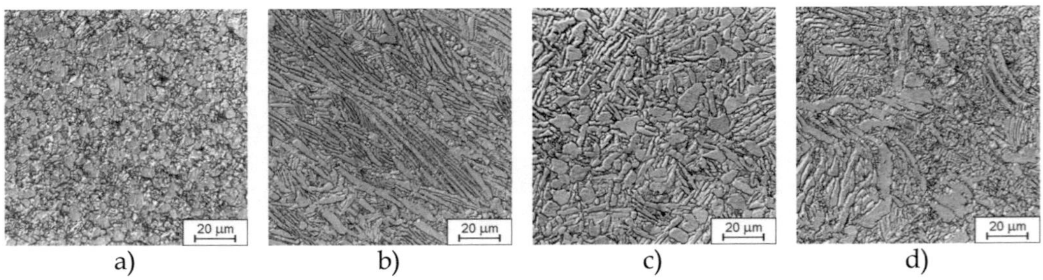

Fig. 23. The microstructure of Ti-6Al-4V alloy after sequential deformation with various dwell time: a) 1s - transverse section, b) 1s - longitudinal section, c) 100s - transverse section, b) 100s - longitudinal section (Kubiak, 2004)

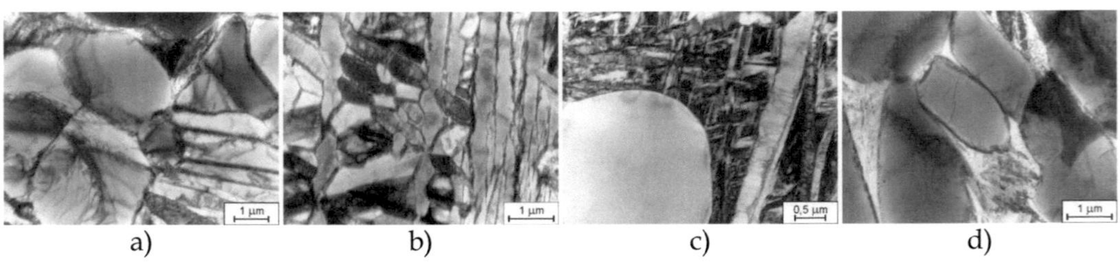

Fig. 24. The microstructure of Ti-6Al-4V alloy after sequential deformation with various dwell time: a) 1s - transverse section, b) 1s - longitudinal section, c) 100s - transverse section, b) 100s - longitudinal section (Kubiak, 2004)

In the 1980s and 1990s deformation maps for CP titanium were calculated on the basis of material constants, describing possible deformation mechanisms operating during processing. In the late 1990s the map of microstructure changes depending on deformation conditions was developed for Ti-6Al-4V alloy (Fig. 25) (Seshacharyulu et al., 1999).

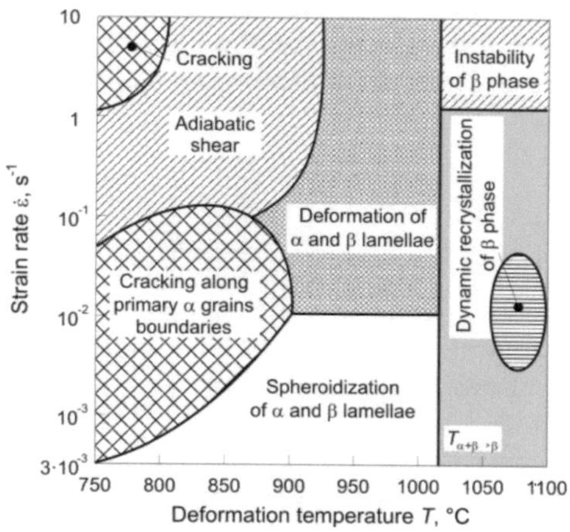

Fig. 25. The map of microstructure changes and phenomena occurring during hot deformation of Ti-6Al-4V alloy (Seshacharyulu et al., 1999)

Technological hot plasticity of α+β titanium alloys, characterized by plastic flow stress and critical strain, depends on morphology and stereological parameters of the phases in the alloy microstructure and deformation conditions (Kubiak, 2004).

The flow curves $\sigma = f(\varepsilon)$ of α+β titanium alloys have similar character. Three stages of flow stress changes can be distinguished, what is characteristic for materials in which dynamic recrystallization occurs (Fig. 26):

- increase up to σ_{pm} value,
- decrease down to σ_{ps} value,
- stabilization at σ_{ps} value.

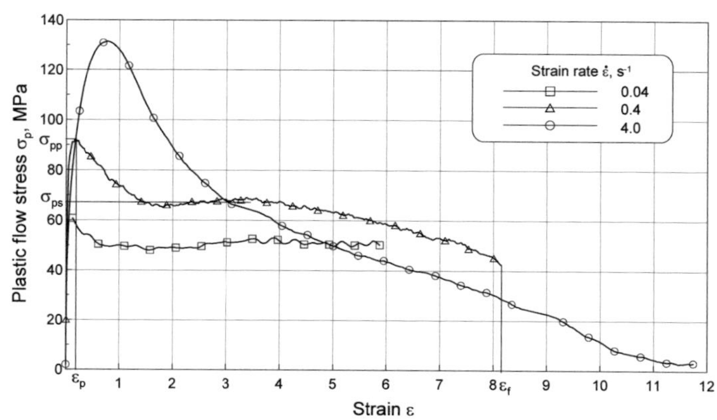

Fig. 26. The effect of strain rate on plastic flow stress at 900°C for Ti-6Al-4V alloy - coarse-grained lamellar microstructure (Kubiak, 2004)

Deformation of Ti-6Al-4V alloy at 1050°C - the range of β phase stability - results in reduction of maximum plastic flow stress σ_{pm} regardless of the initial phase morphology and dispersion. Coarse lamellar microstructure shows the maximum plastic flow stress. Reduction of the size of colonies and lamellae of α and β phases leads to significant decrease in flow stress in comparison with bi-modal and globular microstructure. It was also found that deformation rate have a pronounced effect on the flow stress σ_{pm} and ε_m strain for Ti-6Al-4V alloy. Increase in strain rate results in higher σ_{pm} and ε_m values (Figs 27 and 28).

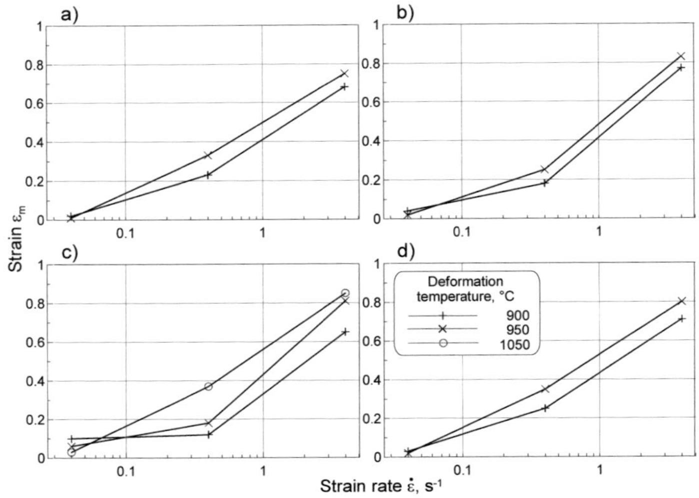

Fig. 27. The dependence of ε_m strain on strain rate, temperature of deformation and microstructure of Ti-6Al-4V alloy: a) bi-modal, b) globular, c) coarse-grained lamellar, d) fine-grained lamellar microstructure (Kubiak, 2004)

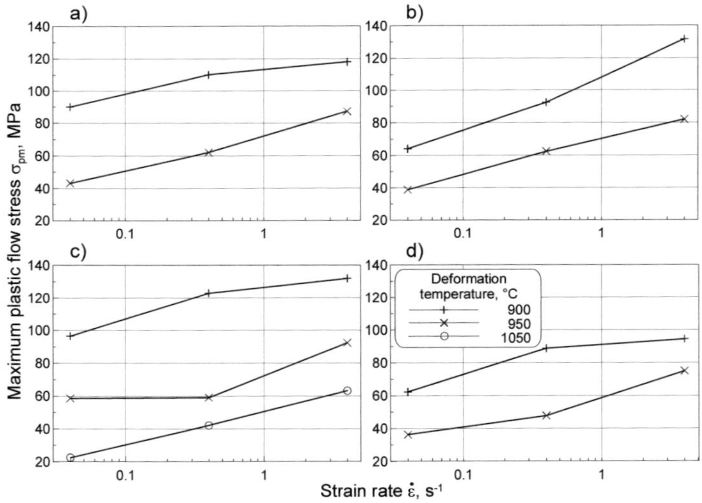

Fig. 28. The dependence of maximum plastic flow stress σ_{pm} on strain rate, temperature of deformation and microstructure of Ti-6Al-4V alloy: a) bi-modal, b) globular, c) coarse-grained lamellar, d) fine-grained lamellar microstructure (Kubiak, 2004)

The influence of microstructure morphology and deformation condition on maximum flow stress σ_{pm} for Ti-6Al-5Mo-5V-2Cr-1Fe alloy is similar to that found for Ti-6Al-4V alloy.

The effect of conditions of heat treatment and degree of plastic deformation in thermomechanical process on development of microstructure and plasticity of Ti-6Al-4V and Ti-6Al-2Mo-2Cr titanium alloys in hot tensile test was also investigated (Motyka & Sieniawski, 2010). On the basis of dilatometric results and previous findings conditions of heat treatment and plastic deformation were defined and two schemes of thermomechanical processing were worked out, denoted TMP-I and TMP-II respectively (Fig. 29).

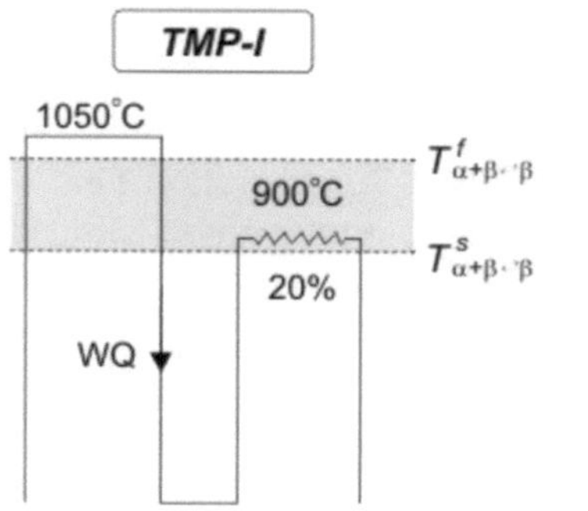

Fig. 29. Schemes of thermomechanical processing of Ti-6Al-4V alloy with forging reduction $\varepsilon \approx 20\%$ (a) and $\varepsilon \approx 50\%$ (b) (WQ - water quenching) (Motyka & Sieniawski, 2010)

Initial microstructure of Ti-6Al-4V alloy was composed of globular, fine α grains and β phase in the form of thin layers separating α grains (Fig. 30a). Quenching of Ti-6Al-4V alloy from the β phase temperature range led to formation of microstructure composed solely of martensitic α′(α″) phase (Fig. 30b). Microstructure after following plastic deformation in the α+β↔β range with forging reduction of about 20% (TMP-I) and 50% (TMP-II) comprised elongated and deformed grains of primary α phase in the matrix of β transformed phase containing fine globular grains of α secondary phase (Fig. 30c,d). Higher degree of initial deformation led to obtaining finer microstructure containing more elongated α grains – f_α = 16 for ε = 20% and 21.1 for ε = 50%. The larger volume fraction of α phase was also found (Tab. 2).

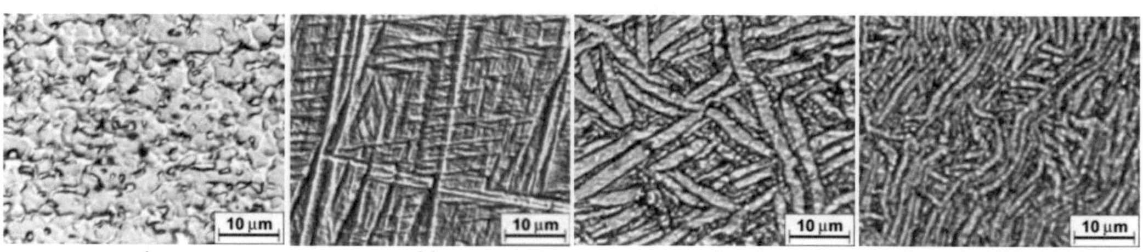

a) b) c) d)

Fig. 30. Microstructure (DIC) of Ti-6Al-4V alloy before thermomechanical processing (a), after quenching from the β phase range (b) and after deformation in the α+β↔β range with forging reduction of 20% (c) and 50% (d) (Motyka & Sieniawski, 2010)

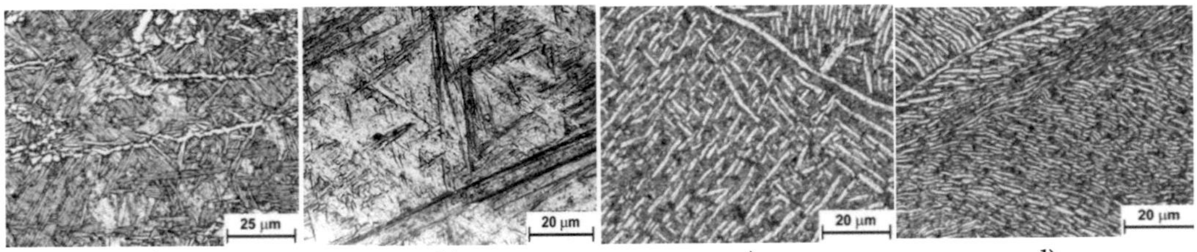

a) b) c) d)

Fig. 31. Microstructure (DIC) of Ti-6Al-2Mo-2Cr alloy before thermomechanical processing (a), after quenching from the β phase range (b) and after deformation in the α+β↔β range with forging reduction of 20% (c) and 50% (d) (Motyka & Sieniawski, 2010)

Initial microstructure of Ti-6Al-2Mo-2Cr alloy was composed of colonies of parallel α-lamellae enclosed in primary β phase grains (Fig. 31a). Solution heat treatment led to formation of microstructure composed of martensitic α′(α″) phase, similarly to Ti-6Al-4V alloy (Fig. 31b). Microstructure after thermomechanical processes (TMP-I and TMP-II) comprised fine, elongated grains of α phase in the matrix of β transformed phase (Figs 31c,d). In contrary to Ti-6Al-4V alloy primary β phase grain boundaries were observed. Higher degree of initial deformation in thermomechanical process led to obtaining finer microstructure and larger volume fraction of α phase (Table 2).

Condition of Ti-6Al-4V alloy	Stereological parameters			
	V_α %	$\bar{a}_\alpha$ µm	$\bar{b}_\alpha$ µm	$\bar{f}_\alpha$
As received	82	4.1	5.3	0.77
TMP-I processed	59	51.3	3.2	16
TMP-II processed	79	23.2	1.1	21.1

Condition of Ti-6Al-2Mo-2Cr alloy	V_α %	$\bar{a}_{\beta prim}$ µm	$\bar{b}_{\beta prim}$ µm	$\bar{f}_{\beta prim}$	R µm	g µm
As received	76	137	42	3.26	12	1
TMP-I processed	34	-	-	-	-	4
TMP-II processed	40	-	-	-	-	1

Table 2. Stereological parameters of microstructure of as-received and thermomechanically processed Ti-6Al-4V and Ti-6Al-2Mo-2Cr alloys; where: V_α - volume fraction of α phase, $\bar{a}_\alpha$ and $\bar{b}_\alpha$ - length of sides of rectangular circumscribed on α grain section, $\bar{f}_\alpha$ - elongation factor of α phase grains, $\bar{a}_{\beta prim}$ and $\bar{b}_{\beta prim}$ - length of sides of rectangular circumscribed on primary β grain section, $\bar{f}_{\beta prim}$ - elongation factor of primary β phase grains, R - size of the colony of parallel α lamellae, g - thickness of α-lamellae (Motyka & Sieniawski, 2010)

TEM examination of Ti-6Al-4V alloy revealed fragmentation of elongated α grains (Fig. 32a) and presence of globular secondary α grains in the β transformed matrix (Fig. 32b) after TMP-I thermomechanical processing. Higher dislocation density in elongated α grains was observed after TMP-II processing (larger forging reduction) (Figs 33a,b).

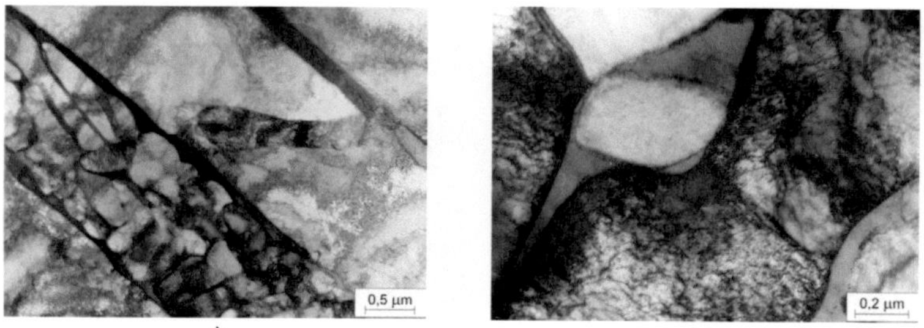

a) b)

Fig. 32. Microstructure (TEM) of Ti-6Al-4V alloy after TMP-I process: fragmentation of α phase (a), globular secondary α phase grain (b) (Motyka & Sieniawski, 2010)

In Ti-6Al-2Mo-2Cr alloy after TMP-I processing dislocations were observed mainly near grain boundaries (Fig. 34a). It was found that the secondary α phase in β transformed matrix occurs in lamellar form (Fig. 34b). Higher degree of deformation in TMP-II process led to higher dislocation density in α phase grains (Fig. 35a) and fragmentation of elongated α grains (Fig. 35b).

In Ti-6Al-2Mo-2Cr alloy higher volume fraction of β phase (Tab. 3) was found than in Ti-6Al-4V alloy which can be explained by higher value of coefficient of β phase stabilisation K_{β}.

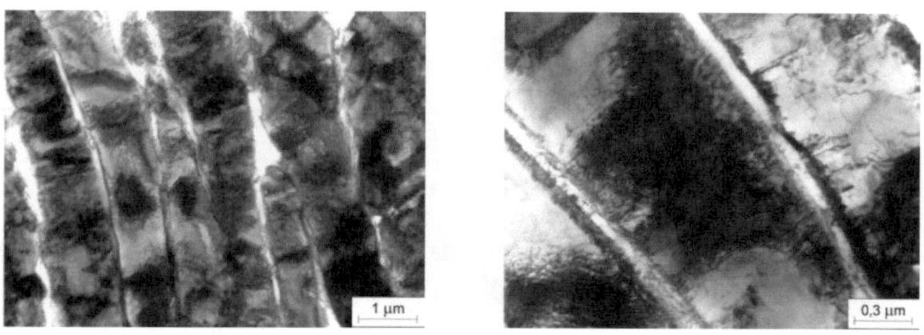

a) b)

Fig. 33. Microstructure (TEM) of Ti-6Al-4V alloy after TMP II process: high dislocation density in α grains (Motyka & Sieniawski, 2010)

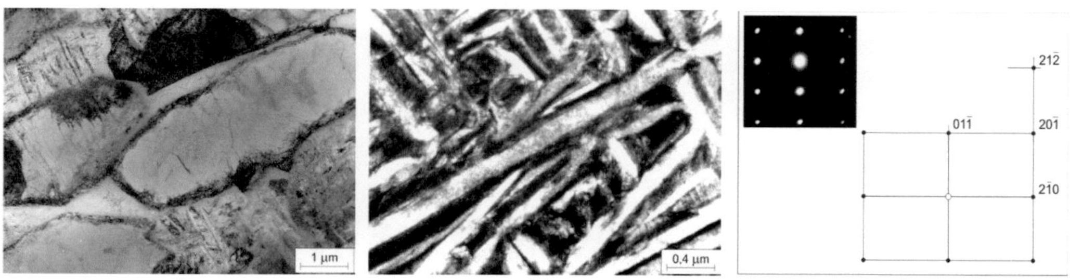

a) b) c)

Fig. 34. Microstructure (TEM) of Ti-6Al-2Mo-2Cr alloy after TMP-II process: dislocations in α grains (a), lamellae precipitations of α grains in β transformed phase (b) with indexing of diffraction (c) (Motyka & Sieniawski, 2010)

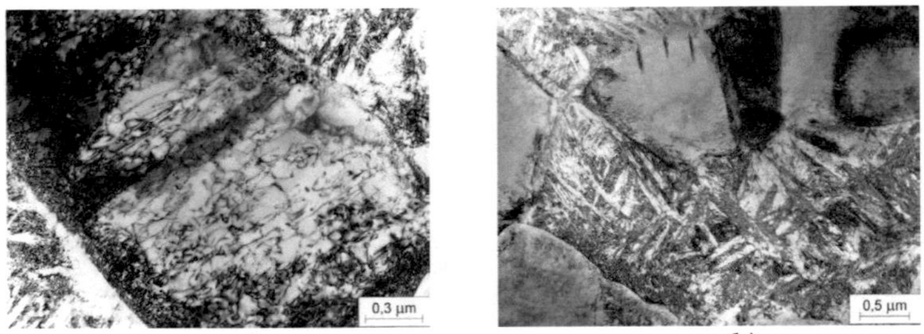

a) b)

Fig. 35. Microstructure (TEM) of Ti-6Al-2Mo-2Cr alloy after TMP-II process: dislocations in α grains (a), precipitations of α grains in β transformed phase (b) (Motyka & Sieniawski, 2010)

Critical temperatures of α+β↔β phase transformation [°C]	Condition of Ti-6Al-4V alloy		
	As-received	TMP-I processing	TMP-II processing
Start of α+β↔β	894	882	912
Finish of α+β↔β	979	976	1009
	Condition of Ti-6Al-2Mo-2Cr alloy		
	As-received	TMP-I processing	TMP-II processing
Start of α+β↔β	803	800	809
Finish of α+β↔β	991	992	1011

Table 3. Critical temperatures of α+β↔β phase transformation of as received and thermomechanically processed two-phase alloys (Motyka & Sieniawski, 2010)

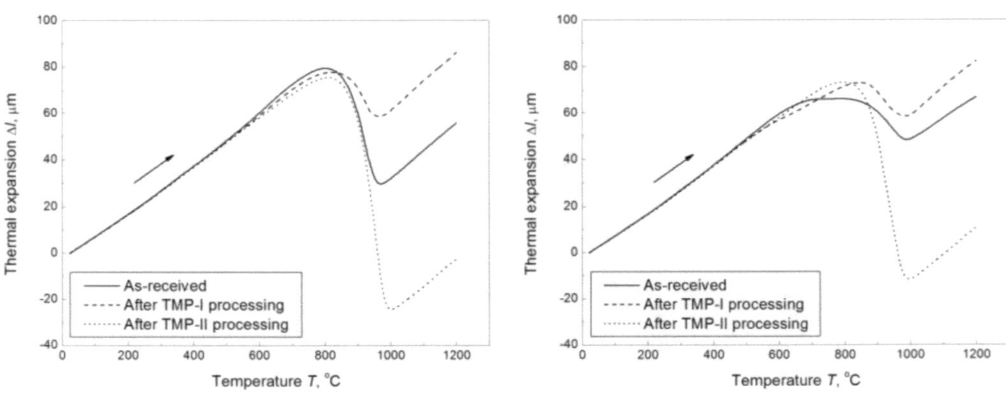

a) b)

Fig. 36. Dilatometric curves of Ti-6Al-4V (a) and Ti-6Al-2Mo-2Cr (b) alloys in as-received state and after thermomechanical processing (Motyka & Sieniawski, 2010)

Dilatometric examination revealed the influence of forging reduction on critical temperatures of $\alpha+\beta\leftrightarrow\beta$ phase transformation. The TMP-II thermomechanical processing with highest strain applied ($\varepsilon\approx50\%$) caused significant increase in finish temperature of $\alpha+\beta\leftrightarrow\beta$ phase transformation in both examined titanium alloys (Tab. 3 and Fig. 36). The temperature range of phase transformation was considerably wider in Ti-6Al-2Mo-2Cr alloy (Tab. 3).

On the basis of tensile tests at 850°C and 925°C on thermomechanically processed Ti-6Al-4V and Ti-6Al-2Mo-2Cr alloys it was found that the maximum flow stress σ_{pm} decreased with growing temperature of deformation but increased with strain rate (Fig. 37). It was found that the maximum flow stress σ_{pm} determined in tensile test is higher at lower test temperature 850°C for the strain rate range applied (Fig. 37). There is no significant effect of degree of initial deformation (forging) of two investigated alloys on σ_{pm} value for both 850°C and 925°C test temperature (Fig. 37).

The relative elongation *A* of hot deformed Ti-6Al-4V and Ti-6Al-2Mo-2Cr titanium alloys decreased with the increasing strain rate $\dot{\varepsilon}$ in the whole range applied (Fig. 38). For strain rate $\dot{\varepsilon}$ above 0.1 the influence of forging reduction ε in thermomechanical processing and tensile test temperature is very slight. Considerable differences are visible for $\dot{\varepsilon} = 1{\cdot}10^{-2}\ s^{-1}$ where the maximum *A* value was achieved for both alloys deformed at 850°C. After thermomechanical processing TMP-II ($\varepsilon \approx 50\%$) alloys exhibit maximum elongations, typical for superplastic deformation (Fig. 38). It seems that higher grain refinement obtained in thermomechanical process enhanced hot plasticity of two-phase titanium alloys deformed with low strain rates. Similar behaviour was observed in previous works on superplasticity of thermomechanically processed Ti-6Al-4V alloy (Motyka, 2007; Motyka & Sieniawski, 2004). It was found that fragmentation and globularization of elongated α phase grains during initial stage of hot deformation restricted grain growth and resulted in higher values of total elongation in tensile test.

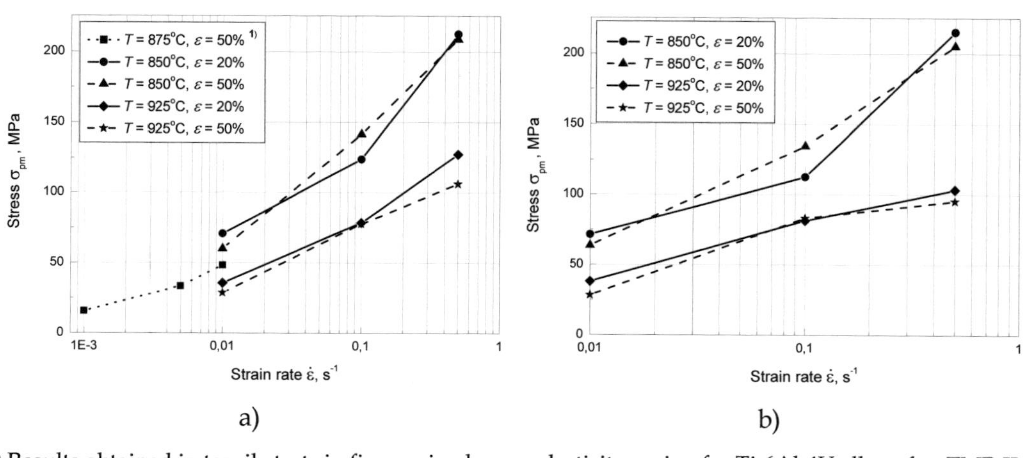

[1]) Results obtained in tensile tests in fine-grained superplasticity region for Ti-6Al-4V alloy after TMP-II processing (Motyka & Sieniawski, 2004)

Fig. 37. The σ_{pm} - $\dot{\varepsilon}$ dependence (on the basis of tensile test) for Ti-6Al-4V (a) and Ti-6Al-2Mo-2Cr (b) alloys after processing TMP-I and TMP-II (Motyka & Sieniawski, 2010)

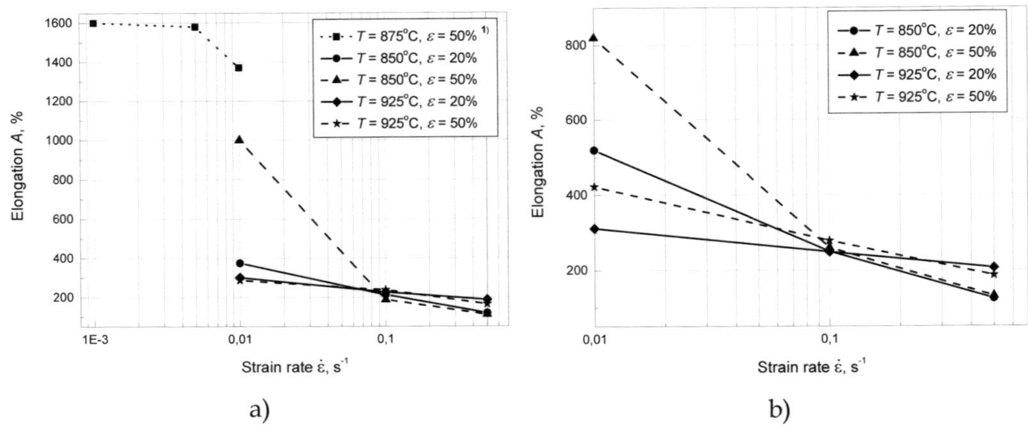

1) Results obtained in tensile tests in fine-grained superplasticity region for Ti-6Al-4V alloy after TMP-II processing (Motyka & Sieniawski, 2004)

Fig. 38. The A - $\dot{\varepsilon}$ dependence (on the basis of tensile test) for Ti-6Al-4V (a) and Ti-6Al-2Mo-2Cr (b) alloys after processing TMP-I and TMP-II (Motyka & Sieniawski, 2010)

4. Superplasticity of titanium alloys

Superplasticity is the ability of polycrystalline materials to exhibit very high value of strain (tensile elongation can be even more than 2000%), appearing in high homologous temperature under exceptionally low stress which is strongly dependent on strain rate. Generally two types of superplasticity are distinguished: fine-structure superplasticity (*FSS*) - considered as an internal structural feature of material and internal-stress superplasticity (*ISS*) caused by special external conditions (e.g. thermal or pressure cycling) generating internal structural transformations that produce high internal stresses independent on external stresses.

FSS phenomenon is observed in isotropic fine-grained metallic materials under special conditions: limited range of low strain rates and temperature above 0.4 T_m. Main features of superplastic deformation are: high value of strain rate sensitivity parameter ($m > 0.3$), lack of strain hardening, equiaxial shape of grains not undergoing changes, conversion of texture during deformation, low activity of lattice dislocations in grains and occurrence of intensive grain boundary sliding (GBS) with associated accommodation mechanisms (Grabski, 1973; Nieh et al., 1997).

One of the titanium alloys which has been extensively studied in aspect of superplasticity is widely used Ti-6Al-4V alloy. Results concerning research on this alloy published in world scientific literature indicate meaningful progress in evaluation and applications of superplasticity in last 30 years. In the beginning of 70's maximum superplastic tensile elongation of Ti-6Al-4V alloy was about 1000% at the strain rate of 10^{-4} s^{-1} (Grabski, 1973), whereas in few last years special thermomechanical methods were developed that enabled doubling the tensile elongation and increasing strain rate by the factor of 100 (Inagaki, 1996) (Table 4).

Alloy	Phase composition	Elongation ε [%]	Grain size $\overline{d}$ [μm]	Temperature T [°C]	Strain rate $\dot{\varepsilon}$ [s^{-1}]
		Two-phase α + β alloys			
Ti-4Al-4Mo-2Sn-0.5Si (*IMI550*)	α + β	2000	4	885	$5 \cdot 10^{-4}$
Ti-4.5Al-3V-2Mo-2Fe (*SP-700*)	α + β	2500	2-3	750	10^{-3}
Ti-5Al-2Sn-4Zr-4Mo-2Cr-1Fe (*β-CEZ*)	α + β	1100	2-3	72	$2 \cdot 10^{-4}$
Ti-6Al-4V	α + β	2100	2	850	10^{-2}
Ti-6Al-2Sn-4Zr-2Mo	α + β	2700	1-2	900	10^{-2}
Ti-6Al-2Sn-4Zr-6Mo	α + β	2200	1-2	750	10^{-2}
Ti-6Al-7Nb (*IMI367*)	α + β	300	6	900	$3 \cdot 10^{-4}$
Ti-6.5Al-3.7Mo-1,5Zr	α + β	640	6-7	600	10^{-4}
Ti-6Al-2Sn-2Zr-2Mo-2Cr-0,15Si	α + β	2000	4	885	$5 \cdot 10^{-4}$
		Intermetallics based alloys			
Ti-24Al-11Nb	α_2 (Ti_3Al) + β	1280	4	970	10^{-3}
Ti-46Al-1Cr-0.2Si	γ (TiAl) + α_2 (Ti_3Al)	380	2-5	1050	10^{-3}
Ti-48Al-2Nb-2Cr	γ (TiAl) + α_2 (Ti_3Al)	350	0.3	800	$8.3 \cdot 10^{-4}$
Ti-50Al	γ (TiAl) + α_2 (Ti_3Al)	250	<5	900-1050	$2 \cdot 10^{-4}$- $8.3 \cdot 10^{-3}$
Ti-10Co-4Al	α + Ti_2Co	1000	0.5	700	$5 \cdot 10^{-2}$
		Titanium matrix composites			
Ti-6Al-4V + 10%TiC	α + TiC	270	5	870	$1.7 \cdot 10^{-4}$
Ti-6Al-4V + 10%TiN	α + TiN	410	5	920	$1.7 \cdot 10^{-4}$

Table 4. Superplastic deformation conditions of selected titanium alloys and titanium matrix composites (Sieniawski & Motyka, 2007)

Relatively new group of superplastic titanium alloys are TiAl or Ti_3Al intermetallics based alloys (Tab. 2). It is well known that intermetallics based alloys have a high relative strength, and good high-temperature creep resistance. Widespread usage of those materials is limited mainly by their low plasticity precluding forming of structural components using conventional plastic working methods. In this case pursuit to obtain fine-grained microstructure enabling superplastic deformation seems to be very promising [Hofmann et al., 1995; Imayev et al., 1999; Kobayashi et al., 1994; Nieh et al., 1997).

Main criterion for superplastic materials is possibility of obtaining fine-grained and equiaxial microstructure. Desired microstructure is most often obtained by conventional plastic working methods coupled with suitable heat treatment and severe plastic deformation methods (i.e. equal-channel angular pressing – ECAP). Superplastic forming (SPF) of titanium alloys is limited by relatively long time and high deformation temperature. It was established that grain refinement causes increase of strain rate and decrease of superplastic deformation temperature (Fig. 39) (Sieniawski & Motyka, 2007).

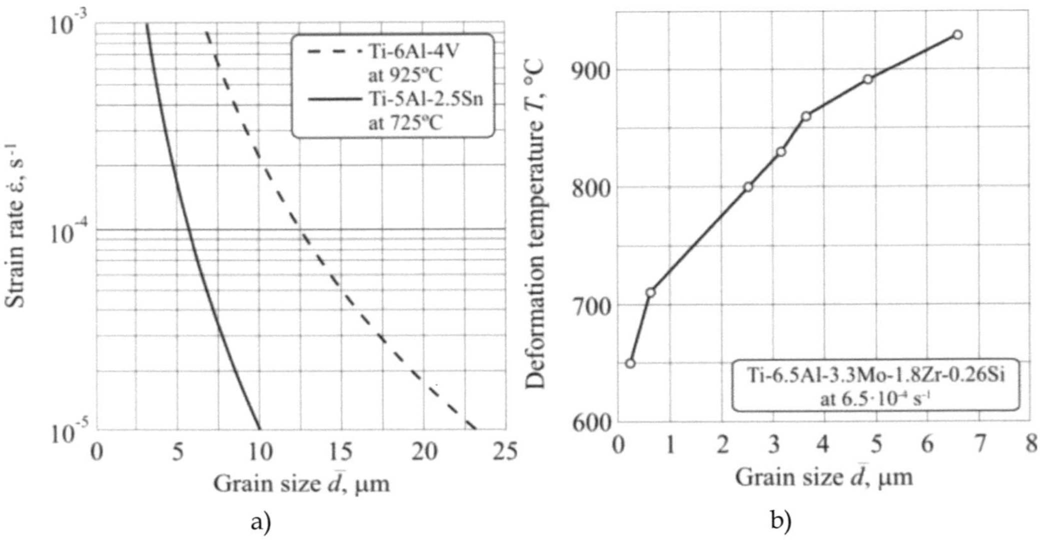

Fig. 39. Effect of grain size on strain rate of superplastic deformation for Ti-6Al-4V and Ti-5Al-2.5Sn alloys (a) and on superplastic deformation temperature for Ti-6.5Al-3.3Mo-1.8Zr-0.26Si alloy (b) (Sieniawski & Motyka, 2007)

Taking into account the mechanism of superplastic deformation equiaxed microstructure favours proceeding of GBS. It was found that in fine grained polycrystalline materials with grains elongated crosswise deformation direction GBS is limited. The main reason is difficulty of deformation accommodation in triple points. Transverse deformation is also related to cavities formation along grain boundaries and precludes superplastic deformation (Nieh et al., 1997). It is emphasised that superplastic deformation does not cause shape changes of equiaxed grains. However, gradual transformation of texture is observed what indicates that GBS plays a crucial role in superplastic deformation (Grabski, 1973; Zelin, 1996).

On the basis of the results of research works conducted at the Department of Materials Science of Rzeszow University of Technology it was found that initial microstructure of superplastic titanium alloy can be different from equiaxed one. High superplasticity was observed in Ti-6Al-4V alloy with microstructure composed of strongly elongated and deformed α grains (Fig. 40a) (Motyka & Sieniawski, 2004). It was established that during heating and first stage of superplastic deformation significant changes of the morphology of phases occur (Fig. 40b) (Motyka, 2007).

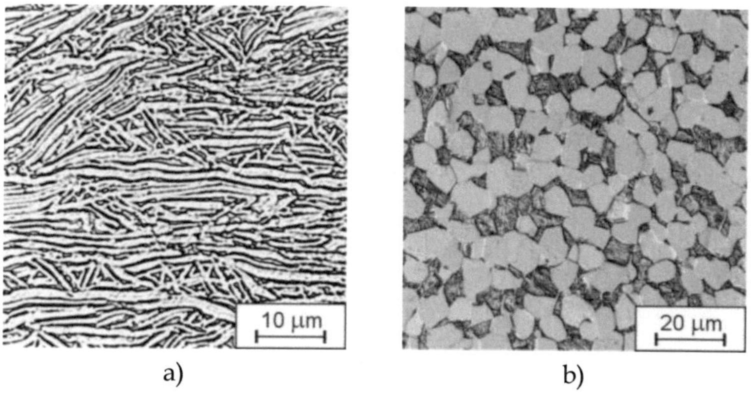

Fig. 40. Microstructure of Ti-6Al-4V alloy before (a) and after superplastic deformation (b) - temperature 850°C and strain rate of 10^{-3} s^{-1} (Sieniawski & Motyka, 2007)

Along with grain size and shape, volume fraction of particular phases in the alloy also affects its superplasticity. Properties of phases in two-phase α+β titanium alloys differ considerably. α phase (hcp) has less slip systems and two order of magnitude lower self-diffusion coefficient than β phase (bcc). These features suggest that in the superplasticity conditions α phase has a higher plasticity than β phase. It was confirmed by results obtained from experiments on superplasticity in Ti-6Al-4V alloy where deformation in α grains was observed. Density of dislocations was found to be very low in β grains (Bylica & Sieniawski, 1985; Inagaki, 1996; Jain et al., 1991, Meier et al., 1991; Nieh et al., 1997).

It was established that increase in volume fraction of β phase in alloy causes decrease of the effect of α grain size (Meier et al., 1991). Maximum values of elongation and strain rate sensitivity factor m as a function of β volume fraction is shown in Figure 41. Increase in relative volume of β phase causes improvement of superplasticity of titanium alloys. The best superplastic properties of two-phase α+β titanium alloys are achieved for 40-50% volume fraction of β phase (Nieh et al., 1997). Whereas similar properties of intermetallics based alloys are possible for about (20-30)% volume fraction of β phase (Kim et al., 1999; Lee et al., 1995).

Superplasticity of titanium alloys depends on relationship between grain growth control and plasticity. β grains are characterized by high diffusivity therefore they grow extremely rapidly at the superplastic deformation temperature which does not favour superplastic flow (Meier et al., 1992). Particular volume fraction of α phase considerably limits β grains growth because in this case long distance diffusion of alloying elements is necessary (e.g. vanadium in β phase). The second phase, besides stabilization of microstructure, influences the rate of grain boundary (α/α, β/β) and phase boundary (α/β) sliding (Inagaki, 1996; Jain et al., 1991, Meier et al., 1991; Nieh et al., 1997). Increase of volume fraction of β phase causes decrease of α/α grain boundary areas and consequently their contribution to deformation by GBS. It is thought that improvement of superplasticity of α+β titanium alloys caused by increase of volume of β phase should be considered in following aspects (Inagaki, 1996): α/β phase boundary sliding, β/β GBS and contribution of other deformation mechanisms.

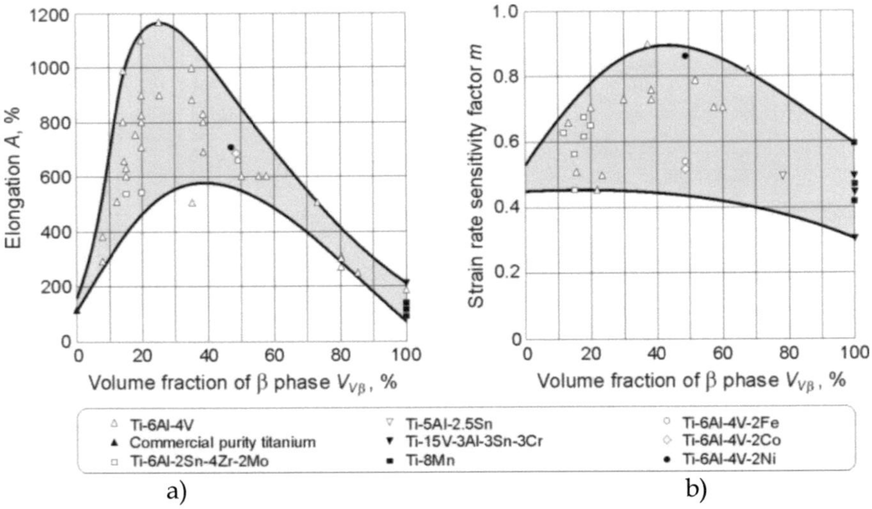

Fig. 41. Effect of volume fraction of β phase on elongation ε (a) and strain rate sensitivity *m* (b) in selected titanium alloys (Nieh et al., 1997)

Most often microstructure of α+β superplastic titanium alloys is composed of α and β grains which have similar size and shape. Interesting results was obtained for Ti-6Al-4V alloy where α grains were separated by thin films of β phase. Superplastic elongation in this case was more than 2000%. Further investigations indicated that during superplastic deformation thin films of β phase coagulated in triple points into larger particles having irregular forms. Thanks to that α/α grain boundaries free of β thin films were formed. It can be expected that sliding along these grain boundaries proceeds easily. However it was revealed that at this stage superplastic deformation is almost completed and deformation within grains becomes dominant deformation mechanism. It seems that α/α grain boundary sliding is not dominant superplastic deformation mechanism. In this case the effect of β phase thin film can be comparable to role of grain boundaries in single phase materials. Slip and shearing in β phase thin film is caused by movement and rotation of neighbouring α grains. Mentioned processes enable accommodation of grain boundary and phase boundary sliding (Inagaki, 1996). Other investigations also indicate accommodative role of β phase, in which substantially higher dislocations density is observed than in α phase grains. It was noticed simultaneously that dislocations density in α phase increases together with decrease in temperature and increase in strain rate of superplastic deformation (Kim et al., 1999; Meier et al., 1991). Superplasticity of titanium alloys with intermetallic phases like Ti-12Co-5Al and Ti-6Co-6Ni-5Al is observed for grain size about 0.5 μm. Particles of Ti_2Co and Ti_2Ni phases (about 27% of volume) advantageously influence the grain refinement and limit grain growth during superplastic deformation (Nieh et al., 1997).

5. Conclusion

Hot plasticity of two-phase titanium alloys strongly depends on values of stereological parameters of microstructure. It is possible to develop appropriate microstructure of these alloys yielding optimum plastic flow stress and critical strain values based on technological plasticity criterion. The value of critical strain depends on microstructure morphology and deformation conditions. The character of dependence of plastic flow stress as well as results of microstructure examination support conclusion that dynamic processes of microstructure recovery take place above the temperature range of $\alpha+\beta\leftrightarrow\beta$ phase transformation.

Thermomechanical processing enables microstructure and hot plasticity development of two-phase $\alpha+\beta$ titanium alloys. Increase in degree of initial deformation (forging) in proposed thermomechanical processing leads to formation of more elongated and refined α grains in tested $\alpha+\beta$ alloys. The most significant effect of degree of initial deformation occurs for the lowest strain rate and lower tensile test temperature used, resulting in considerable rise of elongation *A*.

High superplasticity of the Ti-6Al-4V alloy does not necessarily require equiaxial microstructure. Changes of the morphology of phases during heating and first stage of superplastic deformation enables superplastic behaviour of the alloy with initial microstructure composed of strongly elongated and deformed α grains.

6. References

Balasubramanian, S. & Anand, L. (2002). Plasticity of Initially Textured Hexagonal Polycrystals at High Homologous Temperatures: Application to Titanium, In: *Acta Materialia,* Vol. 50, pp. 133-148, ISSN 1359-6454

Brooks, J.W. (1996). Processing Wrought Nickel and Titanium Superalloys, *Proceedings of Int. Conf. "Thermomechanical Processing in Theory, Modelling and Practice [TMP]2"*, ISBN 916-305-4213, Stockholm, Sweden, September 1996

Bylica, A. & Sieniawski, J. (1985). *Titanium and Its Alloys* (in Polish), PWN, ISBN 83-0105-888-9, Warsaw, Poland

Churchman, A.T. (1955). The Formation and Removal of Twins in Ti at Low and High Temperatures, In: *Journal of the Institute of Metals,* Vol. 83, pp. 39-40

Ding, R., Guo, Z.X. & Wilson A. (2002). Microstructural Evolution of a Ti-6Al-4V Alloy During Thermomechanical Processing, In: *Materials Science and Engineering,* Vol. A327, pp. 233-245, ISSN 0921-5093

Ezugwu, E.O. & Wang, Z.M. (1997). Titanium Alloys and their Machinability - A Review, In: *Journal of Materials Processing Technology,* Vol. 68, pp. 262-274, ISSN 0924-0136

Ghosh, A.K. & Hamilton, C. H. (1979). Mechanical Behaviour and Hardening Characteristics of a Superplastic Ti-6Al-4V Alloy, In: *Metallurgical Transactions,* Vol. 10A, pp. 699-706, ISSN 0026-086X

Glazunov, S. & Kolachev, B.A. (1980). *Physical Metallurgy of Titanium Alloys* (in Russian), Mietallurgiya, Moscow, Russia

Glazunov, S. & Mojsiejew, W.N. (1974). *Structural Titanium Alloys* (in Russian), Mietallurgiya, Moscow, Russia

Grabski, M. (1973). *Nadplastyczność strukturalna metali,* Śląsk, Katowice, Poland

Hadasik, E. (1979). *Analysis of plastic working processes of titanium alloys based on plastometer examinations* (in Polish), PhD Thesis (not published), Katowice, Poland

Hofmann, H., Frommeyer, G. & Herzog, W. (1995) Dislocation Creep Controlled Superplasticity at High Strain Rates in the Ultrafine Grained Quasi-eutectoid Ti-10Co-4Al Alloy, *Proceedings of the Conference „Titanium '95: Science and Technology"*, ISBN 978-186-1250-05-6, Birmingham, United Kingdom

Imayev, V.M., Salishchev G.A., Shagiev, M.R., Kuznetsov, A.V., Imayev, R.M., Senkov, O.N. & Froes F.H. (1999), Low-temperature Superplasticity of Submicrocrystalline Ti-48Al-2Nb-2Cr Alloy Produced by Multiple Forging, *Scripta Materialia*, Vol. 40, No. 2, pp. 183-190, ISSN 1359-6462

Inagaki, H. (1996). Mechanism of Enhanced Superplasticity in Thermomechanically Processed Ti-6Al-4V, *Zeitschrift für Metallkunde*, Vol. 87, No. 3, pp. 179-186, ISSN 0044-3093

Jain, M., Chaturvedi, M.C., Richards, N.L. & Goel, N.C. (1991). Microstructural Characteristics in α Phase During Superplastic Deformation of Ti-6Al-4V, *Materials Science Engineering*, Vol. A145, pp. 205-214, ISSN 0921-5093

Kajbyszew, O.A. & Krajuchijn, W.I. (1967). Properties of Deformation Mechanism of Magnesium Alloy under Hydrodynamic Load (in Russian), In: *Fizika I Mietalowiedienije Materialow*, Vol. 3, No. 24, pp. 553-557

Kim, J.H., Park, C.G., Ha, T.K. & Chang, Y.W. (1999). Microscopic Observation of Superplastic Deformation in a 2-phase Ti_3Al-Nb Alloy, *Materials Science Engineering*, Vol. A269, pp. 197-204, ISSN 0921-5093

Kobayashi, M, Ochiai, S., Funami, K., Ouchi, C. & Suzuki, S. (1994). Superplasticity of Fine TiC and TiN Dispersed Ti-6Al-4V Alloy Composites, *Materials Science Forum*, Vol. 170-172, pp. 549-554, ISSN 0255-5476

Kubiak, K. & Sieniawski J. (1998). Development of the Microstructure and Fatigue Strength of Two Phase Titanium Alloys in the Processes of Forging and Heat Treatment, *Journal of Materials Processing Technology*, Vol. 78, pp. 117-121, ISSN 0924-0136

Kubiak, K. (2004). *Technological Plasticity of Hot Deformed Two-Phase Titanium Alloys* (in Polish), Rzeszow University of Technology Press, ISBN 83-7199-332-3, Rzeszow, Poland

Lee, C.S., Kim, J.S., Lee, Y.T. & Froes F.H. (1995). Superplastic Deformation Behavior of Ti_3Al-based Alloys, *Proceedings of the Conference „Titanium '95: Science and Technology"*, ISBN 978-186-1250-05-6, Birmingham, United Kingdom

Lee, W-S. & Lin, M-T. (1997). The Effects of Strain Rate and Temperature on the Compressive Deformation Behaviour of Ti-6Al-4V Alloy, In: *Journal of Materials Processing Technology*, Vol. 71, pp. 235-246, ISSN 0924-0136

Levine, E.D. (1966). Deformation Mechanisms in Titanium at Low Temperature, In: *Transactions of the Metallurgical Society of AIME*, Vol. 236, pp. 1558-1565, ISSN 0543-5722

Lütjering, G. (1998). Influence of Processing on Microstructure and Mechanical Properties of (α+β) Titanium Alloys, *Materials Science and Engineering*, Vol. A243, (1998), pp. 32-45, ISSN 0921-5093

Mc Quillan, A.D. & Mc Quillan, M.K. (1956). Titanium, In: *Metallurgy of the Rarer Metals*, H. M. Finniston, (Ed.), 1-50, Butterworths Scientific Publication, ISBN 031-742-1379, London, United Kingdom

Meier, M.L., Lesuer, D.R. & Mukherjee A.K. (1991). α Grain Size and β Volume Fraction Aspects of the Superplasticity of Ti-6Al-4V, *Materials Science Engineering*, Vol. A136, pp. 71-78, ISSN 0921-5093

Meier, M.L., Lesuer, D.R. & Mukherjee A.K. (1992). The Effects of the α/β Phase Proportion on the Superplasticity of Ti-6Al-4V and Iron-modified Ti-6Al-4V, *Materials Science Engineering*, Vol. A154, pp. 165-173, ISSN 0921-5093

Motyka, M. & Sieniawski, J. (2004). The Influence of Thermomechanical Process Conditions on Superplastic Behaviour of Ti-6Al-4V Titanium Alloy, *Advances in Manufacturing Science and Technology*, Vol. 28, pp. 31-43, ISSN 0137-4478

Motyka, M. & Sieniawski, J. (2010). The Influence of Initial Plastic Deformation on Microstructure and Hot Plasticity of α+β Titanium Alloys, *Archives of Materials Science and Engineering*, Vol. 41, No. 2, pp. 95-103, ISSN 1897-2764

Motyka, M. (2007). Evaluation of Microstructural Transformation During Superplastic Deformation of Thermomechanically Processed Ti-6Al-4V Alloy, *Advances in Materials Science*, Vol. 7, pp. 95-101, ISSN 1730-2439

Nieh, T.G., Wadsworth, J. & Sherby O.D. (1997). *Superplasticity in Metals and Ceramics*, Cambridge University Press, ISBN 978-052-1020-34-3, Cambridge, United Kingdom

Peters, M., Lütjering, G. & Ziegler, G. (1983). Control of microstructures of (α+β) - titanium alloys, *Zeitschrift für Metallkunde*, Vol. 74, No. 5, pp. 274-282, ISSN 0044-3093

Raub, E. & Röschel, E. (1963). Ruthenium with Titanium and Zirconium Alloys (in German), In: *Zeitschrift für Metallkunde*, Vol. 54, pp. 455-462, ISSN 0044-3093

Sakai, T. (1995). Dynamic Recrystallization Microstructure Under Hot Working Conditions. In: *Journal of Materials Processing Technology*, Vol. 53, No. 1-3, pp. 349-361, ISSN 0924-0136

Semiatin, S.L., Seetharaman, V. & Weiss, I. (1998). Hot Workability of Titanium and Titanium Aluminide Alloys - An Overview, In: *Materials Science and Engineering*, Vol. A243, pp. 1-24, ISSN 0921-5093

Seshacharyulu, T., Medeiros, S.C., Frazier W.G. & Prasad Y.V.R.K. (2002). Microstructural Mechanism During Hot Working of Commercial Grade Ti-6Al-4V with Lamellar Starting Structure, In: *Materials Science and Engineering*, Vol. A325, pp. 112-125, ISSN 0921-5093

Seshacharyulu, T., Medeiros, S.C., Morgan, J.T., Malas, J.C., Frazier, W.G. & Prasad Y.V.R.K. (1999). Hot Deformation Mechanisms in ELI Grade Ti-6Al-4V, *Scripta Materialia*, Vol. 41, pp. 283-288, ISSN 1359-6462

Sheppard, T. & Norley, J. (1988). Deformations Characteristics of Ti-6Al-4V, In: *Materials Science Technology*, Vol. 4, pp. 903-908, ISSN 0267-0836

Sieniawski, J. & Motyka M. (2007). Superplasticity in Titanium Alloys, *Journal of Achievements in Materials and Manufacturing Engineering*, Vol. 24, No. 1, pp. 123-130, ISSN 1734-8412

Wassermann, G. & Grewen, J. (1962). *Texturen Metallischer Werkstoffe,* Springer Verlag, ISBN 354-002-9214, Berlin, Germany

Zelin, M.G. (1996). Processes of Microstructural Evolution During Superplastic Deformation, *Materials Characterization,* Vol. 37, pp. 311-329, ISSN 1044-5803

Zwicker, U. (1974). *Titanium and Titanium Alloys* (in German), Springer Verlag, ISBN 978-354-0052-33-3, Berlin-Heidelberg-New York

6

Machinability of Titanium Alloys in Drilling

Safian Sharif[1], Erween Abd Rahim[2] and Hiroyuki Sasahara[3]
[1]*Universiti Teknologi Malaysia,*
[2]*Universiti Tun Hussein Onn Malaysia,*
[3]*Tokyo University of Agriculture and Technology,*
[1,2]*Malaysia*
[3]*Japan*

1. Introduction

1.1 Drilling technology

Hole making is an essential process in the structural frames of an aircraft and contributes to 40 to 60% of the total material removal operations (Brinksmeier, 1990). This process is commonly divided into short hole or deep hole drilling. Short hole drilling typically covers holes with a small depth to diameter ratio having diameter up to 30 mm and a depth of not more than 5 times the diameter. Meanwhile deep hole drilling caters for holes greater than 30 mm in diameter and the depths are usually greater than 2.5 times the hole diameter. Drilling deeper hole with conventional drills requires pecking method to enable easy flow of the chips out of the hole. Deep hole drilling is more difficult especially when hole straightness is the main concern. Therefore, a usual method is to make a circular cut using a hollow-core cutting tool. This technique allows larger hole diameter to be drilled with lesser power. In addition, holes can be produced in many forms which include through holes or blind holes (Fig. 1). Through hole is one which is drilled completely through the workpiece while a blind hole is drilled only to a certain depth.

A twist drill is fabricated with 3 major parts as shown in Fig. 2. The most important features from the analytical point of view are rake angle, point angle, web thickness, nominal clearance angle, drill diameter, inclination angle and chisel edge angle. The rake angle is usually specified as helix angle at the periphery. The direction of the chip flow is attributed to the point angle. The torque decreases with increasing point angle due to the increase of orthogonal rake angle at each point on the main cutting edges. Furthermore, the thrust force always increases with increasing point angle.

Fig. 3 shows the phases involved in a drilling operation, first is the start and centering phase, second is the full drilling phase and finally the break through phase (Tonshoff et al., 1994). To ensure good surface quality and accuracy of the holes are achieved, the first phase is very important (Fig. 3 (a)) in order to avoid the occurrence of premature wear and breakage of the drill. In this phase, the torque and force on the tool constantly increase. The full drilling phase starts once the main cutting edges are fully engaged (Fig. 3 (b)). The break through phase begins when the drill point breaks through the underside of the work piece and the process is stopped when the drill body passed through the work piece (Fig. 3 (c)).

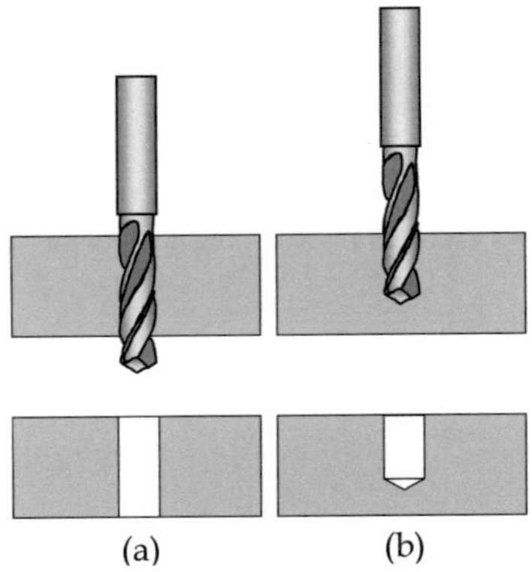

Fig. 1. Type of hole, (a) Through hole and (b) blind hole

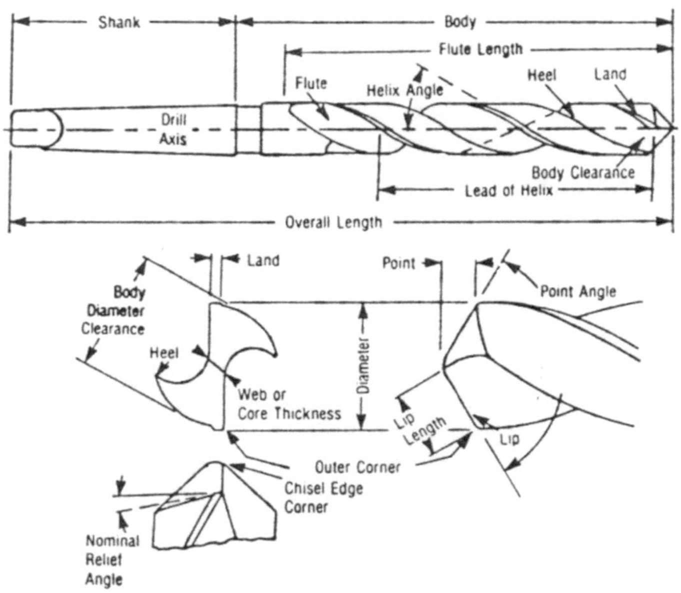

Fig. 2. Drill geometry (Lindberg, 1990)

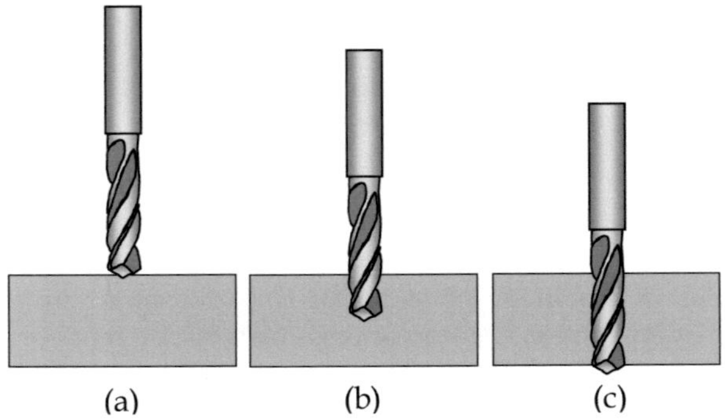

Fig. 3. Drilling phases, (a) centering phase, (b) full drilling phase and (c) break through phase

2. Tool wear in drilling process

Heat generation, pressure, friction and stress distribution are the main contributors of drill wear. The drill wear can be classified into (Kanai et al., 1978): outer corner (w), flank wear (V_b), margin wear (M_w), crater wear (K_M), along with two types of chisel edge wear (C_T and C_M) and chipping at the cutting lips (P_T and P_M). Fig. 4 shows the aforementioned types of wear. Wear starts at the sharp corners of the cutting edges and distributed along the cutting edges until the chisel and drill margin (Schnieder, 2001). Flank wear is considered as one of the criterion to measure the performance of a drill. It occurs due to the friction between the workpiece and the contact area on the clearance surface. However, Kanai et al. (1978) suggested that outer corner wear should be used as the main criteria of tool performance because of the relative ease of measurement and the close relationship between this type of wear and the drill life.

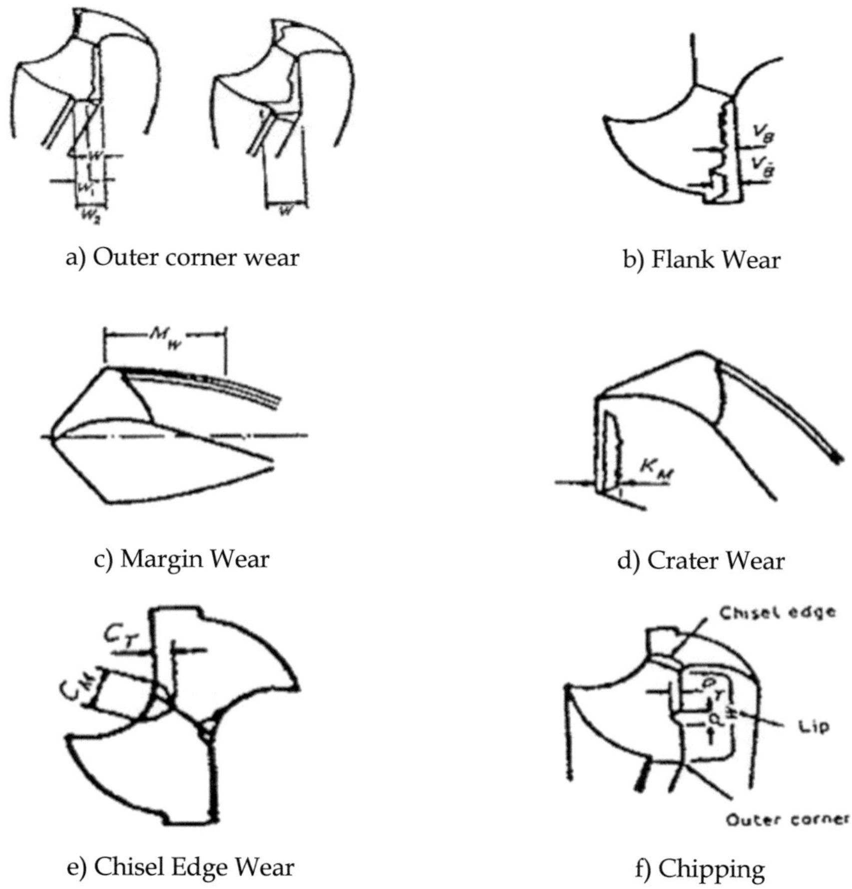

Fig. 4. Types of drill wear (Kanai et al., 1978)

Crater wear was also observed on the rake face of the drill and can be found clearly around the outer corners of the cutting edges (Choudhury & Raju, 2000; Kaldor & Lenz, 1980). According to Dolinsek et al., (2001), wear land behind the cutting edges is less significant as an indicator of tool wear because it depends on the relief angle. They suggested that the drill will be considered damaged once the corner of the drill has been rounded off as shown in Fig. 5. However, Fujise and Ohtani (1998) and Harris et al. (2003) considered the outer corner wear as their tool rejection criteria (Fig. 6). The tools were rejected when the outer corner wear reached 75% of the total margin width. Kaldor and Lenz (1980) also employed the corner wear as the tool life criterion in drilling because of the similar wear behavior of other cutting tools.

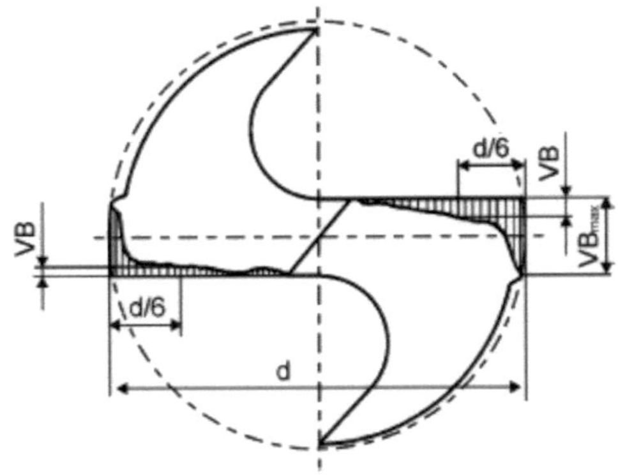

Fig. 5. Location of flank wear land on the drill (Dolinsek et al., 2001)

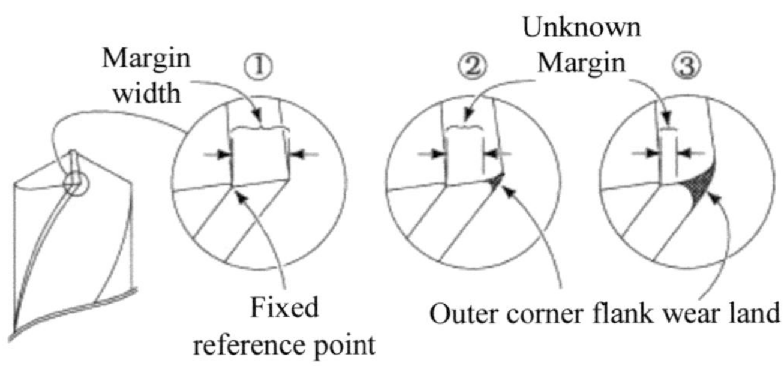

Fig. 6. A method to measure outer corner wear from a fixed reference point (Harris et al., 2003)

Tetsutaro & Zhao (1989) considered that the tool is rejected when the maximum flank wear width, $V_{b,max}$ reached 0.7 mm when drilling plain steel. Wen & Xiao (2000) used to measure the wear width developed on the flank surfaces when they drilled stainless steel. Lin (2002) rejected the tool based on the tool rejection criteria when maximum flank wear land exceeded 0.8 mm, surface roughness value exceeded 5.0 μm, excessive outer corner tearing and chipping of the helix flutes. Choudhury & Raju (2000) have studied the influence of feed and speed on crater wear at different points along the cutting lip in drilling. Ezugwu & Lai (1995) rejected the drill bit when maximum flank wear in excess of 0.38 mm on any of the drill lips, a squeaking noise occurring during machining and fracture or catastrophic failure of the drill. These criteria were used when they investigated the drilling of Inconel 901 using HSS drills.

3. Titanium alloys

Lightweight materials such as titanium alloys are now being constituted in modern aircraft structure especially in jet engine components that are subjected to temperatures up to 1000° C. Titanium alloys possess the best combination of physical and metallurgical properties and have established to be quite attractive as engineering materials due to their high strength-to-weight ratio, low density, excellent corrosion resistance, excellent erosion resistance and low modulus of elasticity (Brewer et al., 1998)

Titanium alloys are classified into groups based on the alloying elements and the resultant predominant room temperature constituent phases. These groups include α alloy, α- β alloy and β alloy. The α alloys can be divided into two types, commercially pure grades of titanium and those with additions of α- stabilizers such as Al and Sn. α alloys are non-heat treatable and are generally very weldable. They have low to medium strength, good notch toughness, reasonably good ductility and possess excellent mechanical properties which offer optimum high temperature creep strength and oxidation resistance (Boyer, 1996; Ezugwu and Wang, 1997). These include alloys such as Ti-3Al-2.5V, Ti-5Al-2.5Sn, Ti-8Al-1Mo-1V and Ti-6Al-2Sn-4Zr-2Mo. A wide variety of application for α alloys includes gas turbine engine casings, air frame skin and structural components and jet engine compressor blades.

Most of the titanium alloys used in the industry contain α- and β- stabilizers. These alloys include Ti-6Al-4V, Ti-6Al-6V-2Sn and Ti-6Al-2Sn-4Zr-6Mo. They are heat treatable and most are weldable especially with the lower β- stabilizer. Their strength levels are medium to high. These alloys possess excellent combination of strength, toughness and corrosion resistance. Typical applications include blades and discs for jet engine turbines and compressors, structural aircraft components and landing gear, chemical process equipment, marine components and surgical implants. Meanwhile, β alloys contain small amounts of α-stabilizing elements as strengtheners and generally weldable, high corrosion resistance and good creep resistance to intermediate temperatures. Additions of vanadium, iron and chromium as stabilizing elements, provide superior hot working characteristics. Ti-10V-2Fe-3Al, Ti-15V-3Cr-3Al-3Sn, Ti-15Mo-2.7Nb-3Al-0.2Si and Ti-3Al-8V-6Cr-4Mo-4Zr are examples of these alloys. Typical applications include airframe components, fasteners, springs, pipe and commercial and consumer products.

4. Machinability of titanium alloys

Research works on the machinability of titanium alloys have been conducted extensively and reviewed comprehensively by several researchers. The increasing demands of titanium alloys with excellent high temperature, mechanical and chemical properties make them more difficult to machine. According to Ezugwu et al. (2003), machinability can be phrased as the difficulty to machine a particular material under a given set of the machining parameters such as cutting speed, feed rate and depth of cut. It can be rated in terms of tool life, surface quality, the reaction of cutting forces and also machining cost per part. Basically, work hardening, low thermal conductivity, abrasiveness, high strength level and high heat generated were the dominant reasons for the difficulty in machining titanium alloys. Heat is the most important factor that needs to be aware of when machining titanium alloys. Excessive heat could damage the cutting tool rapidly. The main sources of heat during machining are from the shear zone, from the tool-chip interface friction and from the tool-workpiece interface friction. However, too much heat is not the only reason associated with tool failures. The lack of rigidity in holding the tool holder with cutting tool and workpiece can also shorten the tool life. Non-rigid setups with vibration or inconsistent cutting pressure and interrupted cuts often cause tool chipping or fracturing. Prolong machining also causes severe chipping and fracture of the tool edge.

5. Performance evaluation in drilling of titanium alloys

Among the various machining processes, drilling can be considerably as the most difficult process in comparison to milling and turning. Many researchers have studied the machinability of titanium alloys in the past, especially in turning and milling operations. Although extensive investigation reports have been published, no considerable progress is being made and reported on the drilling of these alloys.

5.1 Tool wear

Tool wear of cutting tools in metal cutting accounts for a significant portion of the production costs of a component. Tool wear occurs due to the physical and chemical interaction between the cutting tool and workpiece as a result of the removal of small particles of the tool material from the edge of the cutting tool.Tool wear takes place in three stages as shown in Fig. 7 (Vaughn, 1966). Tool wear developed rapidly in the initial stage and then grew uniformly until it reached its limiting value. In the third stage, the tool wear developed rapidly and caused tool failure. Machining beyond this limit will cause catastrophic failures on the tool and usually this should be avoided.

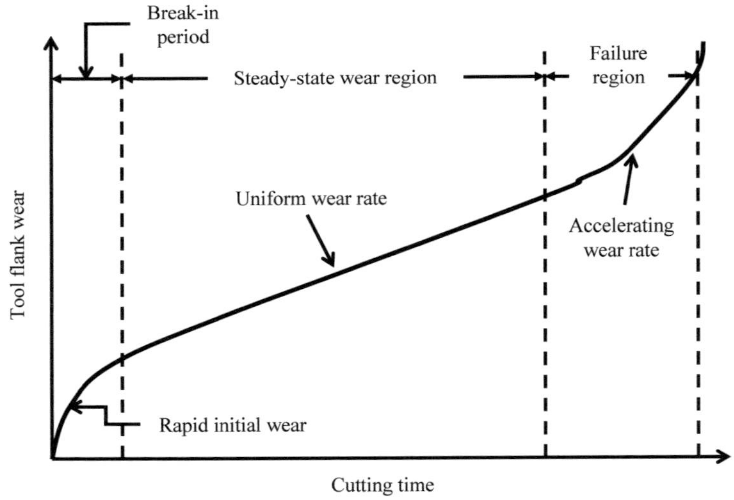

Fig. 7. Typical stages of tool wear in machining (Vaughn, 1966)

The main problem in drilling titanium and its alloys is the rapid wear of the cutting tool. Permissible rates of metal removal are low, in spite of the low cutting forces. The inhibitor in machining titanium alloys are the high temperature generated and the unfavorable temperature distribution in the cutting tool (Ezugwu & Wang, 1997; Vaughn, 1966). Due to the low thermal conductivity of titanium alloys, the temperature on the rake face can be above 900° C even at moderate cutting speed.

Various types of wear can be observed when drilling titanium alloys, namely non-uniform flank wear, excessive chipping and micro-cracking. These types of wear are the dominant tool failure modes when drilling Ti-6Al-4V. The wear occurs along the drill's cutting edges or the flank faces. An increase in cutting speed led to a proportional increase in the flank wear width. The increase of flank wear rate may encourage adherence of workpiece material and may lead to attrition wear and eventually ended up in severe chipping.

During drilling of Ti-48Al-2Mn-2Nb, Mantle and co-workers (Mantle et al. (1995)) found that the workpiece material adhered to the chisel and the cutting edges. The adherence of Ti-48Al-2Mn-2Nb was thinner than Ti-6Al-4V and after verification under the SEM, they concluded that the adhered material was the main contributor to the tool failure. Titanium is highly chemically reactive with the tendency of welding onto the cutting tool during machining. In the beginning, the adhered material may protect the cutting edges from wear as shown in Fig. 8. In this figure, the adhesion occurred mainly at the cutting edge, near the periphery and on the chisel edge. However with prolonged drilling, the adhered material becomes unstable and breaks away from the tool carrying along small amount of tool particles. This situation may lead to severe chipping on the cutting edge.

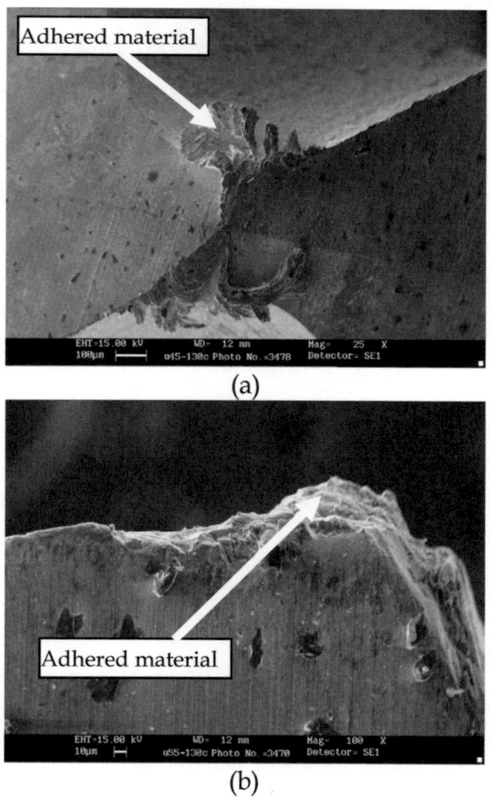

Fig. 8. Adherence of workpiece materials observed at: (a) chisel edge and (b) cutting edge of after drilling Ti-6Al-4V (Rahim, 2005)

Fig. 9 shows a thermal crack on the flank face of the drill (Rahim, 2005). It can be seen that the crack line propagated perpendicularly to the cutting edge. Cracks on the cutting tool and fracture of the entire cutting edge were mainly observed when machining titanium alloys at higher cutting conditions (Ezugwu et al., 2000; Jawaid et al., 2000). Cracks usually originate from the chipped area and gradually propagate along the worn flank face. Chipping at cutting edges is attributes mainly by the generation of cyclic surface stresses during drilling, which may lead to the stress cycling results in the formation of cracks parallel on the cutting edge. The propagation of cracks with prolonged machining, leads to chipping along the cutting edge. Chipping can also occur without the presence of crack formation, especially at the initial stages of the wear progress. If cracks become very numerous, they may join and cause small fragments of the cutting edge to break away.

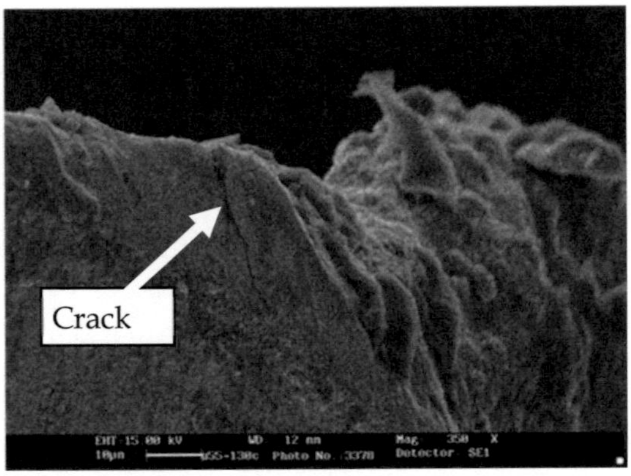

Fig. 9. Crack on the flank face after drilling Ti-6Al-4V for 1 minute at 55 m/min and 0.06 mm/rev (Rahim, 2005)

Cantero et al. (2005) reported on the approach in drilling Ti-6Al-4V under dry condition. Using a 6 mm diameter with TiN coated carbide drill, they recommended that speed and feed rate for drilling of Ti-6Al-4V were 50 m/min and 0.07 mm/rev respectively. Attrition and diffusion were the dominant tool wear mechanisms, especially in the helical flute of drill. With prolonged drilling, these tool wear mechanisms lead to the catastrophic failure of the drill. Attrition wear is a removal of grains or agglomerates of tool material by the adherent chip or workpiece (Dearnley and Grearson, 1986). This could be due to intermittent adhesion between the tool and the workpiece as a result of the irregular chip flow and the breaking of a partially stable built-up edge. When seizure between the tool and the workpiece is broken, small fragments of the tool can be plucked out due to weakening of the binder and transported material via the underside of the chip or by the workpiece. The presence of fatigue during machining operation can initiate cracks and also encourage cracks propagation on the tool.

Furthermore, diffusion wear is associated with the chemical affinity between the tool and workpiece materials under high temperature and pressure during machining of titanium alloys (Hartung and Kramer, 1982; Kramer, 1987). An intimate contact between the tool-workpiece interface at temperature above 800° C provides an ideal environment for diffusion of tool material across the tool-workpiece interface. The EDAX analysis (Fig. 10) confirmed that tool elements (C, Co and W) had diffused into the interface between tool-workpiece during drilling Ti-6Al-4V (Rahim & Sharif, 2009). Diffusion wear is significant at the tool-workpiece interface, especially at high cutting temperature. Due to high chemical reactivity of titanium alloys, carbon reacts readily with titanium. Therefore, the formation of titanium carbide occurred at the interface between the tool and work material.

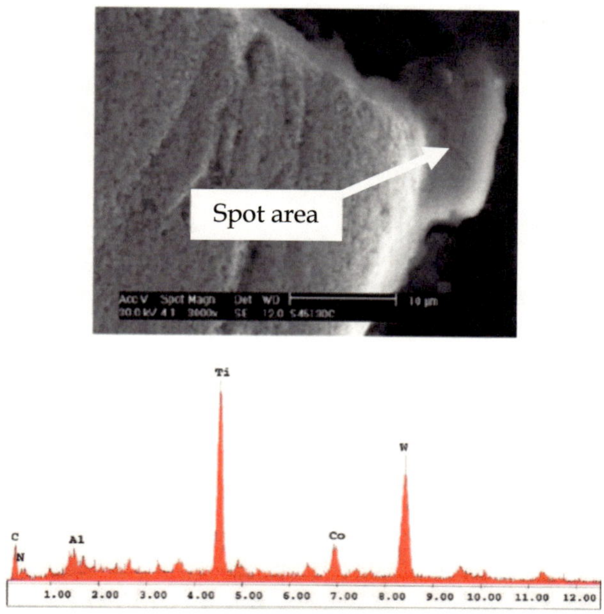

Fig. 10. EDAX Section of worn tool, showing adherent workpiece material on the cutting edge after drilling Ti-6Al-4V for 2 minutes at 45 m/min and 0.06 mm/rev (Rahim & Sharif, 2009)

5.2 Thrust force and torque

Piezoelectric dynamometer is commonly used to measure the cutting force in most machining processes. The various types of the dynamometer depend on the machining process such as turning, milling, drilling or grinding. A three components dynamometer is able to measure the cutting force and feed force, especially in milling and turning operations. Meanwhile, two components dynamometer is normally used in drilling process to measure the thrust force and torque.

Comparison of thrust force against the coolant-lubricant conditions at cutting speed of 60 m/min and feed rate of 0.1 mm/rev is presented in Fig. 11 (Rahim & Sasahara, 2011). It was found that the air blow condition produced the highest thrust force in comparison to the other coolant-lubricant conditions. In contrast, the MQLPO (palm oil using MQL condition) and flood conditions exhibited comparable and the lowest thrust force among the other conditions tested. As expected, the flood condition demonstrated the lowest torque among the other conditions tested as shown in Fig.12. Through the comparison, it was found that air blow did not reduce the drilling torque as much as the other coolant-lubricant conditions. They concluded that the highest value of thrust force and torque for the air blow condition could be attributed to higher amount of friction between tool-chip interface, hence, more heat is generated during the drilling process. Furthermore, Lopez and co-workers found that the cutting force produced by high pressure internal cooling method was lower compared with the external cooling, which has a beneficial effect on workpiece deformation and hole quality (Lopez et al. , 2000).

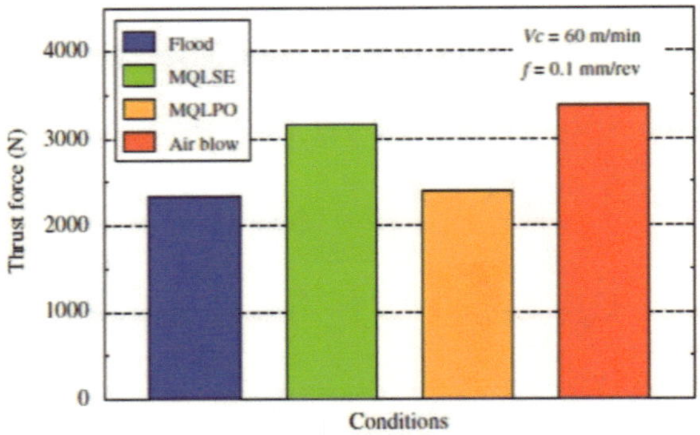

Fig. 11. Thrust force when high speed drilling of Ti-6Al-4V under various coolant-lubricant conditions (Rahim & Sasahara, 2011)

The influence of drilling parameters has been assessed for different material characteristic and properties of titanium alloys (Mantle et al. ,1995). Result shows that the thrust force and torque for Ti-48Al-2Mn-2Nb were greater than Ti-6Al-4V. As shown in Figs. 13 and 14, the thrust force decreases as cutting speed increases (Rahim et al., 2008). At the same time, results also showed that low torque values were obtained at the highest cutting speed. This behavior is attributed to the reduction of the contact area between the tool-workpiece interface and the reduction of specific cutting energy. Moreover, with increase of cutting speed, the cutting temperature increases, subsequently reduced the material hardness. As a result, both the thrust force and torque are reduced. Meanwhile, the thrust force and torque values were significantly increased when the feed rate was increased as shown in Fig. 15 (Rahim & Sasahara, 2011). The thrust force and torque are strongly correlated with the chip thickness, which is associated with the feed rate (Liao et al., 2007). This is because high feed rate results in a larger cross sectional area of the undeformed chip, and, consequently, greater thrust force and torque are produced.

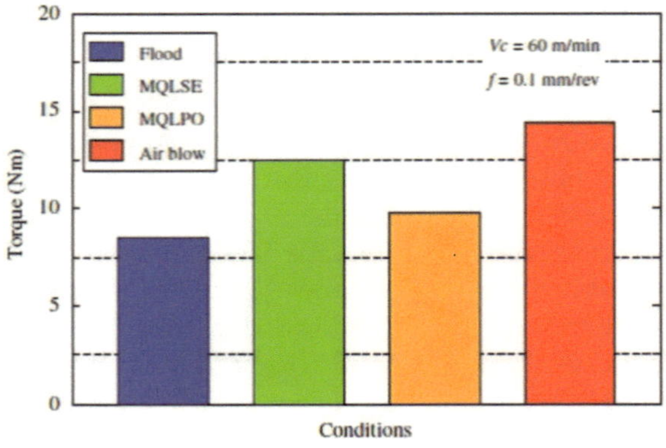

Fig. 12. Torque when high speed drilling of Ti-6Al-4V under various coolant-lubricant conditions (Rahim & Sasahara, 2011)

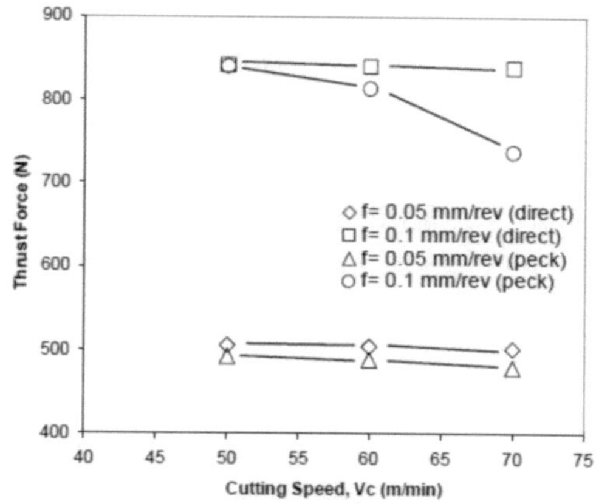

Fig. 13. Comparison of thrust force when drilling Ti-6Al-4V using cemented carbide tool under flood coolant condition (Rahim et al., 2008)

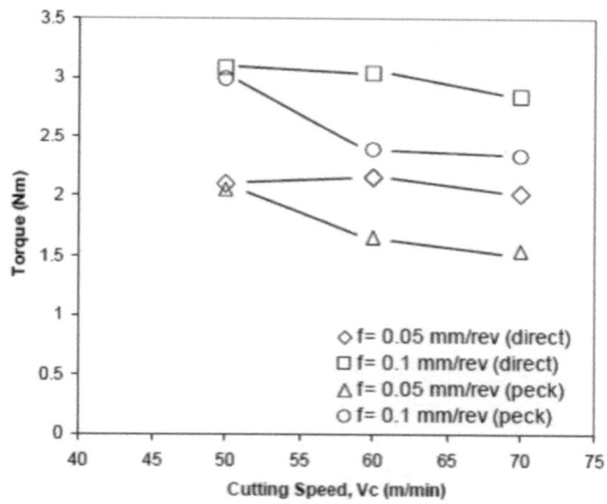

Fig. 14. Comparison of torque when drilling Ti-6Al-4V using cemented carbide tool under flood coolant condition (Rahim et al., 2008)

Some researchers have tested several techniques in drilling of titanium alloys. A step feed drilling or intermittently decelerated feed drilling and vibratory drilling were conducted by Sakurai co-wrokers (Sakurai et al., 1992; Sakurai et al., 1996) to examine the cutting force and cutting characteristic of TiN coated cobalt HSS and oxide treatment nitridized cobalt HSS when drilling Ti-6Al-4V. Results of their study showed that step feed drilling contributed a lower thrust force and torque as compared to continuous conventional drilling. In addition, the thrust force and torque on TiN drills are lower than oxide treatment nitridized drills in both conventional and step feed drilling. As reported by Okamura and co-workers (Okamura et al., 2006), the non-vibration drilling shows a tremendous reduction on thrust force. However, the value tends to decrease once the vibration exceeded 20 kHz. It is believed that the natural frequency of measurement systems does not exceed the vibrating frequency. In another work by Rahim and co-workers (Rahim et al., 2008) showed that pecking drilling method significantly reduces the thrust force and torque.

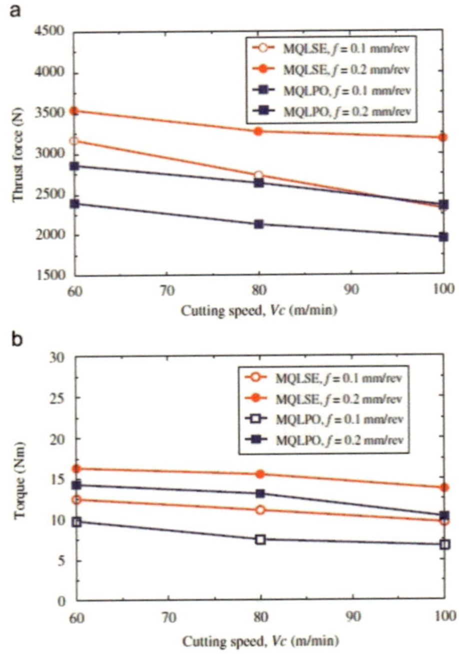

Fig. 15. (a) Thrust force and (b) torque for MQLSE and MQLPO under various cutting speed and feed rate (Rahim & Sasahara, 2011)

5.3 Temperature

Embedded thermocouples were one of the earliest technique used for the estimation of temperatures in various manufacturing and tribological applications. In order to use this technique, particularly in machining, a number of fine deep holes have to be made in the stationary part, namely the workpiece or the cutting tool, and the thermocouples are inserted in different locations in the interior of the part, with some of them as close to the surface as possible. In drilling process, the measurement of temperature by thermocouple wires can be done by embedding the wires in the workpiece and cutting tool as shown in Figs. 16, 17, 18 and 19, respectively. These methods are able to measure the workpiece and cutting tool temperature, especially when drilling titanium alloys.

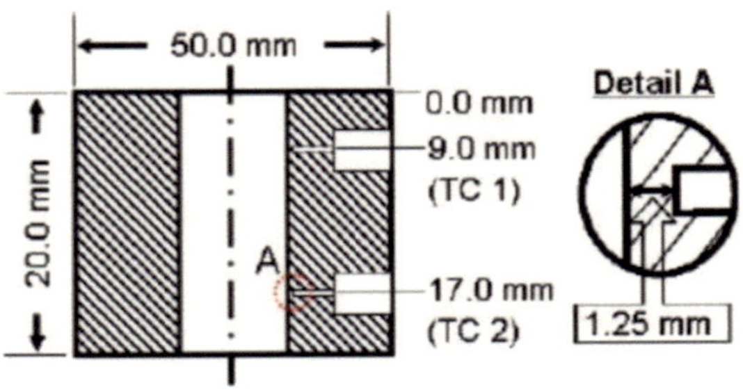

Fig. 16. Thermocouple locations (Rahim & Sasahara, 2010a)

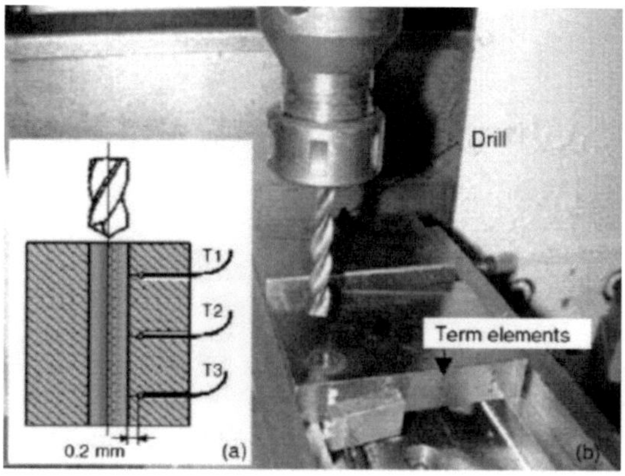

Fig. 17. System for measurement of the temperature in the workpiece (Zeilmann & Weingaertner, 2006)

Cutting speed and feed rate are among the factors that contribute to the variation of temperature during drilling titanium alloys. The cutting temperature increases with the cutting speed. This corresponds with the high cutting energy, deformation strain rate as well as the heat flux (Rahim & Sasahara, 2011). Furthermore, drilling at high feed rate increases the friction and stresses, thus increasing the cutting temperature.

The application of different cooling methods provides a variation in temperature results. For example, the maximum temperatures recorded for drilling with abundant emulsion through the interior of the tool stayed in the range of 22-32% of the values obtained with the application of MQL with an external nozzle as shown in Fig. 20 (Zeilmann & Weingartner, 2006). Comparing drilling with MQL applied with an external nozzle and dry drilling, the values obtained for the second condition were approximately 6% superior, ranging from 455 to 482 °C. Furthermore, flood and MQL conditions recorded a low workpiece temperature in comparison to the air blow condition as shown in Fig. 21 (Rahim & Sasahara, 2011).

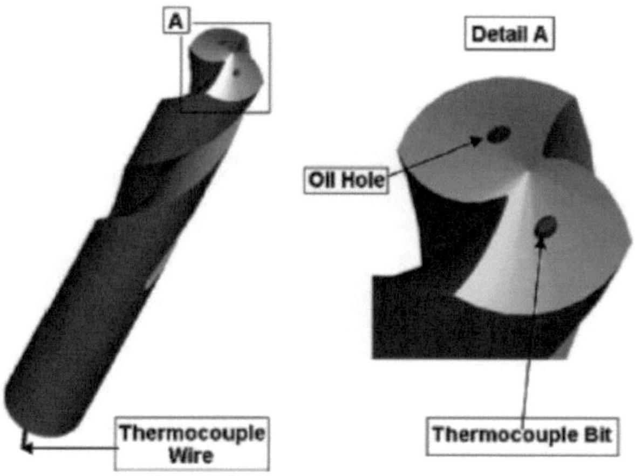

Fig. 18. The thermocouple was inserted through the oil hole of internal coolant carbide drill (Ozcelik & Bagci, 2006)

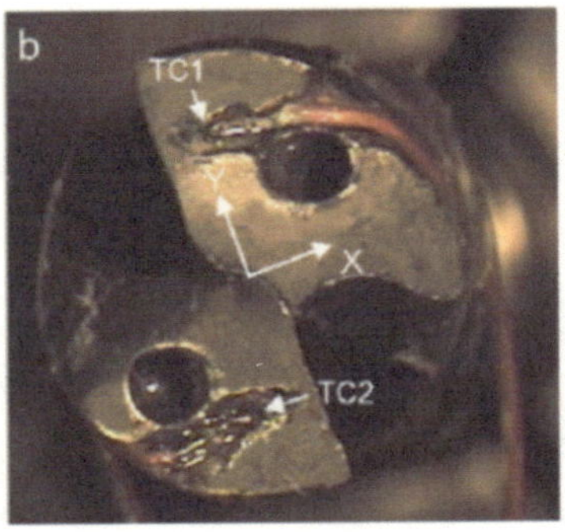

Fig. 19. Top view and coordinates of thermocouple tips on drill flank face (Li & Shih, 2007)

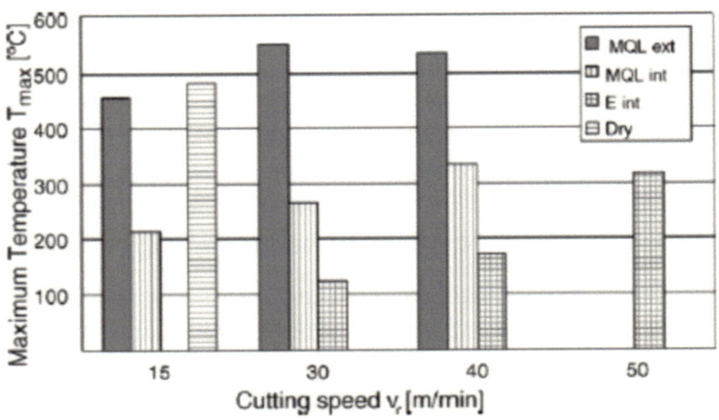

Fig. 20. Maximum temperature in the piece for different cutting fluids conditions when drilling Ti-6Al-4V using grade K10 cemented carbide tool (Zeilmann & Weingartner, 2006)

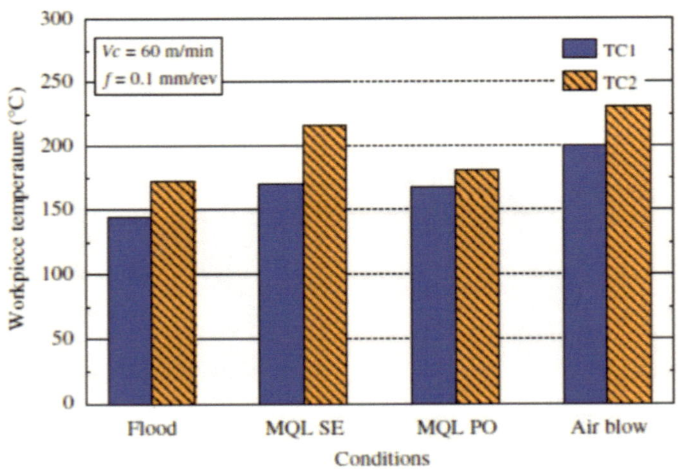

Fig. 21. Maximum workpiece temperature in high speed drilling of Ti-6Al-4V under various coolant-lubricant conditions (Rahim & Sasahara, 2011)

As reported by Pujana and co-workers (Pujana et al., 2009), the cutting temperature was higher when using ultrasonic-assisted drilling in comparison to the non-vibration drilling. In this case, the higher the vibration amplitude, the higher the temperature variations. Okamura and co-authors (Okamura et al., 2006) have designed a low-frequency vibration drilling machine to drill a Ti-6Al-4V. They described the effect of low-frequency vibration drilling on cutting temperature. Results showed that, higher amplitude of 0.24 mm and frequency of 30 Hz exhibited lower cutting temperature as compared to non-vibration drilling.

6. Surface integrity

Surface integrity is defined as the unimpaired or enhanced surface condition of a material resulting from the impact of a controlled manufacturing process (Field and Kahles, 1964). Damaged layer and surface integrity of the finished surface significantly influence the wear resistance, corrosion resistance and fatigue strength of the machined components. Surface integrity produced by metal removal operation can be categorized as geometrical surface integrity and physical surface integrity. To find the impact of the manufacturing process on the material properties both categories effects must be considered. Surface integrity aspects are very important, especially in aerospace industry with respect to the high degree of safety. Surface integrity is concerned primarily on the effect of the machining process on the changes in surface and sub-surface of the component which are categorized as surface roughness, plastic deformation, residual stress and microhardness.

6.1 Surface roughness

There are three essential parameters in a surface roughness; arithmetical mean deviation of the profile (R_a), maximum height of the profile (R_{max}) and height of the profile irregularities in ten points (R_z). It is believed that the higher surface roughness value is responsible for the decrease of the fatigue strength on the machined surface. Significant improvement in surface roughness can be obtained when low feed rate and high cutting speed are employed. However, the response of surface roughness towards cutting speed was less significant when compared to feed rate. Sun and Guo (Sun & Guo, 2009), reported that surface roughness value increased with increase in feed rate and radial depth-of-cut.

Previous study showed that surface roughness value is lower at high cutting speed when drilling Ti-6Al-4V using carbide drills (Sharif & Rahim, 2007). During machining at high cutting speed, the cutting temperature increases due the small contact length between tool-workpiece interfaces. This could be due to the decrease in the value of coefficient of friction, which results in low friction at the tool-workpiece interface. These factors could contribute to the improvement in surface roughness values as shown in Fig. 22 (Rahim & Sasahara, 2010b). In addition, as the cutting speed increases, more heat is generated thus softening the workpiece material, which in turn improves the surface roughness. However, a low cutting speed may lead to the formation of built-up edge and hence deteriorates the machined surface. Investigation revealed that at high feed rate the surface roughness is poor, probably due to the distinct feed marks produced at high feed rate (Rahim & Sasahara, 2010)

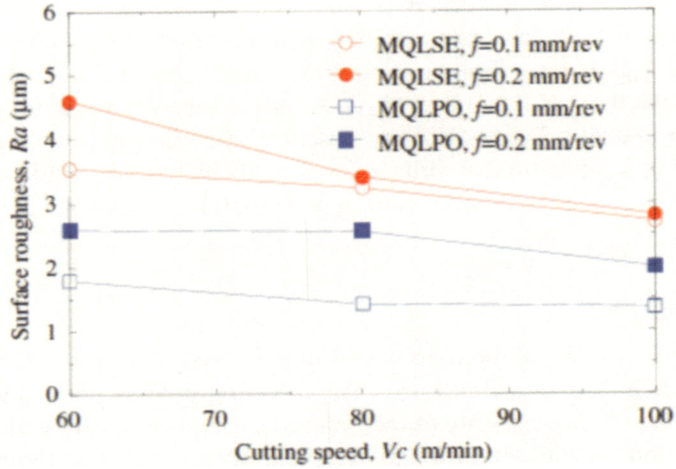

Fig. 22. Comparison of surface roughness level obtained when drilling Ti-6Al-4V using TiAlN coated carbide tool under MQLSE and MQLPO (Rahim & Sasahara, 2010b)

Types of cutting fluid also influence the surface roughness of the machined surface. Under the MQL condition, vegetable oil (MQLPO: palm oil) exhibits better surface roughness than synthetic ester (MQLSE) as shown in Fig. 22 (Rahim & Sasahara, 2010b). It can be suggested that less heat is generated using palm oil thus provided enough time to cool and lubricate the tool-workpiece interface. Apparently, such reduction may attribute to better lubrication and shorter tool-chip contact length during drilling. Moreover, surface roughness measured by peck drilling is far better than conventional drilling method (Rahim et al, 2008).

6.2 Microhardness

The microhardness alterations observed during machining may be due to the effect of thermal, mechanical and chemical reaction. Many researchers believed that the workpiece material is subjected to work hardening and thermal softening effect during machining, especially at high cutting temperature and pressure (Che Haron, 2001; Ginting & Nouari, 2009). When machining titanium alloys, the hardness just beneath the machined surface was found to be softer than the bulk material hardness due to the thermal softening effect. However, when the depth below the machined surface increases, the hardness value starts to increase before reaching its peak value and finally drops gradually to the bulk material hardness as shown in Fig. 23. The increase in hardness value is directly associated with the effect of work hardening. This effect depends on the temperature, cutting time and the mechanism of internal stress relaxation (Ginting and Nouari, 2009).

Fig. 23 shows that the microhardness of the sub-surface at 0.025 mm underneath the machined surface was below the average base material hardness. This indicates that the machined surface experienced thermal softening effect or over aging due to the localized heating during the drilling process (Rahim & Sasahara, 2010b). Turning test of titanium alloys by Haron and Jawaid (Haron & Jawaid, 2005) and Ginting and Nouari (Ginting & Nouari, 2009) have also indicated significant drop of microhardness value near the surface of machined layer. They pointed out that the existence of high cutting temperature and high cutting pressure produced noticeable softening in the surface region.

Fig. 23 also shows that there is hardening layer below the softened layer whose hardness depends on the cutting parameters (i.e cutting speed, feed rate, depth of cut) as well as mechanical and thermal interaction. It was generally observed in the work hardening region that the microhardness increases with increase in cutting speed and feed rate. An increase in microhardness of the surface layer, as a result of high feed rate could be associated by the high rubbing load between the tool and the machined surface and the consequent work hardening effect.

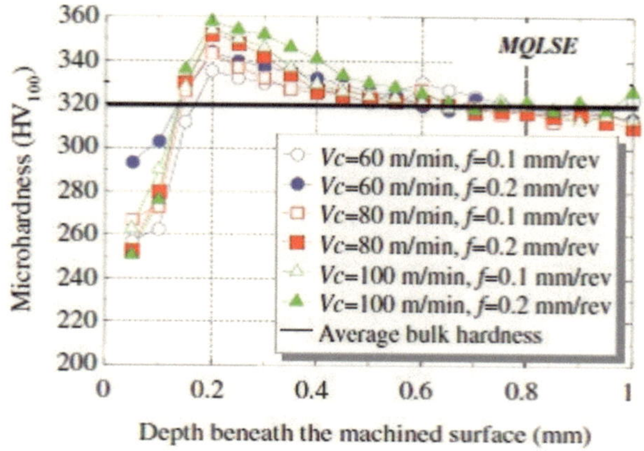

Fig. 23. Sub-surface hardness variations after drilling Ti-6Al-4V using MQLSE (Rahim & Sasahara, 2010b)

In another work by Rahim and Sharif (Rahim & Sharif, 2006), it was reported that the hardness value underneath the drilled surface was higher than the average hardness of the bulk material when drilling Ti-5Al-4V-Mo/Fe (Fig. 24). Meanwhile, a significant changed of microhardness values were also observed underneath the machined surface. It was due to the transformation of beta phase to alpha phase during drilling (Cantero et al., 2005).

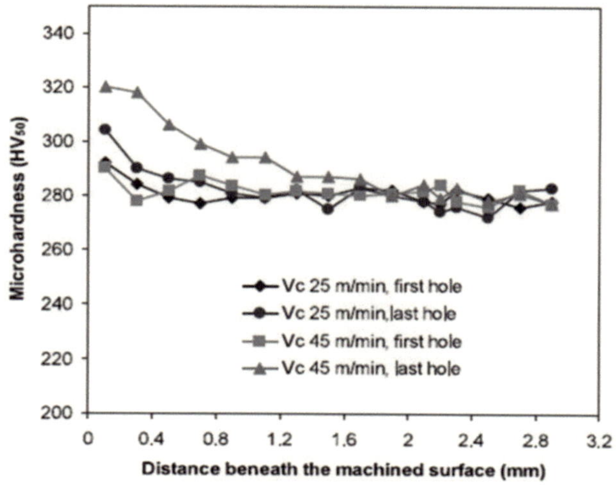

Fig. 24. Microhardness variation beneath the surface produced when drilling Ti-5Al-4V-Mo/Fe (Rahim & Sharif, 2006)

6.3 Sub-surface plastic deformation

It is discernible that the surface and sub-surface of the machined surface are subjected to plastically deform. Sub-surface plastic deformation is particularly due to the effect of large strain, strain rate and temperature. In addition, a freshly cut surface may be burnished by a dull cutting tool, hence work hardened the machined surface. Jeelani and Ramakrishnan (Jeelani & Ramakrishnan, 1983) observed that the machined surface is severely damaged with the plastic flow in the direction of the tool motion. Meanwhile, Velasques and co-workers (Velasques et al., 2010) found that the severe deformation beneath the machined surface is associated with high cutting speed. Sub-surface plastic deformation area can be divided into three zones, namely highly perturbed region, a plastically deformed layer and unaffected zone. Normally, the sub-surface plastic deformations of microstructure of the machined surfaces are examined under the high magnification microscope in etched condition.

In most cases, plastic deformation occurs towards the spindle rotational direction. When drilling titanium alloys at higher cutting speeds and feed rates, a thicker plastic deformation can be observed. At this condition, the temperature between tool-chip interface increases thus sticking friction region occurred. Therefore, the combination of high cutting temperature and sticking friction contributed to the severe and noticeable subsurface plastic deformation (Rahim & Sasahara, 2010).

Fig. 25 shows an evidence of sub-surface plastic deformation when drilling Ti-5Al-4V-Mo/Fe (Rahim & Sharif, 2006). In this figure, the deformation is found to be severe after prolonged drilling. In this case, no white layer especially on the top of the machined surface is observed. The authors stated that high cutting force and temperature are the dominant factors which lead to the severe plastic deformation. Cantero and co-workers (Cantero et al., 2005) also found the same phenomenon and they concluded that the plastic deformations during machining are caused by mechanical forces from the cutting tool acting upon the work-piece. Additional deformation can occur as a consequence of temperature gradients due to localized heating of the machined surface area.

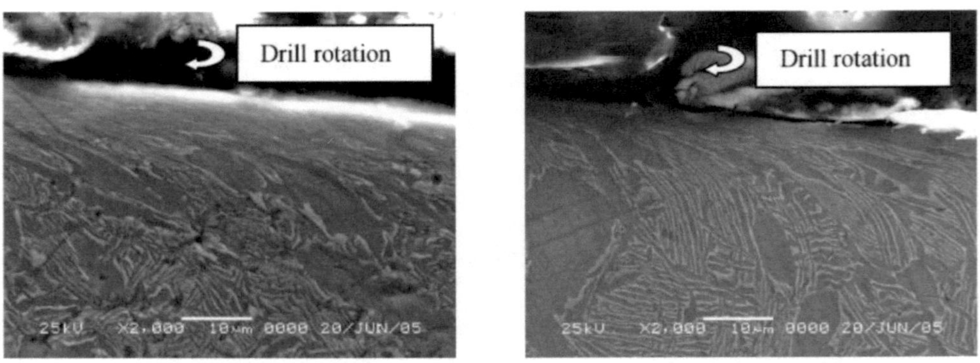

(a) After 7 minutes at 25 m/min (b) After 2 minutes at 45 m/min

Fig. 25. Magnified view of the machined sub-surface when drilling Ti-5Al-4V-Mo/Fe (Rahim & Sharif, 2006)

7. Conclusions

Creditable works have been carried out by many researchers in drilling of titanium alloys which resulted in significant improvements in the productivity of titanium parts. The application of drilling strategies, introduction of newly developed tool geometry and coolant conditions have improved the surface integrity and increased the tool life performance of the cutting tools in several folds. In general the following conclusions can be drawn when drilling titanium alloys:

i. Adhesion, attrition and diffusion are the operating tool wear mechanisms when drilling titanium alloy.
ii. Flank wear, excessive chipping, cracking and tool breakage are the dominant tool failure modes.
iii. The values of thrust force and torque decrease with increase in cutting speed. In contrast, these values increase significantly when the feed rate is increased.
iv. Cutting speed and feed rate significantly affect the surface roughness of the machined surface whereby high cutting speed and low feed rate resulted in the better surface finish.
v. Under various coolant-lubricant conditions, air blow produces higher cutting temperature as compared to other condistions.
vi. The machined surface deteriorates due to the effect of metallurgical changes and surface quality during drilling at high cutting speed, feed rate and under various coolant conditions.

8. Acknowledgments

The authors wish to thank the Ministry of Higher Education of Malaysia and Research Management Center, UTM for their financial supports to this work through the Research University Grant (RUG) funding number Q.J130000.7124.02H43. Special gratitude is also extended to UTHM and TUAT, Japan and Technology and Mitsubishi Materials Corporation for providing the cutting tools.

9. References

Boyer, R. R. (1996). An Overview on the Use of Titanium in the Aerospace Industry. *Material Science & Engineering*, Vol. 213, pp. 103-114, ISSN 0921-5093

Brewer, W. D., Bird, R. K., & Wallace, T. A. (1998). Titanium alloys and Processing for High Speed Aircraft. *Materials Engineering & Science A*, Vol. 243, pp. 299-304, ISSN 0921-5093

Brinksmeier, E. (1990). Prediction of Tool Fracture in Drilling. *Annals of CIRP*, Vol. 39, No. 1, pp. 97-100, ISSN 0007-8506

Cantero, J. L., Tardio, M. M., Canteli, J.A., Marcos, M., & Miguelez, M. H., (2005). Dry drilling of alloy Ti-6Al-4V. *Int. J. Mach. Tools & Manuf.*, Vol. 45, pp. 1246-1255, ISSN 0890-6955

Choudhury, S. K., & Raju, G. (2000). Investigation into Crater Wear in Drilling. *Int. J. Mach. Tools & Manuf.*, Vol. 40, pp. 887-898, ISSN 0890-6955

Dolinsek, S., Sustarsic, B., & Kopac, J. (2001). Wear Mechanism of Cutting Tools in High Speed Cutting Process. *Wear*, Vol. 250, pp. 349-352, ISSN 0043-1648

Ezugwu, E. O., & Lai, C. J. (1995). Failure Mode and Wear Mechanism of M35 High Speed Steel Drills When Machining Inconel 910. *J. Mater. Process. Technol.*, Vol. 49, pp. 295-312, ISSN 0924-0136

Ezugwu, E. O., & Wang, Z. M. (1997). Titanium Alloys and Their Machinability - A Review. *J. Mater. Process. Technol.*, Vol. 68, pp. 262-274, ISSN 0924-0136

Ezugwu, E. O., Wang, Z. M., & Machado, A. R. (2000). Wear of Carbides Tools When Machining Nickel And Titanium Base Alloys. *Tribology Transaction*, Vol. 43, No. 2, pp. 263-268, ISSN 0301-679X

Ezugwu, E. O., Bonney, J., & Yamane, Y. (2003). An Overview of the Machinability of Aeroengine Alloys. *J. Mater. Process. Technol.*, Vol. 134, pp. 233-253, ISSN 0924-0136

Field, D., & Kahles, J.M., (1964). The surface integrity of machined and high strength steels. *DMIC Report*, Vol. 210, pp. 54-77

Fujise, K., & Ohtani, T. (1998). Machinability of Ti-6Al-4V Alloy in drilling with Small Drills. *Proc. 4th International Conference on Progress of Cutting and Grinding*. Urmqi and Torpan, China, October 1998

Ginting, A., & Nouari, M., (2009). Surface integrity of dry machined titanium alloys. *Int. J. Mach. Tools & Manuf.*, Vol. 49, pp. 325-332, ISSN 0890-6955

Harris, S. G., Doyle, E. D., Vlasveld, A. C., Audy, J., & Quick, D. (2003). A Study of the Wear Mechanism of $Ti_{1-x} Al_x N$ and $Ti_{1-x-y} Al_x Cr_y N$ Coated High Speed Steel Twist Drill Under Dry Machining Conditions. *Wear*, Vol. 254, pp. 723-734, ISSN 0043-1648

Hartung, P.D. & Kramer, B.M. (1982). Tool Wear in Machining Titanium. *Annals of CIRP*, Vol. 31, No. 1, pp. 75-80, ISSN 0007-8506

Haron, C. H. C. (2001). Tool Life and Surface Integrity in Turning Titanium Alloy. *J. Mater. Process. Technol.*, Vol. 118, pp. 231-237, ISSN 0924-0136

Haron, C. H. C, & Jawaid, A. (2005). The effect of machining on surface integrity of titanium alloy Ti-6% Al-4% V. *J. Mater. Process. Technol.*, Vol. 166, pp. 188-192, ISSN 0924-0136

Jawaid, A., Sharif, S., & Koksal, S. (2000). Evaluation of Wear Mechanism of Coated Carbide Tools When Face Milling Titanium Alloy. *J. Mater. Process. Technol.*, Vol. 99, pp. 266-274, ISSN 0924-0136

Jeelani, S., & Ramakrishnan, K., (1983). Subsurface plastic deformation in machining 6Al-2Sn-4Zr-2Mo titanium alloy. *Wear* , Vol. 85, pp. 121-130, ISSN 0043-1648

Kaldor, S., & Lenz, E. (1980). Investigation in Tool Life of Twist Drill. *Annals of CIRP*, Vol. 52, pp. 30-35, ISSN 0007-8506

Kanai, M., Fujii, S., Kanda, Y. (1978). Statistical Characteristics of Drill Wear and Drill Life for the Standardized Performance Tests. *Annals of CIRP*, Vol. 27, No. 1, pp. 61-66, ISSN 0007-8506

Kramer, B. M. (1987). On Tool Materials for High Speed Machining. *ASME Journal of Engineering for Industry*, Vol. 109, pp. 87-91, ISSN 0022-1817

Li, R., & Shih, A. J., (2007). Spiral point drill temperature and stress in high-throughput drilling of titanium. *Int. J. Mach. Tools & Manuf.*, Vol. 47, pp. 2005-2017, ISSN 0890-6955

Liao, Y. S., Chen, Y. C., & Lin, H. M., (2007). Feasibility study of the ultrasonic vibration assisted drilling of Inconel superalloy. *Int. J. Mach. Tools & Manuf.*, Vol. 47, pp. 1988-1996, ISSN 0890-6955

Linberg, R. A. (1990). *Processes and Materials of Manufacture*, 4th Edition, Prentice-Hall Inc., ISBN 978-81-203-0663-9, New Jersey, USA

Lopez, L. N., Perez, J., Llorente, J. I., & Sanchez, J. A. (2000). Advanced Cutting Conditions For the Milling of Aeronautical Alloys. *J. Mater. Process. Technol.*, Vol. 100, pp. 1-11, ISSN 0924-0136

Mantle, A. L., Aspinwall, D. K., & Wollenhofer, O. (1995). Twist Drill of Gamma Titanium Aluminade Intermetallics. *Proc. 12th Conference of the Irish Manuf. Committee*, United Kingdom, May 1995

Okamura, K., Sasahara, H., Segawa, T., Tsutsumi, M., (2006). Low-frequency vibration drilling of titanium alloy. *International Journal of Japan Society of Mechanical Engineers, Series C*, Vol. 49, pp. 76-82, ISSN 1347-538X

Ozcelik, B., & Bagci, E. (2006). Experimental and numerical studies on the determination of twist drill temperature in dry drilling: A new approach. Materials and Design, Vol 27, pp. 920-927, ISSN 0261-3069

Pujana, J., Rivero, A., Celaya, A., & Lopez, L.N. (2009). An analysis of ultrasonic-assisted drilling of Ti-6Al-4V. *Int. J. Mach. Tools & Manuf.*, Vol. 49, pp. 500-508, ISSN 0890-6955

Rahim, E. A. (2005), Performance evaluation of uncoated and coated carbide tools when drilling titanium alloy, *Master Thesis*, Universiti Teknologi Malaysia

Rahim, E. A. , & Sharif, S. (2006). Investigation on Tool Life and Surface Integrity when Drilling Ti- 6A1-4V and Ti-5A1-4V-Mo/Fe. *International Journal of Japan Society of Mechanical Engineer, Series C*, Vol. 49, No. 2, pp. 340-345, , ISSN 1347-538X

Rahim, E. A. , Kamdani, K., & Sharif, S. (2008). Performance Evaluation of Uncoated Carbide Tool in High Speed Drilling of Ti6Al4V. *Journal of Advanced Mechanical Design, System and Manufacturing*, Vol. 2, No. 4, pp. 522-531, ISSN 1881-3054

Rahim, E. A., & Sharif, S. (2009). Evaluation of tool wear mechanism of TiAlN coated tools when drilling Ti-6Al-4V. International Journal of Manufacturing Technology and Management, Vol. 17, No. 4, pp. 327-336, ISSN 1741-5195

Rahim, E. A., & Sasahara, H. (2010a). The Effect of MQL Fluids on the Drilling Performance of Nickel-Based Superalloy. *4th International Conference on High Performance Cutting*, Gifu, Japan, October 2010

Rahim, E. A., & Sasahara, H. (2010b). Investigation of Tool Wear and Surface Integrity on MQL Machining of Ti-6Al-4V using Biodegradable Oil. *Proceedings* of the *Institution of Mechanical Engineers*, Part B, Journal of Engineering Manufacture. Accepted to be published

Rahim, E. A. & Sasahara, H. (2011) A Study of the Effect of Palm Oil as MQL Lubricant on High Speed Drilling of Titanium Alloys. *Tribology International*, Vol. 44, pp. 309-317, ISSN 0301-679X

Sakurai, K., Adichi, K., Ogawa, K., & Niba, R., (1992). Drilling of Ti6Al4V. *Journal of Japan Institute of Light Metals*, Vol. 42, No. 7, pp. 389-394, ISSN 1880-8018

Sakurai, K., Adichi, K., Kamekawa, T., Ogawa, K., & Hanasaki, S., (1996). Intermittently Decelerated Feed Drilling of Ti6Al4V Alloy. *Journal of Japan Institute of Light Metals*, Vol. 46, No. 3, pp. 138-143, ISSN 1880-8018

Schneider, G. (2001). Applied Cutting Tool Engineering. *Tooling and Production*, Vol. 66, No. 12, pp. 37-45, ISSN 0040-9243

Sharif, S. & Rahim, E. A. (2007). Performance of Coated and Uncoated Carbide Tools when Drilling of Titanium Alloy - Ti-6Al4V. *J. Mater. Process. Technol.*, Vol. 185, pp. 72-76, ISSN 0924-0136

Sun, J., & Guo, Y.B., (2009). A comprehensive experimental study on surface integrity by end milling Ti-6Al-4V. *J. Mater. Process. Technol.*, Vol. 209, pp. 4036-4042, ISSN 0924-0136

Tetsutaro, H., & Zhao, H. (1989). Study of a High Performance Drill Geometry. *Annals of CIRP*, Vol. 38, No. 1, pp. 87-90, ISSN 0007-8506

Tonshoff, H. K., Spintig, W., & Konig, W. (1994). Machining of Holes Developments in Drilling Technology. *Annals of CIRP*, Vol. 43, No. 2, pp. 551-561, ISSN 0007-8506

Vaughn, R. L. (1966). Modern Metals Machining Technology. *ASME Journal of Engineering for Industry*, pp. 65-71, ISSN 0022-1817

Velasquez, J.D.P., Tidu, A., Bolle, B., Chevrier, P., & Fundenberger, J.J. (2010). Sub-surface and surface analysis of high speed machined Ti-6Al-4V alloy. *Materials Science & Engineering A*, Vol. 527, pp. 2572-2578, ISSN 0921-5093

Wen, C. C., & Xiao, D. L.(2000). Study On the Various Coated Twist Drills For Stainless Steels Drilling. *J. Mater. Process. Technol.*, Vol. 99, pp. 226-230, ISSN 0924-0136

Zeilmann, R.P., & Weingaertner, W.L., (2006). Analysis of temperature during driling of Ti6Al4V with minimal quantity of lubrication. *J. Mater. Process. Technol.*, Vol. 179, pp. 124-127, ISSN 0924-0136

Part 3

Surface Treatments of Titanium Alloys for Biomedical and Other Challenging Applications

7

Chemico-Thermal Treatment of Titanium Alloys – Nitriding

Iryna Pohrelyuk and Viktor Fedirko
Physical-Mechanical Institute of National Academy of Sciences of Ukraine
Ukraine

1. Introduction

Titanium and its alloys are widely used in aircraft, rocket production, shipbuilding, machine industry, chemical and food industry, medicine due to their high specific strength, good corrosion resistance and biological passivity. However, titanium has properties which limit its application as construction material. Particularly, tendency to surface adhesion and galling at friction results in its lowest wear resistance among construction materials. The application of titanium alloys in friction units and in the places of direct contact is impossible without additional surface treatment for higher strength. The corrosion resistance of titanium alloys is not often satisfactory. Therefore titanium alloys also need the additional protection in the aggressive media.

The chemical heat treatment, particularly nitriding, allows to extend the functionality of titanium alloys, enhancing the wear resistance and providing the high anticorrosion characteristics in aggressive media. However, the molecular nitrogen is a reactionless gas as a result of significant bond strength in molecule (ΔH_o=940 kJ/mole). Therefore it is very important to intensify the interaction between titanium and nitrogen and to elaborate the relevant nitriding methods.

2. Basic regularities of high-temperature interaction of titanium alloys with nitrogen

One of the ways to solve the problem of intensification of nitriding of titanium alloys is the high-temperature saturation based on temperature dependence of diffusion coefficient of nitrogen in titanium. Therefore we will consider the basic regularities of nitriding of titanium alloys at high temperatures.

2.1 Kinetic regularities of high-temperature interaction of titanium alloys with nitrogen

Kinetics of nitrogen absorption by titanium alloys was widely studied by both the thermogravimetric analysis, when the mass change after different exposures in nitrogen at constant temperature is fixed and the manometric method, when the change of nitrogen pressure in the closed system is determined. The results of these studies in a wide temperature range (550...1600 ^{o}C) showed that the process of nitrogen absorption by titanium is described by the parabolic dependence. That is the dependence of relative mass increase of the nitrided samples on time can be presented by a function:

$$(\Delta m/S)^2 = K\tau, \quad (1)$$

where $\Delta m = (m_2 - m_1)$ – difference between sample's mass after and before nitriding; S – nitriding surface square; τ – nitriding duration; K – parabolic constant of nitriding rate.

The appreciable deviations from the parabolic law caused by the presence of negligible quantity of oxygen impurities in nitrogen are observed at the initial stage of reaction.

With the rise of temperature during isothermal exposure the intensity of interaction of titanium with nitrogen increases substantially. The parabolic constant of nitriding rate (K) is determined by the tangent of angle of inclination of straight lines of time dependence of square of mass increase. It characterizes quantitatively the relative intensity of saturation process (table 1). Values K increase practically on order with the rise of temperature on one hundred degrees (for example, at 900 ^{o}C $K = 8{,}6\times10^{-12}$ g^2/cm$^4\times$s and at 1000 ^{o}C $K = 3{,}5\times10^{-11}$ g^2/cm$^4\times$s).

Alloy	$K_p = K_o \exp(E/RT)$	
	K_o, g^2/(cm$^4\times$s)	-E, kJ/mole
VT1-0	0,11	214
OT4-1	0,42	229
VT5-1	0,6	248
VT20	0,16	241
PT-7M	0,8	250
OT4	0,02	223
VT6s	1,9	258
VT6	0,9	267
VT23	0,4	252
VT32	0,4	240
VT35	0,1	238

Table 1. Parameters of temperature dependence of parabolic constant of nitriding rate K_p

The temperature dependence of parabolic constant of nitriding rate is described by Arrhenius equation:

$$K = K_o \exp(-E/RT), \quad (2)$$

where E – activation energy of nitriding process, J/mole; K_o – preexponential multiplier, g^2/cm$^4\times$s; R – gas constant; T – temperature. Activation energy of nitriding determined by many researchers has different values for the certain temperature ranges and it increases with the rise of temperature of isothermal exposure as a rule. It allows to assert that process which determines the nitrogen absorption rate is changed with temperature.

The value n in law:

$$(\Delta m/S)^n = K\tau \quad (3)$$

change of mass increase of the nitrided samples of titanium alloys in time determined by treatment of nitriding isotherms of each investigated alloy in the double logarithmical coordinates *ln(Δm/S)-ln τ* (cotangent of angle of inclination of the received straight lines) are near to 2 that corresponds to the parabolic law of interaction (1) as well as for the unalloyed titanium.

The calculated constants (*K, K₀, E*) of the dependences *(1, 2)* at the use of corresponding mathematical models (Matychak et al., 2007, 2008, 2009, 2011) at the predetermined temperature and duration of saturation process allow to forecast the thickness of the phases formed during the nitriding process or determine time and temperature parameters of nitriding at which the thicknesses of the formed layers would be predetermined.

The results of investigation of kinetics of saturation process of titanium alloys by nitrogen testify that alloying influences on nitriding rate (fig. 1).

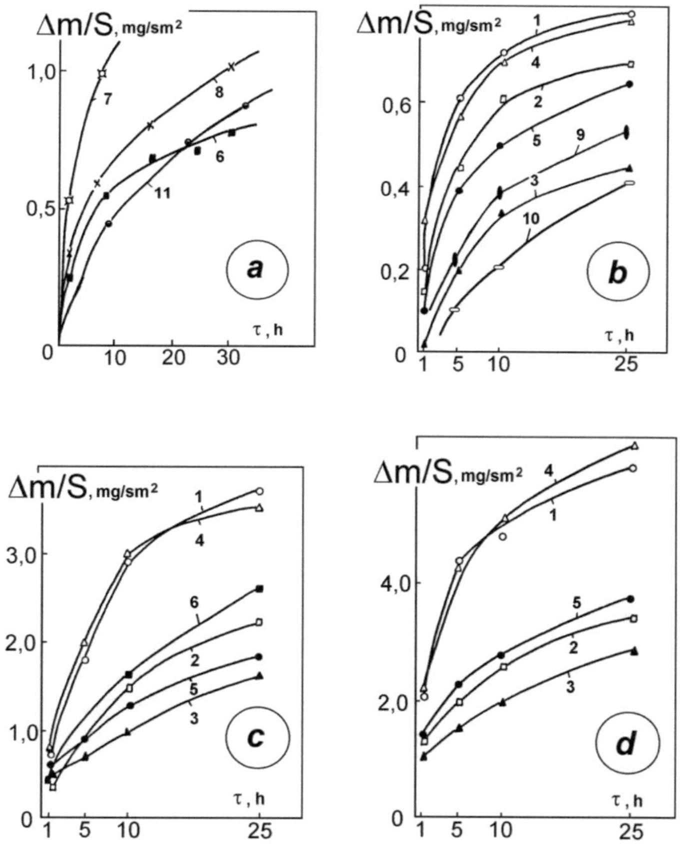

Fig. 1. Kinetics of nitriding of titanium alloys : a - 950 °C; b - 900 °C; c - 1000 °C; d - 1100 °C; 1 - VT1- 0; 2 - PT-7M; 3 - VT5- 1; 4 - OT4- 1; 5 - VT6s; 6 - VT5; 7 - VT1-D; 8 - VT3- 1; 9 - VT6; 10 - VT20; 11 - VT23.

With the rise of temperature of isothermal exposure the influence of alloying elements on the nitriding process increases that confirmed by the increase of distance between the isotherms of nitriding (fig. 1): difference in the mass increase of titanium alloys with different chemical composition at saturation by nitrogen increases.

As opposed to the unalloyed c.p.titanium the nitriding rate of titanium alloys, as a rule, is less. The mechanism allowed the alloying elements to decrease the nitrogen diffusion rate in titanium does not discussed in literature. However, as diffusion in titanium nitrides is interstitial, the influence of alloying elements effects either on the decrease of sizes of interstitial intervals in the titanium lattice or on their filling.

Thus, the process of high-temperature interaction of titanium with nitrogen is described by parabolic dependence which is the result of forming of chemical reaction products - nitrides on the metal surface that slows down the behavior in time. The presence of alloying elements in titanium does not change the process substantially and only slows its. Besides, as a result of heterogeneous reaction titanium–nitrogen the considerable dissolution of gas into the metal with formation of solid solution of nitrogen in α and β titanium (gasing) is observed. Therefore the study of nitriding process by only determination of general mass of absorbed nitrogen which includes both the nitride formation and gasing is incomplete and not quite correct. The differential estimation of contribution of both nitride formation and gasing during nitriding is necessary.

2.2 Features of nitride formation at nitriding of titanium alloys

The well coherent with matrix nitride film of the golden color is formed on the surface of titanium alloys during the isothermal exposure in the nitrogen at temperatures above 800 °C. The film can have different tints of base golden color (from bright to mat) which depends on temperature and duration of nitriding, chemical composition of the nitrides. The film loses the brightness at the temperature behind 900 °C. The thickness of the nitride film and the degree of saturation by nitrogen are stipulated by the time and temperature parameters of nitriding and chemical composition of the saturated material. It allows to assert that the change of colour gamut and reflection power of film depends on its thickness and degree of saturation by nitrogen because titanium nitrides, in particular TiN, is characterized by the wide homogeneity region (27…52 at.%).

Nitride film formed at temperatures below 1000 °C repeats the contours of metallic matrix. In the case of the long exposures and temperatures below 1000 °C and above there are growths of film. On fig. 2a the characteristic topography of surface of the nitrided samples is presented: wavy inequalities forming net on surface, which, most probably, repeats the net of grains boundaries of the material matrix. These formations are most noticeable and reach the large sizes at the nitriding temperatures which are higher than temperature of α⇔β polymorphic transformation (fig. 2b). α⇔β phase transformation during the processes of heating and cooling causes the strain hardening, volume changes and formation of surface relief. The origin of considerable compression stresses at forming of nitride film causes the plastic deformation. It promotes the formation of quantities of inequalities. The surface topography, more or less expressed, is observed after nitriding at temperatures even lower than polymorphic transformation, and not arises after nitriding at the certain temperature. With the rise of nitriding temperature there is the growth of fragments of surface net like the growth of grain of titanium matrix. More active nitride formation on the grains boundaries promotes the forming of surface net, and processes, accompanying α⇔β transformation, assist to enhance the surface relief. Alloying weakens the formation of surface relief owing to the rise of temperature of polymorphic transformation of alloys.

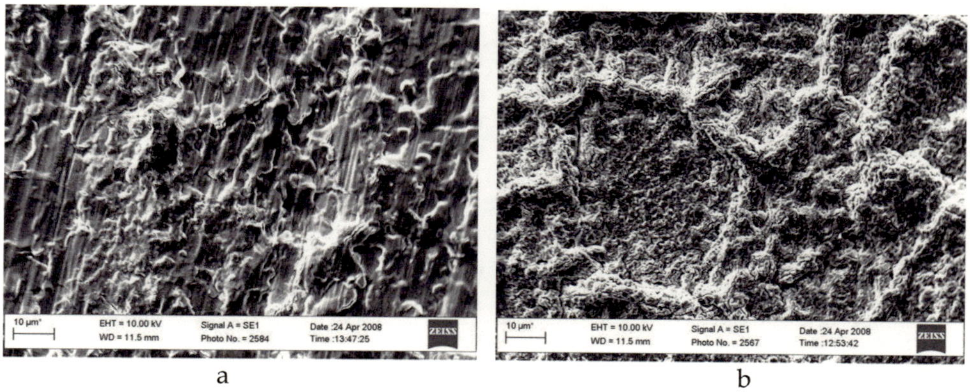

Fig. 2. Surface of nitrided VT1-0 alloy: a – 850 °C, 12h; b – 950 °C, 8h.

The formation of surface relief worsens the quality of nitrided surface with the nitriding temperature rising (fig. 3). For example, nitriding of VT23 alloy at 900°C results in change of surface roughness parameter (R_a) from 0,08 to 0,2 μm. After isothermal exposure in nitrogen at 950°C R_a is 0,4 μm, that is the surface roughness became worse on two classes. The substantial worsening of surface quality during nitriding at high temperatures complicates the obtaining of smooth surface. Therefore the use of nitriding with the purpose to increase the wear resistance of titanium foresees either limitation of process temperature (≤ 900 °C) or additional surface treatment of the nitrided details.

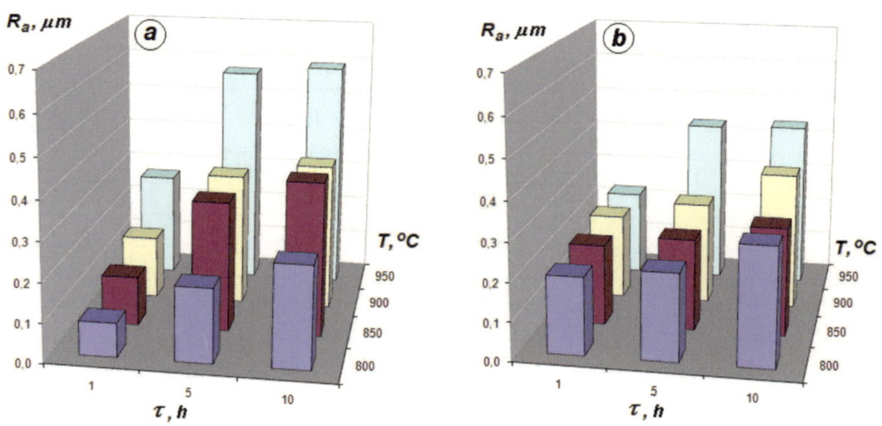

Fig. 3. Surface roughness of VT6 (a) and VT22 (b) titanium alloys after nitriding.

The nitride film consists of only nitrides of base metal – δ-(TiN) and ε-(Ti_2N). The grains of nitride phases have predominating orientations (table 2, 3). It is better expressed for ε-Ti_2N grains, which are mainly oriented on planes [002]. It should be noted that the texture of ε-phase is formed only during the process of nitride film growth (the redistribution of reflexes' intensity is not observed after short-term exposures when nitride film is thin).

Parameters of nitriding	Alloys			
	VT6		VT22	
	$I_{(111)}/I_{(200)}$	$T_{(200)}$	$I_{(111)}/I_{(200)}$	$T_{(200)}$
800 °C, 1 h	-	-	-	-
800 °C, 5 h	1,3113	0,4327	1,0510	0,4876
800 °C, 10 h	1,1015	0,4758	1,0690	0,4893
850 °C, 1 h	1,2053	0,4535	-	-
850 °C, 5 h	1,2047	0,4537	0,9726	0,5069
850 °C, 10 h	0,9864	0,5034	1,0642	0,4845
900 °C, 1 h	-	-	1,105	0,4751
900 °C, 5 h	1,1129	0,4733	1,0209	0,4948
900 °C, 10 h	1,0759	0,4817	0,8165	0,5505
950 °C, 1 h	1,2308	0,4483	1,1349	0,4684
950 °C, 5 h	1,0634	0,4846	0,9272	0,5189
950 °C, 10 h	1,2248	0,4495	0,9865	0,5034

Table 2. The ratio $I_{(111)}/I_{(200)}$ and coefficient of texture plane (200) $T_{(200)}^*$ of TiN_x ($^*T_{(200)}= I_{(200)}/(I_{(200)}+I_{(111)})$ (Hultman et al., 1995)).

Parameters of nitriding	Reflexes Ti_2N (hkl)	Alloys			
		VT6		VT22	
		I, arb. units	$I_{(111)}/I_{(200)}$	I, arb. units	$I_{(111)}/I_{(200)}$
800 °C, 1 h	(111)	254	2,7609	430	2,9861
	(002)	92		144	
800 °C, 5 h	(111)	152	0,4199	268	0,4401
	(002)	362		609	
800 °C, 10 h	(111)	236	0,3758	249	0,3522
	(002)	628		707	
850 °C, 1 h	(111)	265	1,3731	345	1,1616
	(002)	193		297	
850 °C, 5 h	(111)	251	0,5529	139	0,1154
	(002)	454		1205	
850 °C, 10 h	(111)	121	0,1157	-	-
	(002)	1046		2006	
900 °C, 1 h	(111)	316	1,5960	493	4,0410
	(002)	198		122	
900 °C, 5 h	(111)	275	0,6643	406	0,7719
	(002)	414		526	
900 °C, 10 h	(111)	268	0,8845	440	0,7666
	(002)	303		574	
950 °C, 1 h	(111)	259	-	-	-
	(002)	-		91	
950 °C, 5 h	(111)	235	1,5359	216	1,1676
	(002)	153		185	
950 °C, 10 h	(111)	387	3,5833	-	-
	(002)	108		212	

Table 3. Relative intensity of diffraction reflexes (111) and (002) of Ti_2N on the diffraction patterns from VT6, VT22 and T110 alloys after nitriding.

The thin outer layer of golden color consists of TiN and the thick inner layer of white color – Ti_2N. The perceptible growth of TiN is observed only at the long exposures.

Due to the large brittleness of nitride layer the measuring of its quantitative characteristics is complicated. Therefore there are few articles on the kinetics of nitride formation. This explains the substantial spread and certain discordance of the results received by different researchers.

The alloying elements of titanium alloys does not participate in the process of nitride formation. According to the thermodynamic activity of elements in relation to nitrogen (fig. 4), except of formation of titanium nitrides, zirconium nitrides are formed probably at nitriding of titanium alloys. The information about their formation is not found in literature, although at nitriding of alloys with 3...4 % Al, 8…12 % Zr, 1,2...2,6 %V the formation of phase (Ti, Zr)N with the lattice parameter of 0,4283 nm was fixed (Kiparisov & Levinskiy, 1972).

Alloying of titanium influences on the depth of nitride layer. In according with the recent results the nitrided area in (α+β)-alloys is less than in α- and pseudo-α-alloys (fig. 5).

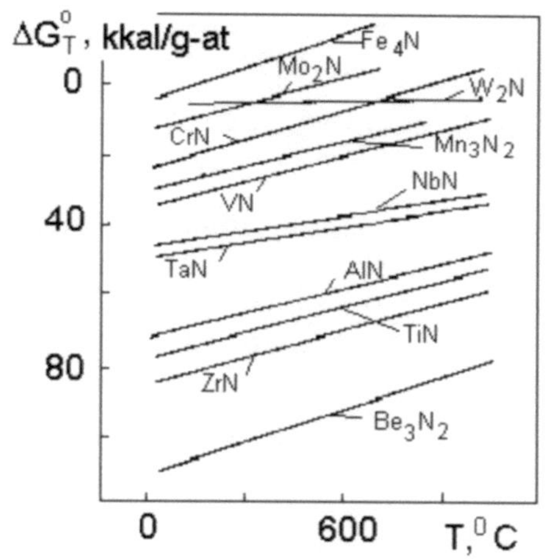

Fig. 4. Change of Gibbs thermodynamic potential ΔG_T^o on 1 g-at of titanium depending on temperature for the reactions of formation of some nitrides (Kiparisov & Levinskiy, 1972).

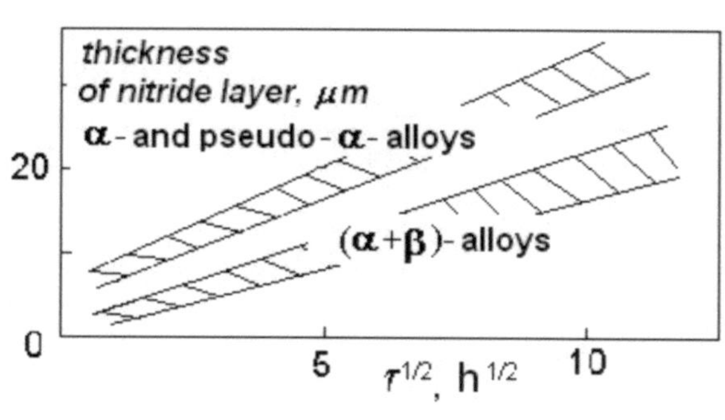

Fig. 5. Thickness of nitride layer on titanium alloys depending on duration of nitriding at 1000 °C.

The beginning of active nitride formation influences on the surface microhardness which depends on the nitrided material and ranges in 4,5…7,0 GPa. The rise of temperature leads to the increase of surface microhardness due to the activating of nitride formation (fig. 6).

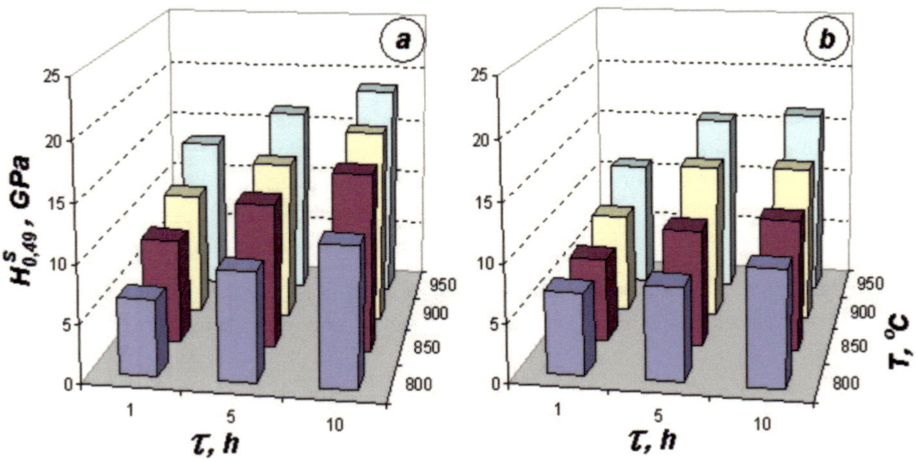

Fig. 6. Dependence of surface microhardness of VT6 (a) and VT22 (b) alloys on duration and temperature of nitriding.

Thus, during the nitriding the growth of nitride film has columnar character (mainly on the grain boundaries). It assists in the rising of surface relief that worsens the surface quality at higher nitriding temperature. During the growth of nitrides the texture Ti_2N is formed mainly on plane [002]. The change of thickness of nitride film with duration is described by parabolic dependence, thus the thickness of nitride film on β- and (α+β)-titanium alloys is less than on α-alloys. The texture growth increases with the rise of nitriding temperature resulting to the increase of imperfectness and heterogeneity of nitride film.

2.3 Regularities of formation of gas saturated layer – Morphology of the nitrided layers

At the high-temperature interaction of titanium alloys with nitrogen, except for formation of nitride area on the metal surface, nitrogen diffuses into the alloy, dissolves and forms the area with increased microhardness, so-called gas-saturated area. This area is identified as α-titanium with the increased lattice parameters (solid solution of nitrogen in α-titanium). The grains of α- solid solution are oriented mainly on plane [211]. The layers with the higher nitrogen concentration are characterized by the texture of α-solid solution.

The process of gasing of titanium alloys is connected with the process of nitride formation. Surface microhardness (H^s_μ), strengthening level of gas-saturated layer ($H_\mu = f(\ell)$) and its depth (ℓ) are determined by time and temperature parameters of nitriding and depend both on chemical and phase composition of alloys. The surface layer of α- and pseudo-α-titanium alloys is the most strengthened (fig. 7). This effect weakens considerably at transition to (α+β)- and, especially, β-alloys. Considerably bigger depth of gas-saturated layer of β-alloys as compared to α-alloys is determined by the values of nitrogen diffusion coefficients in α-

and β-phases of titanium. For pseudo-α- and (α+β)-alloys when the saturation process occurs at low (≤ 850 °C) temperatures and undurable exposures (≤ 5 h) the depth of gas-saturated layer is decreased with the increase of coefficient of β- stabilization of alloy. High temperatures and long exposures assist in the increase of depth of gas-saturated layer with the increase of coefficient of β-stabilization.

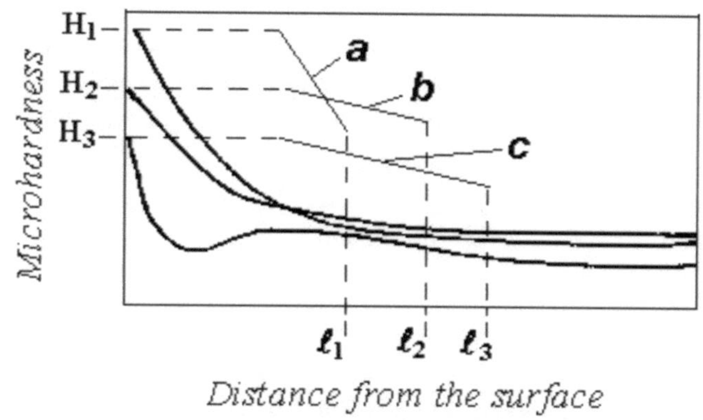

Fig. 7. Distribution of microhardness through cross section of surface layers after nitriding of α- and pseudo-α- (a), (α+β)- (b) and β- (c) titanium alloys at temperatures of $T > T_{\alpha\Leftrightarrow\beta}$.

The morphology of gas-saturated layers after nitriding depends on temperature and metal's phase composition. Let's consider the influence of these factors on the morphology of the gas-saturated layer.

At the temperatures of nitriding below the polymorphic transformation the morphology of gas-saturated layer does not depend on phase composition of alloys (fig. 8a). The layer consists of two parts. The first part contains α-grains with high microhardness due to their strengthening by nitrogen (*α-solid solution stabilized by nitrogen*). The solubility of nitrogen in

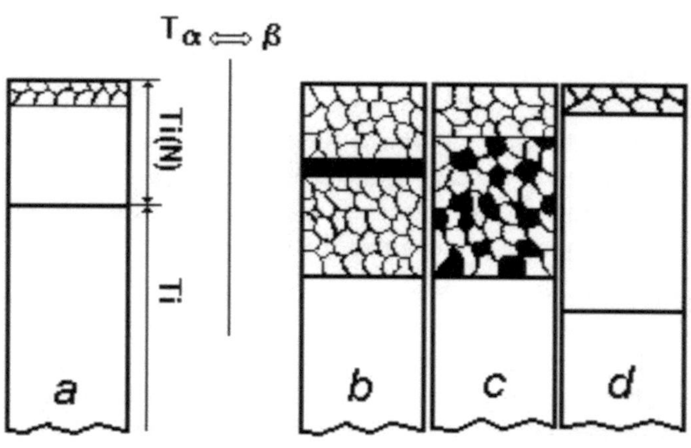

Fig. 8. Structure of gas-saturated layer of α- and pseudo-α- (a, b), (α+β)- (a, c) and β- (d) titanium alloys nitrided at the temperatures of $T< T_{\alpha\Leftrightarrow\beta}$ (a) and $T > T_{\alpha\Leftrightarrow\beta}$ (b-d).

α-titanium is high. Based on the titanium - nitrogen phase diagram, the deformation of lattice at the dissolution of nitrogen is considerable, which results to the significant increase of microhardness of this part of gas-saturated layer. The thickness of this part of gas-saturated layer increases with nitriding time and temperature parameters and connects with the redistribution of alloying elements. The second part of gas-saturated layer is not nearly differ from the alloy matrix. The level of saturation by nitrogen of this part is less substantially than the first part. The microhardness gradient and the redistribution of alloying elements are insignificant. The thickness of gas-saturated area fixed by metallographic method does not give the real depth of nitrogen penetration into the alloy's matrix and as a usual is less than that determined by measuring the microhardness.

At the nitriding temperatures higher than temperature of polymorphic transformation the gas-saturated layer also consists of two parts. However the structure of every part is determined by phase composition of nitrided alloy (fig. 8, b-d). The gas-saturated layers of α-, pseudo-α- and (α+β)-titanium alloys are separated by the phase boundary from the matrix which at the nitriding temperature was β-titanium. During the cooling β-phase is decomposed but the boundary fixed at the high temperature maintains at the room temperature as well as the structure of α-phase formed in result of α⇔β transformation. α-phase which stabilized by nitrogen and α⇔β transformed phase but saturated by nitrogen have significant differences. The morphology of gas-saturated layer of β-titanium alloys does not almost depend on the nitriding temperature. That is caused by the absence of both polymorphic transformations of matrix and part of gas-saturated layer below the level of α-phase stabilizing.

The first part of gas-saturated layer formed at the temperatures of (α + β) - β - area is α-phase (grains of α-solid solution of nitrogen in titanium). The high microhardness of this part is caused by large solubility of nitrogen in α-titanium (21,5 at.% in α-titanium against 0,95 at.% in β-titanium at 1000 °C). It should be noted that microhardness of this part of gas-saturated layer for α- and pseudo-α-titanium alloys exceeds insignificantly the same layer for (α + β)- and β-alloys (18...7 GPa against 10...5 GPa). It is determined by the different nitrogen solubility in α- and β-phases of titanium. The second part of gas-saturated layer consists of metal β → α transformed and enriched by nitrogen. It is separated from the first part by the phase boundary. For α- and pseudo-α-titanium alloys this boundary is detected metallography as a dark band of high etching. For α- and pseudo-α-titanium alloys this part is α-grain of smaller size but with the increased degree of etching compared with α-structure of the first part. For (α + β)-alloys it is mainly α-phase (α-plates) in β-transformed structure (mixture of α- and β-phases). The second part of gas-saturated layer of (α + β)-titanium alloys is often called "the transition area" between the gas-saturated layer and alloy's matrix because of the sharp structural difference as compared to the first part.

With the rise of nitriding temperature and duration the size of the second part of gas-saturated layer decreases, and the size of the first part increases. In addition, there is a coarsening of the structural components of both parts, and also a change of phase correlation in the direction of increasing the quantity of α-phase. The typical microhardness redistribution through the gas-saturated layer of α- and pseudo-α-titanium alloys is shown in fig. 7. It represents the gradient of nitrogen concentration from the surface into the matrix. Some tendency to stabilization of microhardness in the second part of gas-saturated layer for (α+β)-alloy can be explained by different nitrogen solubility in α- and β-phases of titanium.

The structure of gas-saturated layer of β-titanium alloys does not depend on the nitriding temperature and is analogical to the structure of alloys nitrided below temperature of α⇔β transformation when the structural difference between the alloy's matrix and second part of gas-saturated layer is absent (fig. 8). The first part is α-phase stabilized by nitrogen with the boundaries decorated by initial β-grains.

For β-alloys the microhardness distribution of gas-saturated layer has the original regularity (fig. 8): the curve passes through a minimum on the boundary of the part of gas-saturated layer stabilized by nitrogen.

For β-titanium alloys the strengthening of surface layers is significantly lower than for alloys of other structural classes (microhardness distribution curves are in the region of lower hardness values) (fig. 7).

Thus, the basic characteristics of gas-saturated layer i.e depth and degree of strengthening of surface layers (surface microhardness, hardness redistribution), depend on the phase composition of nitrided material. The most strengthening of surface layers is proper for α- and pseudo-α-titanium alloys and substantially decreases at the transition to (α+β)- and, especially, to β-alloys. The depth of gas-saturated layer of β-alloys is considerably larger than depth of gas-saturated layer of α-alloys. The morphology of gas-saturated layer of titanium alloys depends on nitriding temperature and phase composition of the nitrided alloy. For β-alloys the morphology of gas-saturated layer does not depend on the nitriding temperature and thus is identical to alloys with other structures nitrided in α-area.

2.4 Redistribution of alloying elements

The strengthened surface layer consists of nitride and gas-saturated area. As it was shown above, it is the result of the high-temperature interaction of titanium with nitrogen. This interaction is accompanied with the redistribution of alloying elements in alloy's surface layers. Let's consider the general regularities of alloying elements redistribution at nitriding of titanium alloys.

The alloying elements of titanium alloys are categorized as α- (Al), β- (Mn, V, Mo, Cr, Fe, Nb, Si, W) - stabilizers and neutral reinforcers (Zr, Sn). During the saturation of Ti-alloys by nitrogen there is the redistribution of alloying elements between the nitrided layer and matrix as well as in the gas-saturated layer.

The increase of electron concentration during the nitrogen dissolution leads to the decrease of solubility of alloying elements in titanium due to the limited solubility and also because the formation of continuous series of solid solutions. It assists in redistribution of alloying elements in the surface layers of titanium alloys: their separation from solid solution and diffusion into titanium matrix.

Thus, at thermodiffusion saturation by nitrogen there is diffusion of elements separated from solid solution into the alloy (fig. 9). The intensity of this diffusion is determined by the solubility and diffusion mobility of alloying elements.

The alloying elements have different solubility and diffusion coefficients in α- and β-modifications of titanium. According to the calculations, taking into account the diffusion constants, the diffusion mobility of alloying elements is decreased in a sequence Fe→Mn→Zr→Cr→Al→Sn→Nb→V→Mo. With regard to solubility, then the solubility of zirconium is unlimited, and tin and aluminum are characterized by high solubility in α-titanium. Vanadium, molybdenum and niobium are less soluble in α-titanium but dissolve indefinitely in β-titanium. Iron, chromium and manganese are limitedly solubles in β-titanium and solubility in α-titanium is small. Iron, manganese and chromium are redistributed the most actively in surface layers because their solubility is minimal and diffusion mobility is the highest. The solubility of zirconium, aluminum and tin in α-titanium is high and diffusion mobility is low. Therefore, the substantial redistribution of these elements does not occur. Molybdenum, vanadium and niobium redistribute more active than zirconium, aluminum and tin but more weaker than iron, manganese and chromium.

Except for diffusion, there is the concentration of β-stabilizing elements separated from solid solution of nitrogen in α-titanium on the boundary of nitride - gas-saturated areas, even with high diffusion constants. These effects are caused by no occupied bonds between atoms located on the phases' interface and possible anomalous value of electron concentration in these areas.

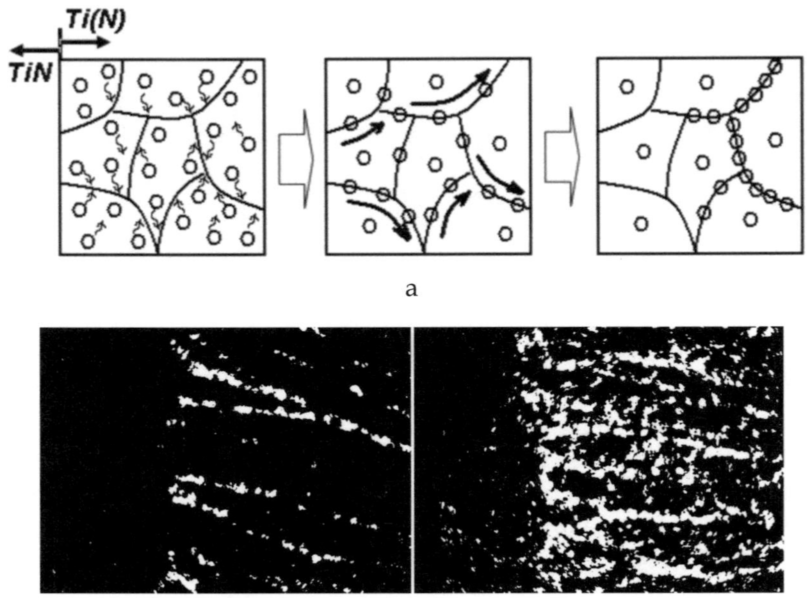

Fig. 9. Scheme of redistribution of alloying elements in the gas-saturated layer of titanium alloys at nitriding (a) and images of surface layers of OT4-1 (b) and VT6s (c) alloys in the characteristic rays K_αMn and K_αV (1100 °C, 1 h).

Among these elements the special attention deserves aluminum. Since aluminum is α-stabilizer with high affinity to nitrogen (ΔH^o_{298} is -318,0 and -335,0 kJ/mole for AlN and TiN respectively), the increase of electron concentration of alloy during stabilization of hexagonal close-packed lattice of solid solution of nitrogen in α-titanium does not influence on the solubility of aluminum and does not assist its diffusion. Releasing of lattice energy and energy stability of system is achieved by the redistribution of other alloying elements. The similar selective redistribution, not so clear expressed, is observed for the systems, in which solubility and diffusion mobility of alloying elements differs significantly (for example, zirconium and molybdenum, tin and iron etc.). Near the interface gas (nitrogen) - metal and afterwards near nitride - gas-saturated layer the areas (clusters) with the high concentration of aluminum are formed (fig. 10). It is possible that the redistribution and coagulation of aluminum will be over by establishing of short range ordering completing by decomposition of solid solution with formation of superstructure of α_2-phase (Ti_3Al) type.

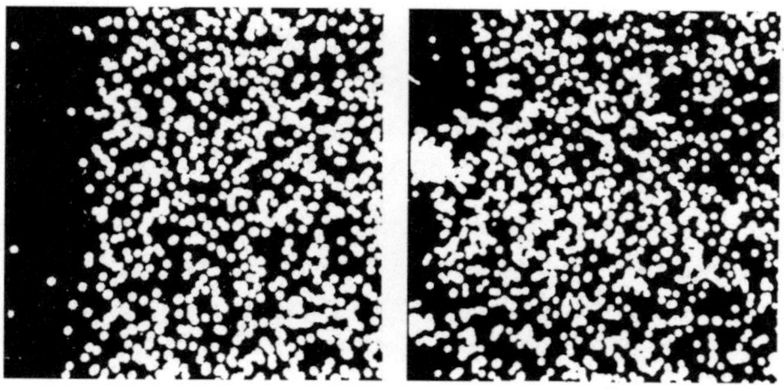

a b

Fig. 10. Image of surface layers of OT4-1 alloy in the characteristic rays $K_\alpha Al$: a – 900°C, 100 h; b – 950 °C, 8 h.

With the rise of temperature of isothermal exposure the diffusion of alloying elements is activated because the diffusion coefficients increase. The active motion of alloying elements along grain boundaries leads to their loosening, nitrogen diffusion becomes accelerated and nitriding rate increases.

The morphology of gas-saturated layer of titanium alloys is connected with the redistribution of alloying elements in the surface layers. There is the active diffusion of alloying elements from the most enriched by nitrogen part of layer. In that part with nitrogen concentration the redistribution is negligible: alloying elements are redistributed between α- and β-phases of titanium. Aluminum (α-stabilizer) and, as a rule, the neutral reinforcers (zirconium, tin) are located in α-phase of titanium, β-stabilizers enrich β-phase.

Thus, the basic constituents of alloying elements redistribution at nitriding of titanium alloys are as follows: 1) separation of alloying elements from the hexagonal close-packed lattice of nitrogen solid solution in α-titanium; 2) diffusion of alloying elements from the surface into the alloy's matrix.

The first process is controlled by the solubility and the second – by the diffusion constants.

In result of alloying elements redistribution, as a rule: 1) a concentration of aluminum increases near the boundary nitride – gas-saturated areas; 2) the surface layers are depleted by β-stabilizing elements.

The redistribution of alloying elements causes the structural and phase changes in the surface layers of alloys, determining the morphology of the nitrided layer.

In spite of the fact that at the increase of temperature the degree of surface strengthening increases continuously (thickness of both nitrided layer and its constituents, surface microhardness and gradient of nitrogen concentration on the cross section of surface layers increase), the use of temperature as the factor of intensification of saturation process has substantial limitations. In particular, the perceptible structural and concentration heterogeneity (on nitrogen and alloying elements) of surface layers caused by the active diffusion processes, and also the inconvertible grain growth of titanium matrix at the saturation temperatures of $(\alpha+\beta)$ - β- areas result in the substantial decrease of fatigue life, plasticity of the nitrided details. The heightened requirements to these characteristics cause the limitation on the saturation temperature (α-area) that does not always provides the provide level of surface strengthening ($H_{\mu}^{s} \geq 6...8$ GPa; $\ell \geq 100$ µm). Moreover, with the increase of temperature the brittleness of nitrided layer increases catastrophically. In the result of thick nitride film forming the surface quality of the nitrided layer becomes worse (surface roughness increases, imperfection and heterogeneity of nitride film grow because the effect of growth texture increases). It influences negatively on the wear- and corrosion resistance of the nitrided details.

At present, it is actual to find out other factors of intensification which allow to provide the effective surface strengthening at lower nitriding temperatures and exclude the negative consequences of the influence of high nitriding temperatures on surface quality and level of mechanical characteristics.

3. Nitriding of titanium alloys at thermocycling

One of the ways to weaken the negative consequences of the high-temperature nitriding is to decrease the time of processing at high temperatures. It can be attained by nitriding in the conditions of thermocycling.

As opposed to the standard methods of the chemical heat treatment there are the additional sources of the influence on the structure at thermocycling. They are inherent only to the process of continuous change of temperature: phase transformations, gradient of temperature, thermal (volume) and interphase tensions caused by the difference of thermophysical characteristics of the phases. The accumulation of structural changes in the material leads to forming, moving and annihilation of point and linear defects, redistribution of distributions, forming of low-angle boundaries, migration of low-angle boundaries with absorption of defects, migration of grain boundaries between recrystallized grains with their coarsening at the simultaneous decrease of grain boundary and surface energies, by the redistribution of alloying elements etc. It results to the increase of mobility of impurity atoms and the acceleration of diffusion processes.

For titanium alloys the thermocyclic treatment is considered as a way to achieve of such structural changes which improve the level of mechanical properties. To estimate the intensity of saturation process in these conditions is impossible due to the absence of such investigations for titanium, although the interaction of steels, aluminum and nickel alloys with gases (carbon, nitrogen) at the cyclic change of temperature, pressure and gas composition is well studied.

Let's consider the regularities of interaction of titanium alloys with nitrogen in the conditions of thermocycling.

Nitriding is intensified at the cyclic change of temperature. It influences on the rise of mass increase of the samples, surface microhardness and depth of the nitrided layer as compared to the isothermal exposure at the middle temperature of thermal cycle (fig. 11).

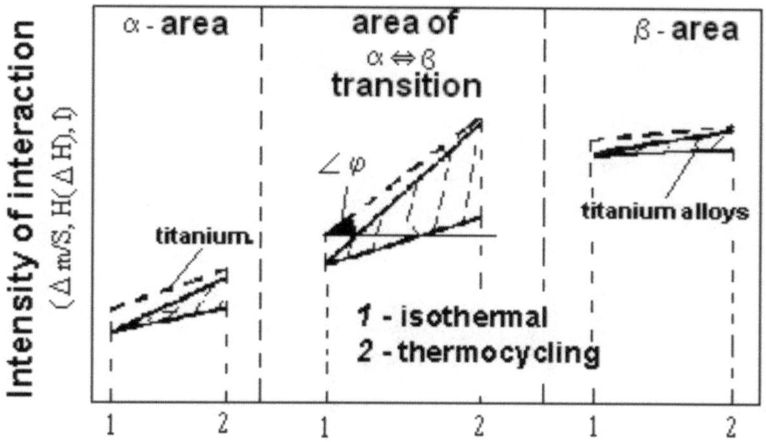

Fig. 11. Influence of thermocycling on intensity of interaction of titanium alloys with nitrogen in different temperature intervals.

The efficiency of thermocycling at nitriding depends on the parameters of thermocyclic treatment (amplitude of thermal cycle, frequency of thermocycling) and increases with their rising.

The maximal effect of thermocycling is proper for the temperature range of polymorphic transformation and correlates with the clear expressed effect of volume strengthening which increase at the rise of amplitude and frequency of thermocycling. Alloying, as a factor of intensification on interaction of titanium alloys with gas media, does not change generally the influence of thermocycling but changes only its intensity (slope of curves φ) depending on participating of certain alloying element in the forming of defect structure.

The morphology of nitrided layer thermocycling as well as after the isothermal conditions is the thick nitride film (≥ 1 μm) and gas-saturated area. The difference from the isothermal nitriding is that the increase of degree of imperfectness of surface layers at thermocycling leads to the forming of surface nitride films with considerable deviation from stoichiometry, mainly the deficit of nonmetal component. Therefore the lattice parameter of TiN after isothermal saturation with the rise of temperature is decreased significantly while the cyclic change of temperature assists to reach the reverse dependence: lattice parameter of TiN increases (fig. 12).

Taking into account the dependence of the lattice parameter of titanium mononitride on nitrogen content in the homogeneity region, the observed regularities allow to suppose that at nitriding in the conditions of thermocycling nitride with considerable deviation from stoichiometry with the deficit on nitrogen is formed on the surface. With displacement of temperature range of thermocycling into the range of lower temperatures the deviation from stoichiometry of surface nitride increases.

It should be noted that forming of these nitride layers gives new possibilities in surface engineering of titanium alloys, in particular, at the complex modification of surface layers by the interstitial elements (Pohreliuk et al., 2007; Pohrelyuk et al., 2009, 2011; Yaskiv et al., 2011; Fedirko et al., 2009).

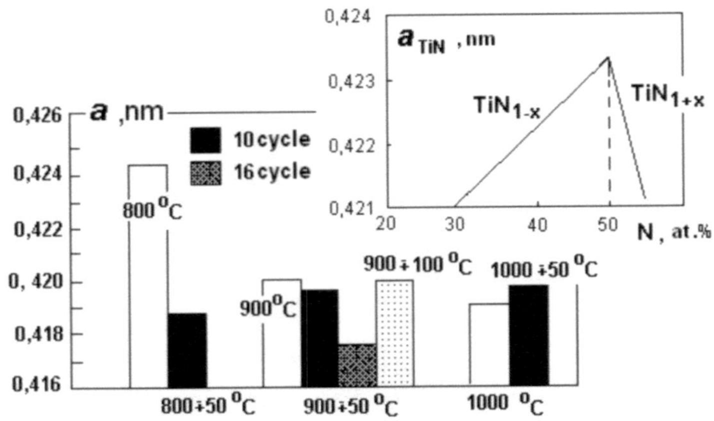

Fig. 12. Lattice parameters of TiN after the isothermal nitriding and nitriding in the conditions of thermocycling. Change of lattice parameter of TiN depending on the content of nitrogen in the homogeneity region (Goldschmidt, 1967).

Thus, at thermocycling the nitriding process of titanium is intensified and reaches the maximum during processing in area of $\alpha \Leftrightarrow \beta$ transition. The efficiency of application of thermocycling at nitriding depends on the parameters of thermocyclic treatment and rises with their increase. The temperature range of thermocycling determines the character of the surface strengthening. The nonstoichiometric nitride films with the deficit of nonmetal component are formed on the titanium surface.

The strength characteristics of titanium are improved after nitriding at thermocycling. The highest strengthening effect is observed at cycling in the area of temperatures of polymorphic transformation of titanium alloys and enhances with the increase of both amplitude and frequency of thermocycling.

4. Influence of initial deformation texture on nitriding of titanium alloys

The intensification of nitriding at thermocycling is based on the structural changes in material. The same changes in the structure of material is possible to provide before thermal heat treatment, for example, using material with deformation texture. Such approach is based on the dependence of diffusion constants on predominating crystallographic orientation (texture).

In practice the metallic materials are used, as a rule, in the polycrystalline state. Although all grains in homogeneous metal have the identical crystalline structure, however they differ in the mutual crystallographic orientation of axes. The analogue of crystallographic orientation of plane in monocrystal for polycrystal is the predominating orientation of grains (texture). One of the basic technological processes causing the formation of crystallographic texture, is plastic deformation. Formation of texture at plastic deformation occurs in the result of crystallographic planes turning in the process of sliding and twinning. In titanium the deformation occurs by sliding on the systems $\{10\bar{1}0\}\langle 11\bar{2}0\rangle$, $\{10\bar{1}1\}\langle 11\bar{2}0\rangle$ and $(0001)\langle 11\bar{2}0\rangle$ (critical shear stress is minimal for plane $\{10\bar{1}0\}$ and maximal for basal plane) and by twinning on planes $\{10\bar{1}2\}$, $\{11\bar{2}2\}$ and $\{11\bar{2}\ \}$ and causes corresponding deformation texture.

The crystallographic texture of titanium alloys depends on the chemical composition (alloying) of metal, degree of deformation, temperature and method of rolling, thickness of semi-finished rolled products, presence of gas-saturated layer etc. Heating of textured material also allows to change the texture (recrystallization, polygonization annealing, polymorphic transformation). Therefore in practice there is a great number of methods to operate the crystallographic texture of titanium alloys allowung to form texture with set-up parameters.

Having established the correlative dependences between crystallographic texture and processes of interaction of titanium alloys with gas media, it is possible to use the texture factor for operating of the intensity of physical and chemical processes in gas - metal system, and, consequently, to influence on improved characteristics of construction material.

Let's illustrate the influence of texture on the nitriding of titanium alloys.

At the gasing of samples with t_1 - "base" $(0001)[10\bar{1}0]$ (fraction of orientations 44 %) deformation texture (the plane of base of hexagonal close-packed lattice is parallel to the rolling plane) the rate of the increase of nitrogen concentration in titanium is higher than for samples with t_2 - "prismatic" $(10\bar{1}0)[11\bar{2}0]$ (fraction of orientations 50 %) texture that assists to form strengthened layers with different parameters (depth of area, surface hardness, gradient of hardness). During nitride formation with "base" texture the density of nucleation centers of nitride phases is larger and time to formation of continuous surface films is less than for the samples with "prismatic" texture (the plane of prism of hexagonal close-packed lattice is parallel to the rolling plane), that assists to form nitride films of different thickness. That is crystallographic texture of titanium alloys influences on the conditions of mass transfer on the gas - metal boundary and diffusion mobility of nitrogen. A schematically influence of crystallographic texture on the processes of interaction of titanium with nitrogen at different phase-boundary conditions on the boundary gas - metal is presented on fig. 13.

Thus, the application of texture factor allows to influence on the intensity of nitride formation and gasing, changing the relation between the dimensions of nitride and gas-saturated areas. Forming by preprocess of the primary crystallographic orientation ("base" or "prismatic" properly), it is possible to provide either higher level of surface strengthening or larger depth of nitrogen penetration in matrix.

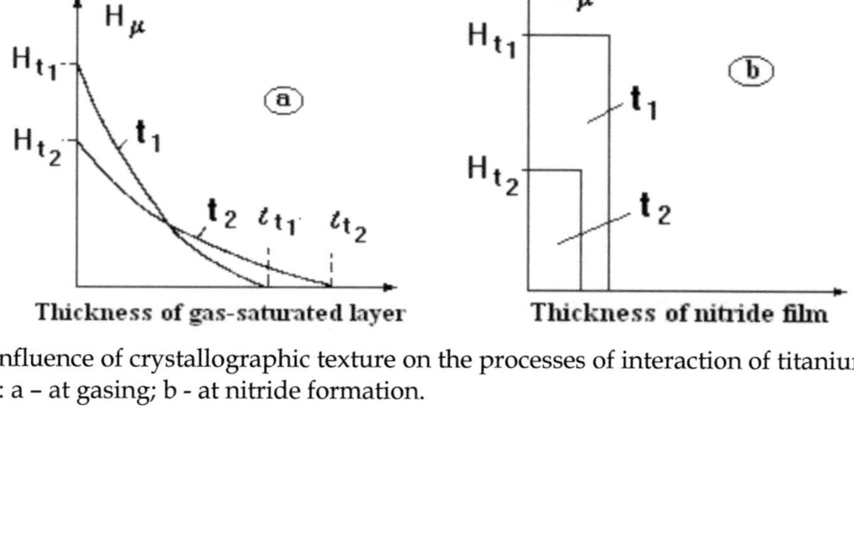

Fig. 13. Influence of crystallographic texture on the processes of interaction of titanium with nitrogen: a – at gasing; b - at nitride formation.

The intensification of process at the use of the considered approach allows to decrease the temperature of treatment, and, consequently, to weaken the negative consequences of the influence of high saturation temperatures on the quality of surface and level of mechanical characteristics.

5. Use of elements of vacuum technology at nitriding

The analysis of results of nitriding of titanium alloys at high temperatures showed that the major reason which decelerates the diffusion of nitrogen into the matrix is the forming of thick nitride film with nitrogen diffusion coefficient less on 2 - 4 orders of magnitude than in matrix (D_{TiN}^{N} = 3,76×10^{-12} cm^2/s; $D_{\alpha\text{-}Ti}^{N}$ = 1,29×10^{-10} cm^2/s; $D_{\beta\text{-}Ti}^{N}$ = 3,92×10^{-8} cm^2/s at 950 ^{o}C). Another reason is the presence of oxide films formed at technological operations of details' manufacturing and their heating to nitriding temperature due to the presence of oxygen impurities in nitrogen. Therefore, the possible ways of intensification of nitriding is to provide the corresponding conditions that allows to : 1) increase the nitrogen diffusion coefficient in nitride film or in general prevent its formation on the initial stages of nitriding; 2) favour the dissociation of existing oxide films and prevent the formation of new ones.

Let's consider some variants of realization of the above approaches.

5.1 Nitriding in rarefied dynamic nitrogen atmosphere

The parabolic character of kinetics of high-temperature interaction of titanium with nitrogen is caused by forming of nitride film on the surface. The amount of nitrogen diffused through nitride layer during its growth is decreased constantly preventing to penetration of nitrogen into the metal.

The calculated nitrogen diffusion rate in titanium and rate of nitrogen supply to the metal surface testify that under the certain conditions even all nitrogen molecules which get on surface can not be sufficient to provide the maximal flux of nitrogen atoms from surface into matrix. Except of it, not all nitrogen molecules, contacting with the surface of metal, interact with surface. In this case the processes connected with supply of nitrogen to the gas - metal reaction area become limiting. It allows to control the maximal nitrogen concentration on titanium surface and thus to provide the necessary nitrogen concentration for nitride formation. The absence of nitride film on the surface removes the diffusion barrier, and, consequently, penetration of nitrogen into titanium matrix intensifies. That is, the nitrogen partial pressure becomes the factor of intensification of nitriding process (fig. 14).

With the decrease of nitrogen partial pressure it is possible to provide the conditions when beginning of nitride film forming is shifted in time, that is at corresponding duration the nitride film is in general absent or its thickness is too thin. Thus, in the certain interval of nitrogen partial pressure the area of solid solution of nitrogen in α-titanium on the surface is formed, providing the more uniform distribution of hardness in the diffusion layer and increasing the depth of nitrogen penetration.

Let's consider the general tendencies in the processes of gasing and nitride formation at nitriding of titanium alloys in the rarefied dynamic nitrogen medium.

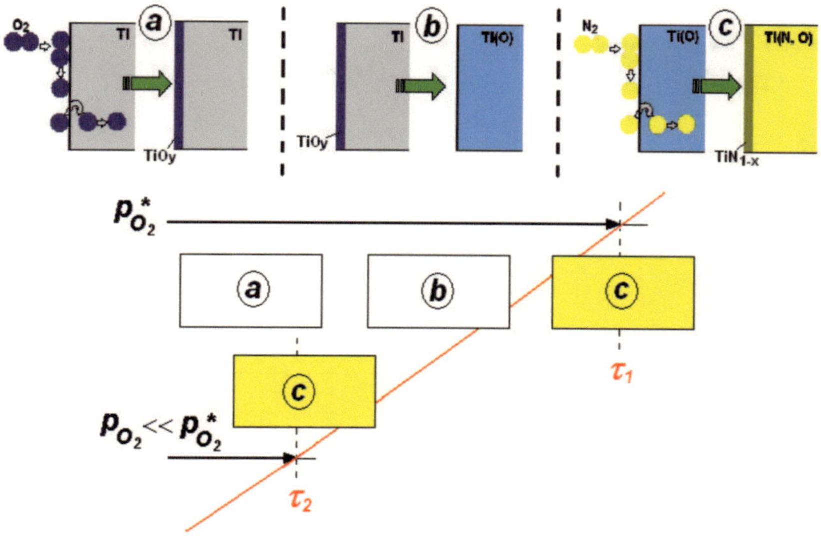

Fig. 14. Stages of nitriding of titanium alloys in nitrogen containing oxygen (a, b, c) and possibility of the intensification of process at the decrease of oxygen partial pressure.

The decrease of pressure from 10^5 to 100 Pa (gas flow rate 0,03 l/min) with the rise of mass increase of samples causes the increase of depth of nitrided layer and the significant decrease of thickness of nitride film. With the increase of gas rarefaction to 10 Pa the nitride thickness is stabilized and the depth of nitrogen penetration into titanium is decreased. With the decrease of gas flow rate on one order of magnitude, the mass increase and depth of nitrided layer increase, and thickness of nitride film decreases. This effect is similar to the decrease of gas partial pressure. At the decrease of pressure to 0,1…1 Pa in order to intensify the nitriding it is necessary to change the nitrogen flow rate. Thus, with the decrease of nitrogen flow rate in the range of 0,03…0,003 l/min the growth of depth of nitrided layer slows down.

The observed regularities and general tendencies in the processes of saturation of titanium alloys in the rarefied dynamic nitrogen medium indicate that in the interval of rarefaction 0,1…10 Pa at the gas flow rate 0,03..0,003 l/min (specific leakage rate $7\times10^{-2}…7\times10^{-4}$ Pa$\times$s^{-1}) the kinetics of nitriding becomes receptive to the processes connected with supply of nitrogen to the gas - metal reaction area.

The analysis of the results on the influence of nitrogen partial pressure and nitrogen supply rate on the mass increase of samples, surface strengthening (surface microhardness), depth of nitrided layer testifies that the providing of the indicated gas-dynamic parameters of gas medium allows the dynamic equilibrium between adsorbed and diffused nitrogen into the titanium matrix to be maintained in certain time interval. In such conditions the nitride film is not formed on the surface and the strengthened area is the solid solution of nitrogen in α-titanium. In due course, in the result of forming of diffusion layer and increase of nitrogen concentration on the gas - metal boundary to the necessary level for nitride formation, that corresponds t*, titanium nitride is fixed continually on titanium alloys (fig. 15).

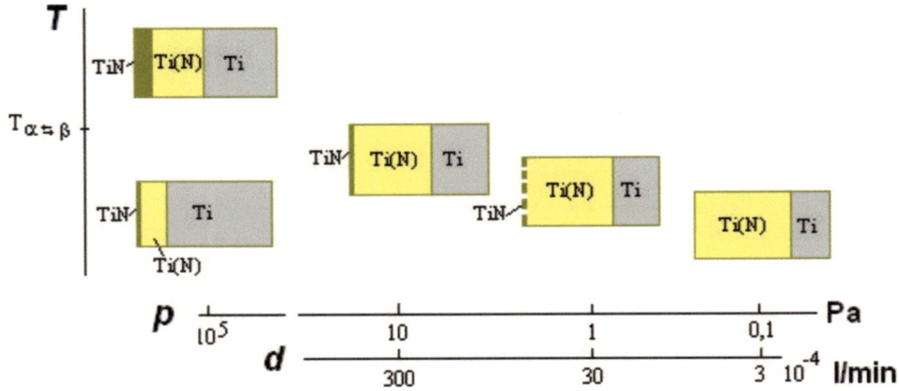

Fig. 15. Influence of temperature (T) and gas-dynamic parameters of nitrogen (p, d) on the phase-structural state of surface layers of titanium alloys.

At nitriding of titanium alloys with conservation of general tendencies the corresponding correctives in the process of saturation contributes the redistribution of alloying elements that influences on the absolute values of characteristics of the nitrided layers.

Nitriding in the rarefied nitrogen as compared to saturation in nitrogen of atmospheric pressure decreases the gradient of nitrogen concentration on the cross section of surface layers and increases the depth of penetration of nitrogen (in 1,3…2,3 time) and decreasing surface strengthening (fig. 16).

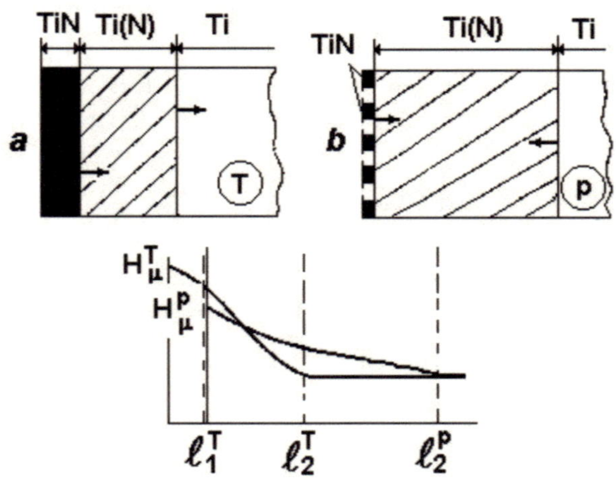

Fig. 16. Saturation temperature (a) and nitrogen partial pressure (b) as factors of the intensification of nitriding process of titanium alloys (arrows are direction of motion of phase boundaries in the areas of the identified parameters as factors of intensification).

As at lowering of nitrogen partial pressure the process of thermodiffusion saturation of titanium alloys intensifies, it makes possible to lower the nitriding temperature and decrease the process duration. The possibility to provide the sufficient surface strengthening at the temperatures of α-area excludes the negative aspects connected with high saturation temperatures, that is the quality of surface rises and the level of mechanical characteristics increases.

Thus, with lowering of nitrogen partial pressure the nitride formation is suppressed. The absence of nitride film on the surface or substantially less its thickness weaken the diffusion barrier and penetration of nitrogen into titanium matrix intensifies. Nitriding in the rarefied dynamic nitrogen condition as compared to nitrogen of atmospheric pressure provides more uniform redistribution of hardness through the diffusion layer and more significant depth of nitrogen penetration.

5.2 Some methods of nitriding improving

The vacuum technology is widely used in practice of thermal heat treatment, including at nitriding. A brief annealing in vacuum (800 °C, 2 h) before nitriding and final treatment of detail surface is recommended to conduct for distressing and prevention of warping (Samsonov & Epik, 1973). To decrease the brittleness of diffusion layer and increase the plasticity of alloys after nitriding on 10…15 % regardless of method it is recommended to conduct the additional annealing of details in vacuum at rarefaction of 4×10^{-2} Pa during 2 h at 800 °C (Kiparisov & Levinskiy, 1972).

The vacuum annealing in this case is the separate technological process. However, such technological process can be used with better efficiency when applying the vacuum technology in nitriding, that is, realizing vacuum annealing not as the separate process but as the element of nitriding process. It allows to considerably shorten and simplify the treatment as in this case the additional processes of heating and cooling are not necessary (fig. 17a) and, consequently, to improve substantially its productivity.

The treatment of titanium alloys in vacuum of 0,1..10 mPa when oxygen partial pressure is about 0,001…0,01 mPa excludes the possibility of formation of surface oxide films. Moreover, at the such oxygen partial pressures and corresponding time and temperature parameters it is possible to provide the conditions for dissociation and dissolution of natural oxide films before nitriding. The surface is activated and, as a result, at the subsequent inflow of nitrogen the adsorption and diffusion processes have been intensified. The upper limit of pressure of residual gases in vacuum is caused by the intensive processes of sublimation, contributing vacuum embittering of alloys's surface, and the lower one - by the active processes of oxidation and oxygen saturation.

That is, providing before the nitrogen supply in the reaction furnace the pressure of the residual gases of 0,1…10 mPa provides the necessary conditions for the intensification of nitriding. At the rarefaction of 0,1…1 mPa and temperatures above 600..700 °C the dissolution of oxide films is began and the effective removal of internal tensions is occurred that determines the lower boundary of temperatures of vacuum treatment. To receive the optimal complex of the mechanical characteristics the use of temperatures above the temperature of polymorphic transformation is undesirable. This determines the upper boundary of temperatures range of vacuum treatment.

The use of vacuum technology elements before nitriding in above indicated temperatures interval assists in the increase of saturation of surface layers with nitrogen, depth of strengthened area, surface quality and more higher temperature of exposure in vacuum. With the increase of temperature the more smooth redistribution of hardness throughout surface layers is achieved.

On this stage the heating in vacuum provides the conditions which exclude the additional oxidation of titanium surface and intensifies the nitriding. The vacuum medium (~1 mPa) during the heating activates the surface in the result of dissociation of oxide films. In result, the supply of nitrogen even of technical purity (with oxygen content to 0,01 % vol.) at the saturation temperatures of 750…850 °C allows to realize the quality nitriding. Providing of low pressure of residual gases of vacuum before supply of nitrogen into the reaction furnace does not require the long-duration isothermal exposure (≤ 2 h). Only heating in vacuum causes the positive result because it excludes the forming of oxide film at heating.

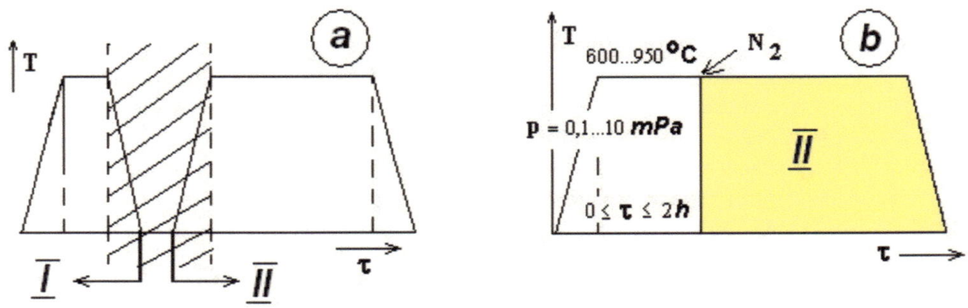

Fig. 17. Efficiency of the use of vacuum technology elements at nitriding of titanium alloys: I - vacuum treatment; II - nitriding.

Thus, the proposed vacuum technology before nitriding (fig. 17b) assists the intensification of thermodiffusion saturation of titanium alloys with nitrogen, allows to decrease the purity requirements to nitrogen for oxygen impurities at high-temperature treatment (> 850 °C) and to realize nitriding at the temperatures of 750..850 °C.

Thus, at the use of one or another intensification factor it is possible to influence on the constituents of nitriding process - nitride formation and gasing. At thermocycling as well as at the high saturation temperatures both nitride formation and gasing intensify. At the use of vacuum technology elements (lowering of nitrogen partial pressure, heating before nitriding and exposure in vacuum) the nitride formation is slowed down but gasing is activated.

At the same time the nitrided layers formed on titanium alloys are not limited by single variant of structure (thick nitride film (≥1 µm) and gas-saturated area). Depending on the conditions of nitriding by the control of intensity of physical and chemical processes on the boundary gas - metal it is possible to form various phase-structural states of surface layers of the nitrided titanium (fig. 18) which allows to change the level of surface strengthening (surface hardness, depth of penetration of nitrogen, distribution of microhardness on the cross section of surface layers) in the wide range, to control by thickness, continuity, composition stoichiometry and content of oxygen impurities, and, consequently, to realize the surface engineering of titanium alloys at nitriding according to the requirements of exploitation.

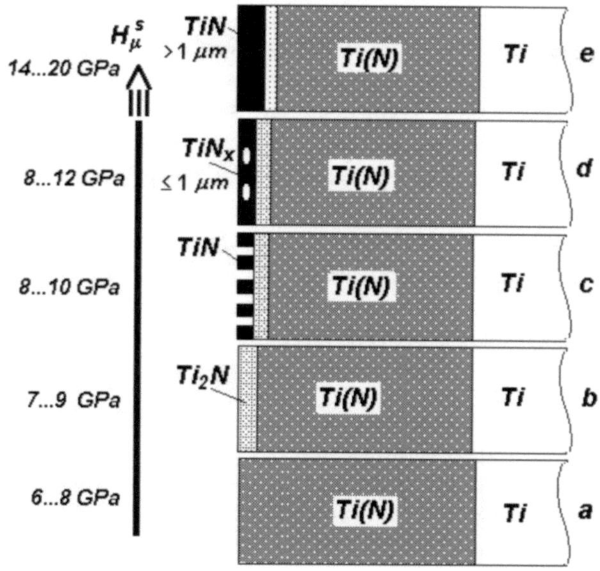

Fig. 18. The phase-structural state of surface layers of nitrided titanium alloys: a – gas-saturated area without nitride film; b, d – thin nitride film and gas-saturated area; c - nitride islands and gas-saturated area; e – thick nitride film and gas-saturated area.

Thus, the change of thermokinetic parameters of saturation, the use of vacuum technology elements and corresponding initial deformation texture allow to intensify the process of nitriding of titanium alloys in molecular nitrogen.

6. Surface engineering of titanium alloys by nitriding for corrosion protection in aqueous solutions of inorganic acids

Thermodiffusion coatings, including nitride ones, protect the titanium alloys against corrosion by combining of covering and electrochemical mechanisms (Chukalovskaya et al., 1993). The covering mechanism is being realized by making of barrier layer on the metal-medium border and thus it depends on its dimension. The electrochemical mechanism is defined by electrochemical characteristics of contacting surface and thus it causes the tendency of system to disturb the balance. In other words, it leads to reactions between surface layer ions and medium. Thus, protective properties of coatings depend on their dimension and structural characteristics (such as uniformity, relief, amount of oxygen impurities). In aggressive medium every mechanism brings in own contribution in protection. The effective combination of these mechanisms is mandatory criterion to ensure the high protective properties of nitride layers. The high saturation temperature (950 °C) provides the high-quality of nitriding in commercially pure nitrogen medium. However, the increasing of nitride coating saturation by nitrogen as well as the increasing of coating dimension due to saturation temperature rising do not lead to the improving of protective properties, but quite the contrary these processes lead to the decrease of these properties due to roughness and defectiveness raising. The saturation temperature determines the changes of surface relief. The nitride film forming at the temperature lower then the temperature of polymorphic transformation, only follows the material's matrix geometry. Nevertheless, at the temperatures higher than polymorphic transformation temperature, the relief fragments, such as burrs, form a grid and thus a roughness has been developed. The roughness is in 0,2…0,3 μm more than after nitriding at 850 °C. The activity of nitride-forming on the grain boundaries at high temperatures assists the relief forming. Processes following $\alpha \leftrightarrow \beta$ transformation, such as deformation strengthening and three-dimensional changes, can only enhance the relief. It should be noted that the plastic deformation has certain influence. The plastic deformation is caused by significant residual stresses during the thick nitride film forming (Rolinski, 1988). The influence of temperature on the surface roughness is essential while the influence of isothermal duration is no significant because in this case the enhancing of surface relief is minimum. For instance, the roughness R_a after different durations (5 h and 10 h) is close: 1,06 and 1,09 μm.

The investigations of influence of temperature-time parameters of nitriding on the nitride coatings dimension reveal that their thickness rises when the temperature and isothermal duration increase. For instance, after nitriding at the 950 °C thickness increases up to 3…4 μm with duration change from 5 to 10 h as well as from 4 to 5 μm at the duration 5 h but with the temperature increasing (from 850 °C to 950 °C) and from 5 to 6 μm at duration 10 h.

At the same time low-temperature nitriding (lower than 950 °C) does not provide the forming of high quality phase composition since the surface becomes dark gold colors. XRD measurements show TiN and Ti_2N reflexes as well as rutile TiO_2 ones. It determines by high thermodynamic relationships between titanium and oxygen, because the active interaction each other begins at 200…300 °C while with nitrogen – at 500…600 °C. Thus the surface oxidation takes place before nitride-forming. Oxide films dissolve and dissociate only at the high temperature (upper than 850 °C).

Therefore, temperature-time and gaseous-dynamic parameters determine the dimension, quality and phase composition of nitride coating. The oxygen partial pressure and saturation temperature determine the purity of nitride coating. The roughness of nitride coating depends on saturation temperature. The thickness of nitride layer is grown with increasing of temperature and duration. Moreover the influence of temperature is more sufficient.

Since every part of nitride structure brings in own contribution in protection against corrosion, it can be possible to optimize the morphology of nitride coatings by manipulating of above-mentioned parameters to achieve the highest protective properties.

The aqueous solutions of inorganic acids dissolve the titanium and its alloys very actively (Kolotyrkin et al., 1982; Gorynin & Chechulin, 1990; Kelly, 1979). To prevent a significant corrosion losses the nitrides, carbides and borides coatings are formed, e.g. in chloric and sulphuric acids corrosion rate of nitrided titanium alloys decreases in hundred times (Kiparisov & Levinskiy, 1972; Tomashov et al., 1985; Fedirko et al., 1998). In the same time according to some studies, the corrosion mechanism has differences in chloric and sulphuric acids (Kolotyrkin et al., 1982; Brynza & Fedash, 1972; Sukhotin et al., 1990). It indicates that protective coatings must have a different structure and phase composition for use in these acids.

The corrosion of nitride coatings passes by the parabolic dependence. At first, the nitride layers are being dissolving and then the oxidation layers takes place. During the corrosion, the corrosion cracking of nitride films has been occurred and then the oxidation and dissolution processes have being following.

Change of quality, morphology and phase composition of nitride coatings influence on their protective properties in chloric and sulphuric acids.

In 30 % HCl the increasing of thickness of nitride film improves the anticorrosion protection: kinetic curves of mass losses of nitrided specimens of significant thickness lie below. When thickness, caused by change of saturation temperature, increases the corrosion losses of light thin nitride films (5 μm) formed at 900 °C are less in 1,2-1,4 times. Another words, the thickenings of nitride films caused by longer isothermal duration improve their protective properties, whereas when it caused by increasing of temperature of saturation - decrease. Losses of nitride films have been increased in 1,5 times (Fig. 19). Obviously, it connected with forming of surface relief. To confirm this assumption the influence of coating thickness was excluded. For that the nitride coatings of different roughness but similar thickness (10 μm) have been forming. It has been achieved by nitriding at 950 °C for 10 h and at 1000 °C for 6 h. It was shown that increasing of roughness is been accompanied by increasing of corrosion rate near in 2 times (Fig. 20).

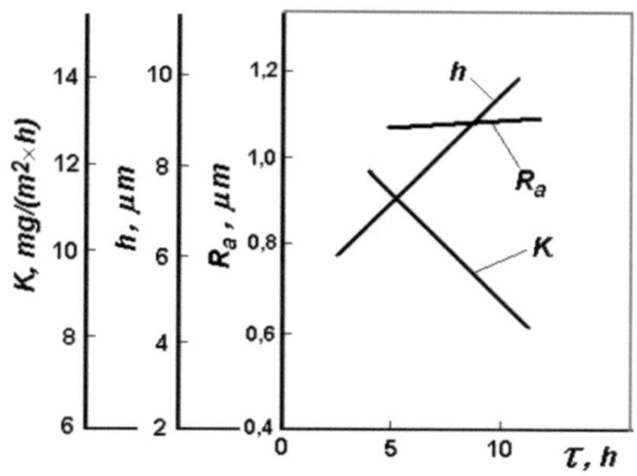

Fig. 19. Dependence of roughness (R_a, μm), nitride coating thickness (h, μm) and corrosion rate (K, mg/m^2×h) in 30% aqueous solution of chloric acid on nitriding duration at 950 °C.

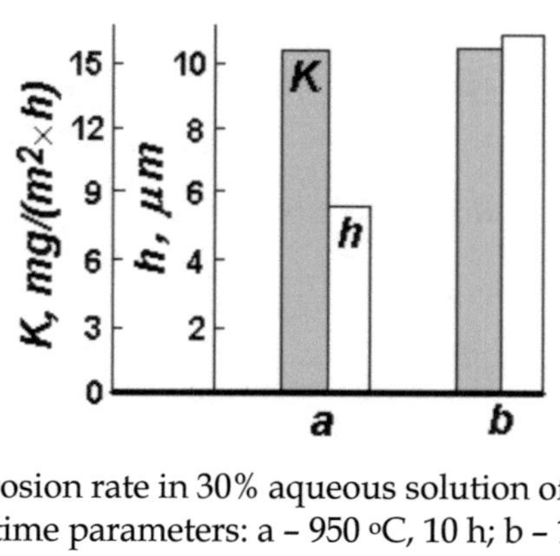

Fig. 20. Dependence of corrosion rate in 30% aqueous solution of chloric acid on the nitriding temperature and time parameters: a – 950 °C, 10 h; b – 1000 °C, 6h.

Electrochemical measurements in 30 % HCl confirm above-mentioned regularities. The forms of anodic curves of nitride coating are similar to the nontreated cases. Nevertheless, the electrochemical values of nitride coating with lower roughness are better: corrosion potential (E_{cor}) becomes higher (+0,08 V versus +0,06 V) and corrosion current density (i_{cor}) becomes lower from 1,0 up to 0,2 A/m^2 at the active dissolution and from 3,0 up to 0,8 A/m^2 at the passive state (fig. 21). Decreasing of the growth of corrosion current density on the different intervals of anodic curve is obviously connected with the forming of oxide films of different composition during the polarization (Gorbachov, 1983).

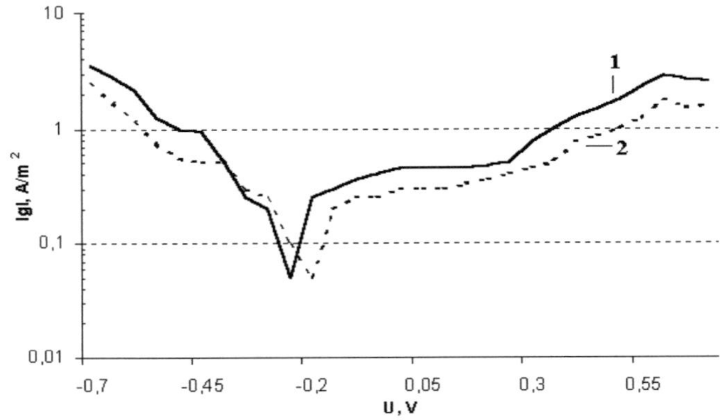

Fig. 21. Polarization curves of nitrided titanium in 30% aqueous solution of chloric acid (1 - 950 °C, 5 h; 2 - 850 °C, 5 h).

In 30% HCl the oxygen impurities or oxides including in nitride layers after the nitriding at 850 °C increase the corrosion rate (Table 4). For instance, the corrosion potential of free of oxygen nitride coating is more positive than oxynitride coating ones (Fig. 21). The corrosion current density decreases up to 10^{-1} A/m^2. The decreasing of anodic current density on the nitride coating in comparison with oxynitride ones indicates about a big braking of dissolution processes and confirms their advantage in protective properties. Cathode polarization passes by the hydrogen polarization mechanism. At the cathode polarization the nitride coating has lower current densities of cathode hydrogen depolarization. It indicates about the increasing of protection against electron conduction at the cathode depolarization of hydrogen. It is obviously that the hydrogen pickup of nitride coating is decrease sufficiently at that it decreases the hydrogen degradation of ones.

Coating technique	Morphology of coating	Corrosion rate, *K*, *mg/m²xh*	
		30% HCl	80% H_2SO_4
Heating and cooling in nitrogen, 950 °C, 5 h	Nitride coating with large surface relief	5,2	12,0
Heating and cooling in nitrogen, 850 °C, 5 h	Multiphase coating (mixture of nitrides and oxides or oxynitrides)	9,7	7,0

Table 4. Corrosion rate of nitride coatings of different morphology in aqueous solutions of inorganic acids

The mass losses investigations of corrosion processes confirm the high protective properties of nitride coating without oxygen impurities (Table 4).

In 80% H_2SO_4 protective properties of nitride coating are changed: nitride layer without oxygen impurities is characterized by the lower corrosion resistance than oxynitride layer, e.g. more negative corrosion potential and higher current density at anodic dissolution (Fig. 22). Since the values of overstresses in both cases are similar the increasing of anodic characteristics of oxynitride coating should be related to addition modification of nitride coating by oxygen.

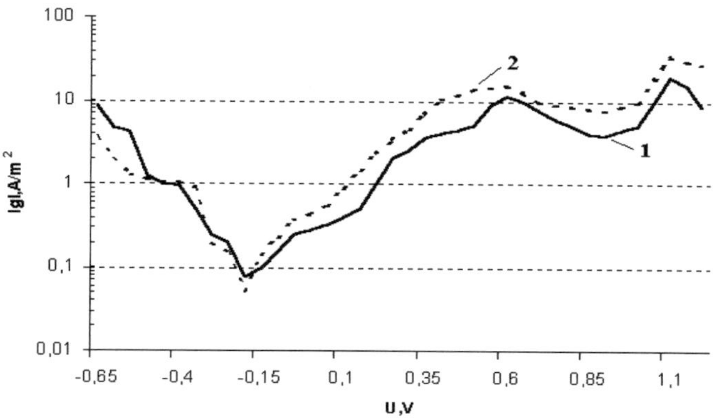

Fig. 22. Polarization curves of oxynitride (1) and nitride (2) coatings on the titanium surface in 80% aqueous solution of sulphuric acid.

Indeed, the kinetic curves of mass losses of nitride coating without oxygen impurities are situated higher and corrosion rate is in 3…10 times bigger than the oxynitride layer ones. It indicates about the good protection properties of oxynitride coatings. The positive role of oxygen impurities at the providing of protective properties of nitride coatings in sulphuric acid should be explained by different activity of chlorine- and sulphate-iones in passivation processes.

Differences in protective properties of nitride coatings in aqueous solutions of sulphuric and chloric acids indicate about necessity of differential approach to a protection against corrosion of titanium. Nitride coatings are more effective for use in chloric acid whereas oxynitride coatings are more effective in sulphuric acid.

Oxygen impurities in nitride coatings increase the corrosion losses in chloric acid whereas it decreases the corrosion dissolution in sulphuric acid.

Decreasing of nitride surface roughness at the simultaneous retaining of big thickness improves the protection properties in inorganic acids. More effective way to decrease the roughness is the decreasing of temperature of saturation. Increasing of thickness due to the large duration increases the protection properties but due to higher temperature - decreases ones.

7. Influence of nitriding and oxidation on the wear of titanium alloys

The methods of thermodiffusion surface hardening of titanium alloys and, specifically, the procedures of thermal oxidation and nitriding have serious advantages over the other available methods under service conditions excluding the application of high contact stresses (>4 MPa) (Gorynin & Chechulin, 1990; Nazarenko et al., 1998). They are technologically simple, guarantee the reliable physical and chemical characteristics of treated surfaces, and do not require any additional technological procedures. Unlike the application of coatings, in this case, we have no problems connected with the adhesion between the hardened layers and the matrix, porosity, and, hence, with the sensitivity to aggressive media.

Let's analyze the wear resistance of titanium after thermal oxidation and nitriding.

To increase the output and efficiency of the technological processes of surface hardening, it is necessary to find out the ways of their intensification because the procedure of thermodiffusion saturation of titanium alloys, e.g., with interstitial impurities (oxygen, nitrogen, carbon, and boron), requires high temperatures and long duration. At present, the problem of intensification of the thermodiffusion saturation of titanium alloys with oxygen finds its solution in new technologies of thermal oxidation. Both the duration of process of getting the desired thickness and degree of hardening of the diffusion layer and its temperature can be decreased by applying of the procedures of boiling-bed and vacuum oxidation.

The thermodiffusion saturation of titanium alloys with oxygen from the boiling bed (650-800 °C, 4-7 h) is accelerated due to the activation of the surface of workpieces in contact (friction) with sand particles, which intensifies the processes of absorption and adsorption (Zavarov et al., 1985). In this case, we observe the formation of a hard (with a surface microhardness of 6.0-8.5GPa depending on the alloy) wear-resistant diffusion layer consisting of a 3-7-μm-thick film of titanium dioxide (in the rutile modification) and an interstitial solid solution of oxygen in titanium with a thickness of 20-70 μm.

However, the non-uniform boiling of sand, its insufficient degree of dispersion and deviations from the required temperature conditions quite often lead to the formation of cavities on the titanium surface and exfoliation of the surface layer, which means that the corresponding workpieces must be rejected.

More stable results are attained in the process of thermal oxidation (700-1050 °C, 0.3-7 h) of titanium alloys in a vacuum (~ 0,1 Pa). In this case, the surface of workpieces is saturated with residual gases of vacuum media (oxygen, nitrogen, and carbon), which results in the formation of a hard wear-resistant layer consisting of two areas: a complex compound of titanium oxides, nitrides, and carbides and an interstitial solid solution of these elements in the matrix.

The procedure of nitriding of titanium alloys is carried out at temperatures of about 950 °C for 15-30 h either under the atmospheric pressure (10^5 Pa) or in rarefied nitrogen ($\leq 10^2$ Pa) (Gorynin & Chechulin, 1990). Long periods of holding at high temperatures result in the irreversible growth of grains in the titanium matrix accompanied by the formation of brittle surface layers and, hence, in a pronounced deterioration of the mechanical characteristics of nitrided workpieces. The characteristics of plasticity and fatigue life of the material prove to be especially sensitive to high-temperature treatment (Table 5). Thus, after nitriding in the indicated mode, the plasticity of unalloyed VT1-0 titanium becomes in 2.3 times lower than after oxidation in a vacuum. At the same time, its fatigue life decreases by a factor of 3-6 depending on the level of strains (Table 5, rows 2 and 3).

No	Treatment	Alloy	σ_B, MПa	δ, %	Fatigue life, cycles	
					ε = 1,0 %	ε = 1,5 %
1	No treatment	VT1-0	450	34,5	2500	900
		OT4-1	790	19,3	10100	2200
		BT6s	1040	15,3	33700	6500
2	Oxidation	VT1-0	374	21,7	1000	300
		OT4-1	729	5,9	2600	250
		BT6s	975	7,2	11700	2000
3	Nitriding	VT1-0	395	9,3	300	50
		OT4-1	735	5,1	2700	230
		BT6s	950	4,8	4200	1500
4	Nitriding in rarefied nitrogen	VT1-0	369	32,9	4000	770
		OT4-1	740	18,4	3700	670
		VT6s	943	14,2	13200	3100

Table 5. Mechanical characteristics of titanium alloys.

Thus, the energy consumption, productivity, and the general level of the attained mechanical characteristics of workpieces for the procedure of oxidation are better than for nitriding. At the same time, the process of nitriding enables to set the higher degrees of surface hardening than oxidation due to the difference between the ionic radii of the interstitial elements (0,148 nm for nitrogen and 0,136 nm for oxygen). Moreover, titanium nitrides are characterized by much smaller Pilling-Bedworth ratios (1,1 for TiN and 1,7 for TiO_2), coefficients of thermal expansion and residual stresses in the surface layer of nitrides, which excludes the possibility of violation of the continuity of the formed film (Strafford, 1979). In addition, the corrosion resistance of nitrided layer is higher than the corrosion resistance of the oxidized layer. Thus, the intensification of nitriding makes it possible to decrease the temperature and time of treatment and preserve the characteristics of toughness of the base material. This enables us to recommend this type of treatment as an efficient and cost-effective method of surface hardening.

The diffusion coefficient of nitrogen in titanium nitride is in 2-4 times lower than in titanium. Indeed, at 850 °C we have $D^{N}_{TiN} = 3{,}95 \cdot 10^{-13}$ cm^2×s^{-1}, $D^{N}_{\alpha\text{-}Ti} = 1{,}81 \cdot 10^{-11}$ cm^2×s^{-1}, and $D^{N}_{\beta\text{-}Ti} = 9{,}03 \cdot 10^{-9}$ cm^2×s^{-1}.

As the nitrogen partial pressure decreases due to the weakening of the barrier effect of the nitride film or its complete vanishing, the process of nitriding intensifies. However, a vacuum of 10^2 Pa is insufficient for a positive effect attained only if the pressure of the dynamic atmosphere of nitrogen is as low as 0,1-1 Pa for a feed rate of the gas of 0,003 l/min (Fedirko & Pohrelyuk, 1995). The elements of vacuum technology (heating in a vacuum and preliminary vacuum annealing) also intensify the thermodynamic saturation of titanium with nitrogen.

The low pressure of residual gases of rarefied atmosphere (~ 10 mPa) prior to the delivery of nitrogen in the stage of heating and, for a short period of time (≤ 2 h) at the saturation temperature, activates the titanium surface and promotes the dissociation of oxide films.

These technological procedures decrease temperature and time of treatment down to 800-900 °C and 5-10 h and, hence, significantly improve the mechanical characteristics of nitrided samples (Table 5, rows 2 and 4). Thus, the plasticity and fatigue life of nitrided VT1-0 (c.p. titanium) are, respectively, in 1,5 and 2,6-4,0 (depending on the level of strains) times higher than in the case of oxidation.

The outlined procedure of nitriding leads to the formation of a hardened layer consisting of a $TiN + Ti_2N$ nitride film (≤ 1 μm) and a deep diffusion layer (100-180 μm).

Let's compare the wear resistance of the oxidized (in a vacuum of 0,1 Pa at 850 °C for 5 h) and nitrided [heating to 850 °C in a vacuum of 10 mPa, creation and maintenance of a dynamic atmosphere of nitrogen (1 Pa, 0.003 l/min) for 5 h, and cooling in nitrogen] titanium. The surface microhardness of nitrided and oxidized layers was 7,9 and 5,1 GPa and the thickness of hardened layer was 100 and 55 μm, respectively.

Wear tests were carried out for the case of boundary sliding friction with lubrication with a AMG-10 hydraulic fluid in an SMTs-2 friction-testing machine by using the disk-shoe mating scheme. The contact load was as large as 1 and 2 MPa. The counter body was made of bronze. The sliding velocity was equal to 0,6 m/s. The friction path was equal to 10 km. To make the contact area of the mating bodies not lower than 90%, the friction couple was run in on a path of 200 m. We analyzed the mass losses of the treated specimens and counter bodies, the friction coefficients, and wear depending on the lengths of the basic friction paths (1, 2, 2, 2, and 3 km).

It was discovered that the wear resistance of nitrided titanium is quite high. After testing, the mass losses of the disk for the indicated contact loads did not exceed 1 mg (Fig. 23, curves 1 and 2). The influence of the load is noticeable (i.e., for 2 MPa, the mass losses are higher) only for the first 1.5 h of operation of the friction couple (this corresponds to a friction path of about 3 km). After this, the mass losses under loads of 1 and 2 MPa are practically equal.

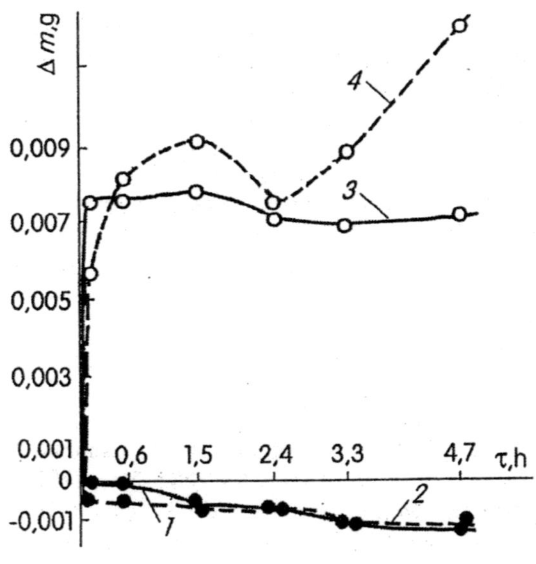

Fig. 23. Kinetics of mass changes of nitrided (1, 2) and oxidized (3, 4) disks of VT1-0 titanium in the process of friction with BrAZh9-4l bronze: 1, 3 – 1 MPa; 2, 4 – 2 MPa.

The process nitriding of titanium alloys is accompanied by the formation of a typical surface pattern, which worsens the quality of the surfaces of the workpieces (the parameter R_a increases by 0,4-0,7 mm). During the friction, the microproasperities (elements of the pattern) are separated, move into the contact zone, and finally, penetrate into the relatively soft counterbody. These hard particles of nitrides play the role of an abrasive substance and make scratches of relatively small depth in the nitrided surface. Thus, to enhance the characteristics of wear resistance, it is necessary to improve the quality of the nitrided surface, which, in turn, depends on the temperature of nitriding (the lower is the saturation temperature, the lower the roughness of the surface) (Fedirko & Pohrelyuk, 1995).

The nitriding noticeably increases the antifriction characteristics of titanium in a friction couple with bronze. The friction coefficient equal to 0,18 is stable and independent of contact pressure.

The process of friction of the oxidized disk is accompanied both by the wear of its surface layers and the process of transfer of small pieces of soft bronze to the oxidized surface. This is connected with the adhesion of bronze particles to the disk leading to the formation of unstable secondary structures, which are destroyed and removed in the course of friction, and the process is repeated again. As a result, the time dependence of the mass of the oxidized disk is not monotonic (Fig. 23, curves 3 and 4). Under a load of 1 MPa, the processes of mass transfer of bronze and its fracture are practically balanced. As a result of fitting of the mating surfaces, the friction coefficient decreases from 0,23 to 0,14 and then stabilizes. Under a load of 2 MPa, the mass transfer of bronze is predominant and, in the course of time, the oxidized surface is more and more intensely rubbed with bronze. The friction coefficient is unstable and varies from 0,24 in the stage of fitting to 0,22-0,21 in the stationary mode. In aggressive media, the appearance of bronze on the oxidized titanium surface leads to the formation of galvanic couples and, thus, promotes the corrosion processes.

The mass increment of the oxidized disk exceeds the degree of wear of the nitrided surface by one or even two orders of magnitude and the wear of the counterbodies is of the same order. Moreover, the wear of the counterbody is more intense in couples with the nitrided and oxidized disks and, thus, determines the wear of the friction couples.

In the friction couple of nitrided titanium with bronze, the wear of bronze exceeds the wear of the nitrided disk by more than three orders of magnitude. Under a load of 1 MPa, the mass loss of the shoe is a monotonically increasing function of time (Fig. 24, curve 1). Under a load of 2 MPa, this process becomes in 1,4-1,6 times more intense (curve 2). For a friction couple of oxidized titanium with bronze, the increase in the load leads to a more pronounced increase in the wear of the shoe (by a factor of 3,4-3,8) (Fig. 24, curves 3 and 4). Moreover, under a contact pressure of 1 MPa, the mass losses of the shoe in the process of friction with the nitrided disk are in 1,9 times greater than for the oxidized disk but, under a pressure of 2 MPa, the situation changes and the mass losses of the shoe coupled with the oxidized disk are in 1,6 times greater than for the nitrided disk.

The mass losses of the friction couples and their elements after different parts of the basic friction path have their own regularities. Let's now consider the wear of the counterbody in the process of friction against the nitrided and oxidized disks.

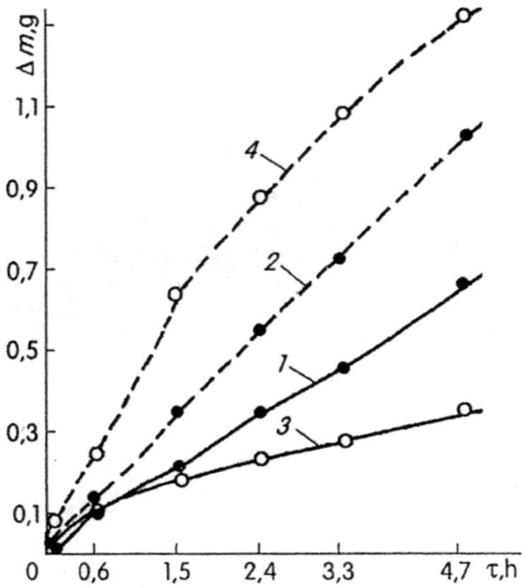

Fig. 24. Kinetics of the mass losses of the counterbody of BrAZh9-4l bronze in the process of friction with nitrided (1, 2) and oxidized (3, 4) disks of VT1-0 titanium: 1, 3 – 1 MPa; 2, 4 – 2 MPa.

In the former case, the wear of the bronze shoe after fitting and the first reference section (200 and 1000 m) first increases, then stabilizes at a level of 0,12 g (0,21 g) under a contact load of 1 MPa (2 MPa) and, finally, begins to increase again. In the latter case, the wear of the bronze is different for different sections of the friction path and under different stresses. Thus, under a load of 1 MPa, the mass loss of the bronze shoe slightly increases after fitting and then stabilizes. Only at the very end of the tests, we observe a weak tendency to increase in the mass losses. Under a load of 2 MPa, the wear of bronze significantly increases after fitting, then decreases and, finally, stabilizes.

It seems possible that the intense wear of the counterbody in the process of friction against the nitrided or oxidized titanium surface is explained by the hydrogenation of bronze as a result of tribodestruction of the lubricant (Goldfain et al., 1977). This observation is confirmed by the formation of bronze powder, which may be caused by the dispersion of hydrated bronze.

Thus, as compared with recommended nitriding, the application of oxidation as a method for increasing the wear resistance of titanium alloys is reasonable only under low contact stresses (≤ 1 MPa). Under higher contact stresses, the processes of oxidation and nitriding are characterized by practically equal levels of energy consumption and productivity but the tribological and mechanical characteristics of nitrided titanium are better than the corresponding characteristics of oxidized titanium.

8. References

Brynza A.P. & Fedash V.P. (1972). About passivation mechanism of titanium in solutions of sulphuric and chloric acids, In: Noviy konstrukcyonniy material- titanium [in Russian], pp. 179-183, Nauka, Moscow

Chukalovskaya T.V., Myedova I.L., Tomashov N.D. at al. (1993). Corrosion properties and electrochemical behavior of nitride layers on titanium surface in sulphuric acid. Zaschita metallov, No. 2, pp. 223- 230 (in Russian)

Fedirko V. M. & Pohrelyuk I. M. (1995). Nitriding of Titanium and Its Alloys [in Ukrainian], Naukova Dumka, Kiev

Fedirko V.M, Pohrelyuk I.M. & Yaskiv O.I. (1998). Corrosion resistance of nitrided titanium alloys in aqueous solutions of hydrochloric acid. Materials Science, Vol. 34, No. 1., pp. 119–121

Fedirko V.M., Pohrelyuk I.M. & Yaskiv O.I. (2009). Thermodiffusion multicomponent saturation of titanium alloys [in Ukrainian], Naukova Dumka, Kiev

Goldfain V. I., Zuev A. M., Kablukov A. G. et al. (1977). Influence of friction and hydrogenation on the wear of titanium alloys, In: Investigation of Hydrogen-Induced Wear [in Russian], pp. 7-80, Nauka, Moscow

Goldschmidt H. J. (1967). Interstitial alloys, Plenum Press, New York

Gorbachov A. K. (1983). Thermodynamics of reductive-oxidative processes in TiN - H2O system. Zaschita metallov, Vol. 19, No. 2., pp. 253– 256 (in Russian)

Gorynin I.V. & Chechulin B.B. (1990). Titanium in mechanical engineering [in Russian], Mechanical engineering, Moscow

Hultman L., Sundgren J.E., Greene J.E., Bergstrom D.R. & Petrov I. (1995). J.Appl. Phys., Vol. 78., p. 5395

Kelly E.J. (1979). Anodic dissolution and passivation of titanium in acidic media. III. Chloride solutions. J. Electrochem. Soc., Vol. 126, pp. 2064-2075

Kiparisov S.S. & Levinskiy Yu. V. (1972). Nitriding of refractory metals [in Russian], Metallurgiya, Moscow

Kolotyrkin Ya.N., Novakovskiy V.N., Kuznetsova Ye. G. at al. (1982). Corrosion behavior of titanium in technological media of chemical industry [in Russian], NIITEChIM, Moscow

Matychak Ya., Fedirko V., Prytula A. & Pohrelyuk I. (2007). Modeling of diffusion saturation of titanium by interstitial elements under rarefied atmospheres. Defect and Diffusion Forum, Vol. 261-262, pp. 47-54

Matychak Ya., Fedirko V., Pohrelyuk I., Yaskiv O. & Tkachuk O. (2008). Modelling of diffusion saturation of (α+β) titanium alloys by nitrogen under rarefied medium. Defect and Diffusion Forum, Vol. 277, pp. 29-34

Matychak Ya. S., Pohrelyuk I. M. & Fedirko V. M. (2009). Thermodiffusion saturation of α-titanium with nitrogen from a rarefied atmosphere. Materials Science, Vol. 45, No. 1, pp. 72-83, ISSN: 1068-820X

Matychak Ya. S., Pohrelyuk I. M. & Fedirko V. M. (2011). Kinetic features of the process of nitriding of (α + β)-titanium alloys, Materials Science, Vol. 46, No. 5, pp. 660-668, ISSN: 1068-820X

Nazarenko P. V., Polishchuk I. E. & Molyar A. G. (1998). Tribological properties of coatings on titanium alloys. Fiz.-Khim. Mekh. Mater., Vol. 34, No. 2, pp. 55-62 (in Ukrainian)

Pohreliuk I., Yaskiv O. & Fedirko V. (2007). Formation of carbonitride coatings on titanium through thermochemical treatment from carbon-nitrogen-oxygen-containing media. JOM Journal of the Minerals, Metals and Materials Society, Vol. 59, No. 6, pp. 32–37

Pohrelyuk I., Yaskiv O., Tkachuk O. & Lee Dong Bok. (2009). Formation of oxynitride layers on titanium alloys by gas diffusion treatment. Metals and Materials International, Vol. 15, No 6, pp. 949–953., ISSN: 1598-9623

Pohrelyuk I. M., Tkachuk O. V. & Proskurnyak R. V. (2011). Corrosion resistance of the nitrated Ti-6Al-4V titanium alloy in 0,9% NaCl, JOM Journal of the Minerals, Metals and Materials Society, Vol. 63, No. 6, pp. 35-40, ISSN: 1047-4838

Rolinski E. (1988). Effect of nitriding on the surface structure of titanium. J. Less-Common Metals, Vol. 141, No. 1., pp. L11 - L14.

Samsonov G.V., Epik A.P. (1973). Refractory coatings [in Russian], Metallurgiya, Moscow

Strafford K. N. (1979). A comparison of the high temperature nitridation and oxidation behaviour of metals. Corn Sci., Vol. 19, No. 1, pp. 49-62

Sukhotin V.N., Lygin S.A., Sunguchayeva I.A. & Gramanschikov V.A. (1990). Passivation of titanium in solutions of chloric acid. Zaschita metallov, Vol. 26, No. 1., pp. 128-131 (in Russian)

Tomashov N.D., Chukalovskaya T.V., Myedova I.L. & Yegorov F.F. (1985). Corrosion and anodic behaviour of carbide, nitride and boride of titanium in sulphuric and phosphoric acids. Zaschyta myetallov, No. 5., pp. 682-688 (in Russian)

Yaskiv O., Pohrelyuk I., Fedirko V.M., Tkachuk O. & Lee Dong-Bok. (2011). Formation of oxynitrides on titanium alloys by gas diffusion treatment. Thin Solid Films, - Vol. 519, No19, pp. 6508 – 6514

Zavarov A. S., Baskakov A. P., & S. V. Grachev. (1985). Thermochemical Boiling-Bed Treatment [in Russian], Mashinostroenie, Moscow

8

Anodic Layer Formation on Titanium and Its Alloys for Biomedical Applications

Elzbieta Krasicka-Cydzik
University of Zielona Gora
Poland

1. Introduction

Properties of the oxide layers on titanium and its implant alloys can be tailored to desired applications by anodizing parameters. Electrochemical oxidation in various electrolytes and different polarization regimes may shape the morphology, structure and chemical composition of oxide layers to enhance the use of titanium materials in electronics, photovoltaic and medicine. Phosphate electrolytes play specific role in the anodizing process. Besides forming compact barrier layer they enable also to form porous and nanostructural oxide layers enriched with phosphates, which enhance their bioactivity.

The formation of anodic layers: thick or thin, compact or porous, gel-like and nanostructural on titanium and its alloys Ti6Al4V and Ti6Al7Nb in phosphoric acid solutions of different concentrations is described in this charter. Basing on morphological and chemical composition analysis (SEM, XPS) as well as on the electrochemical examination the influence of electrolyte composition on enrichment of surface oxide layers with phosphates and fluorides, enhancing their bioactivity, is presented. Studies to use Ti/titania systems as the platforms of the electrochemical biosensors to detect H_2O_2 and glucose proved the opportunity to use the nanotubular titania material as a platform for the 2rd generation biosensors.

2. Oxide layers on titanium and its implant alloys formed in H_3PO_4 solutions

Anodic films formed on titanium and its alloys are of great interest due to the industrial applications of metal covered with oxide layers of various and unique properties [1-7]. These layers have been investigated extensively by many authors [8-11]. Thick oxide layers on titanium, obtained by anodizing, provide improved resistance to local corrosion [12]. Anodizing can result in the adsorption and incorporation of inorganic and organic, biologically important species, e.g. phosphate ions, into the oxide layer. Such surface layers, desirable for medical implants, are not only corrosion resistant in a biological environment, but also compatible with tissue response [13-15]. Anodizing titanium and its alloys has been investigated in a wide range of parameters [16-20], which include also the participation of the electrolyte components, e.g. anions, in the formation of anodic films [21, 22].

At anodizing oxide layers are formed according to the following reaction [17, 23]:

$$n\,Me + m\,H_2O \rightarrow Me_nO_m + 2\,m\,H^+ + 2m\,e^- \qquad (1)$$

Studies in this field [24, 25] have shown that, phosphate ions can be incorporated into the anodic layer on titanium and Ti-6Al-4V, and in turn stimulate the formation of the bio-compatible hydroxyapatite [26].

Anodizing in phosphate solutions exhibit some advantages over other acid and base electrolytes. First of all less corrosive attack of phosphoric acid on titanium and its alloys, when compared with other acidic media, is related to the strong adsorption of phosphate anions on the surface [27,28]. Although unalloyed titanium is resistant [11] to naturally aerated pure solutions of phosphoric acid up to 30 % wt. concentration (~3.6 M) in phosphoric acid of lower concentrations (0.5-4 M), mainly non-dissociated acid molecules and H_2PO4^- of phosphate ions exist [29,30] and they exhibit a strong affinity or complexing power towards most metal cations.

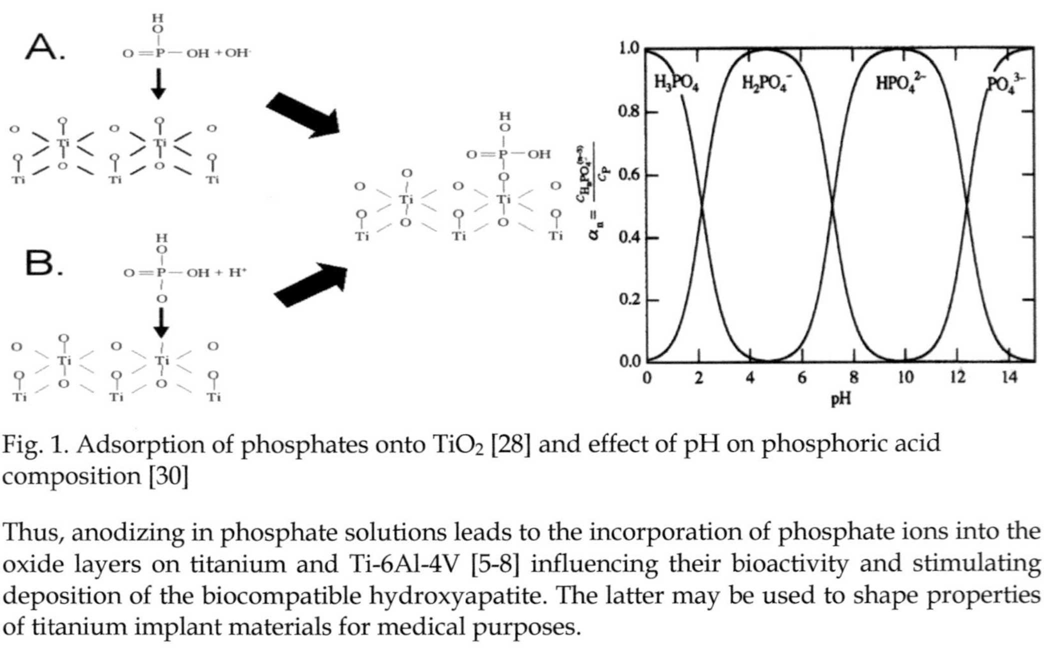

Fig. 1. Adsorption of phosphates onto TiO_2 [28] and effect of pH on phosphoric acid composition [30]

Thus, anodizing in phosphate solutions leads to the incorporation of phosphate ions into the oxide layers on titanium and Ti-6Al-4V [5-8] influencing their bioactivity and stimulating deposition of the biocompatible hydroxyapatite. The latter may be used to shape properties of titanium implant materials for medical purposes.

2.1 Thin porous anodic layers

Anodic layers on pure Ti and its alloys Ti6Al4V ELI and Ti6Al7Nb (ASTM F136-84), alloys in the annealed condition) can be formed by anodizing carried out at ambient pressure and room temperature in non-deaerated electrolyte solutions of 0.5 M H_3PO_4. Both techniques, the galvanostatic at anodizing current density values varied in the range of 0.1-0.5 Am-2 and the potentiostatic at up to 60V [22] are used. The oxide layers, 30-120nm thick, enriched with phosphorus are formed in these conditions. With mechanical and chemical pre-treatment applied to titanium and its alloys: Ti6Al4V ELI and Ti6Al7Nb (Timet Ltd, UK) [12-20], the layers obtained at 60V in phosphoric acid are golden and porous [11, 24-26] (Fig. 2) and they show very stable values of currents in passive region up to 1 V (SCE) (Fig.3) [11]. At other polarization parameters however layers of different thickness and colorization are formed [11]. Due to the presence of phosphates in the anodic layers they are highly bioactive in comparison to oxides formed in other electrolytes. Just after 9 days in the SBF solution (Simulated Body Fluid) they are covered with hydroxyapatite deposists (Fig. 2) [27].

a b c

Fig. 2. Porous titania layers on Ti formed at 0.5 A/m^2 in 0.5 M H_3PO_4 [22] (a,b) and HAp particles on anodic layer after 9 days in SBF solution [21, 25, 27] (c), JEOL JLM 5600 EDS

The bilayer structure of compact oxide covered with HAp particles can be demonstrated also in the impedance tests (Fig. 3). The first time constant in Bode diagrams in range of the high and intermediate frequencies confirmed the high R_t resistance of the barrier layer covering the anodized metal, thus giving the evidence of its high corrosion resistance. The second time constant corresponds to porous layer above the barrier one. Its lack in case of the Ti6Al4V ELI indicates the different characteristics of the coating on this implant material.

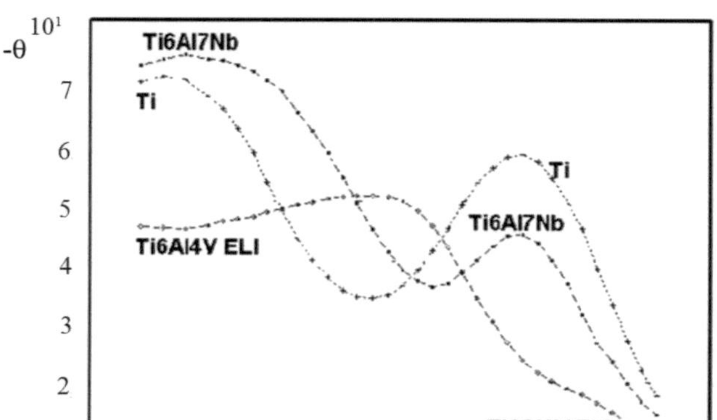

Fig. 3. EIS spectra recorded after 9 days in SBF solution [21, 25, 27] for anodic layers on Ti, Ti6Al4V and Ti6Al7Nb formed at 0.5 A/m^2 in 0.5 M H_3PO_4 [22]

To investigate the effect of phosphoric acid concentration (0.5 - 4 M) on the anodising of titanium and its alloys the galvanostatic and potentiodynamic techniques have been applied [11,19]. Particularly, the galvanostatic method with low current densities, up to 0.6 Am^{-2}, applied in order to minimize side effects (ie. oxygen evolution), and more importantly to determine processes responsible for the growth of the oxide layer on the anodised metal at the early stages of its formation, allowed to observe the abnormal behaviour of titanium at anodizing. In Fig. 4 the results presenting the minimum rates of potential growth at initial stages of galvanostatic anodising in 2 M H_3PO_4 solutions are shown.

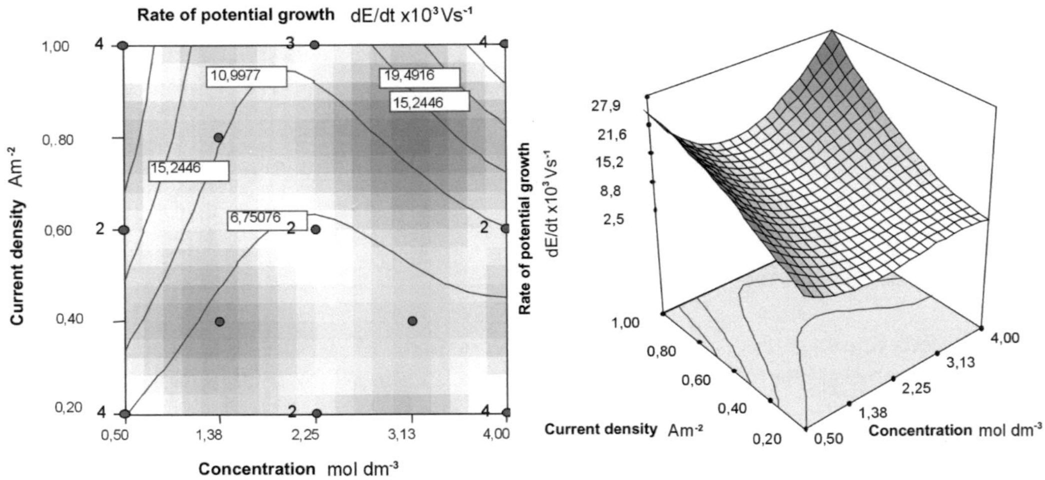

Fig. 4. Surface response for the investigation of the effect of H_3PO_4 concentration and polarization current on the rate of potential growth dE/dt at galvanostatic anodizing of titanium [11]

Also the anodic polarization curves for titanium in electrolytes of different concentration show various shapes and different slopes in active-passive region (Fig.5).Potentiodynamic control at a comparable rate in 2 M H_3PO_4 applied to titanium and two of its implant alloys, Ti6Al4V ELI and Ti6Al7Nb, revealed a shift in corrosion potential toward the anodic direction with the lowest current densities in the passive region. This was possibly due to the effect of adsorption of phosphate ions onto the surface layer.

2.2 Gel like layers on anodic titania

Active-passive transition of titanium in phosphoric acid solutions of 0.5-4 M [11,19] reveals that the growth of anodic layer is affected proportionally by the applied anodic potential, but shows the unusual influence of electrolyte concentration. Under galvanostatic conditions at low current densities (0.1-0.6 Am^{-2}) the slope of dE/dt shows the minimum at the concentration ~2 M H_3PO_4 (Fig. 4), which is resulted due to a coating of an oxide film by an additional gel-like layer during anodizing, [28, 29], similar to the one observed in other media on aluminum.

It was found that the active–passive transition was a process in which an inhibiting effect of phosphate ions on a dissolution of oxide layer was observed during anodizing [29]. Typical examples of voltage *vs* time transients (dE/dt) during the growth of the galvanostatic anodic oxide film on titanium for the current density of 0.5 A/m^2 (Fig. 5a) show that the continuous linear growth of potential to the steady state, demonstrates the lowest value in 2

M H_3PO_4. Polarization curves (Fig.5b), show that after an initial range of cathodic depassivation, samples reach the corrosion potential E_{cor}. Then an active–passive transition is observed with passivating currents the order of a few microamperes, which are typically observed during the passivation of titanium and its alloys. Anodic curves for 0.5 M and 3 M H_3PO_4 solutions (Fig.5b), illustrate quasi-passive behaviour in active-passive transitions, while curves for 1 and 2 M H_3PO_4, having the higher corrosion potential E_{cor}, do not show linear dependence in this potential region. The differences in anodic Tafel slopes are accompanied by a shift of the E_{cor} value in the positive direction by ~0.15V with the increase of H_3PO_4 concentration to 2 M (Fig. 5b).

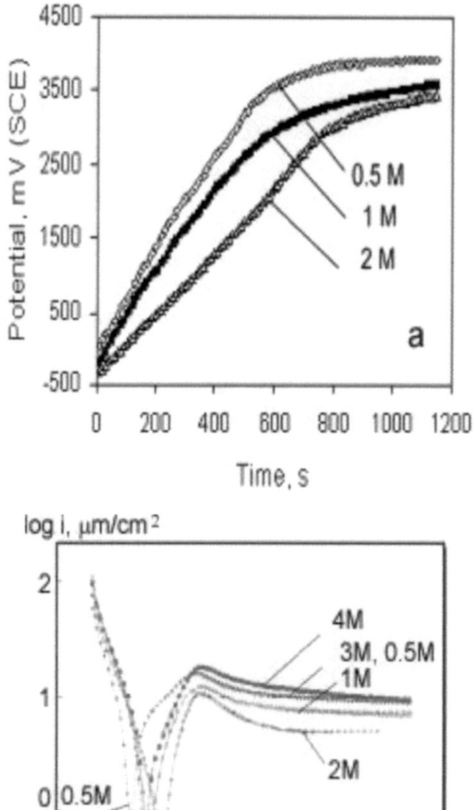

Fig. 5. Corrosion potential values Ecor of Ti anodised at 0.5 A/m2 in 0.5-2 M H3PO4 in SBF solution [11,29] (a) and active-passive transition regions of polarization curves for titanium (scan 3 mV/s)in 0.5-4 M H3PO4 (b)

SEM/EDS examination confirm the evidence of two-layered surface film. Such layers presented in Fig. 6, show the whole surface covered by a gel-like layer of $H_3PO_4 \times 0.5H_2O$.

The SEM/EDS examinations reveal that thin films of anodic titania oxide are covered by gel-like layer with crystalline phosphates nuclei inside. Phosphates deposits are few in layers formed in 0.5 M H_3PO_4, but numerous and uniformly dispersed in a surface oxide of sample anodised in 2 M H_3PO_4. However, the oxide and phosphates are covered with the additional layer consisting of 76.3±3.6 wt.% of phosphorus and 23.7±1.5wt.% of oxygen (Fig.6a and 6b).

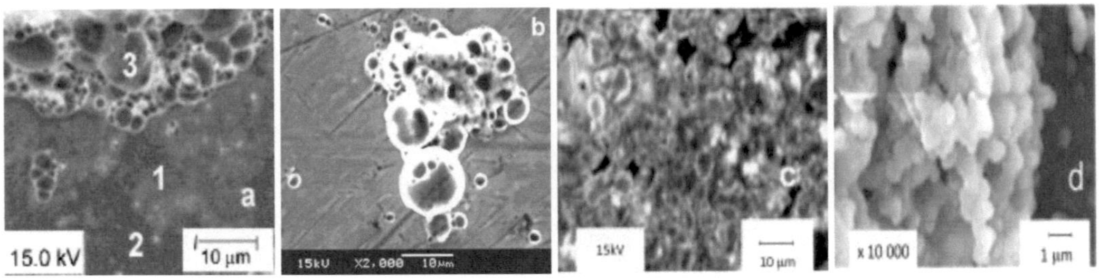

Fig. 6. SEM micrographs of titanium surface anodized at 0.5 A/m² in a,b) 0.5 M, c) 2 M H_3PO_4 [29], and d) HAp particles on anodic layer after 9 days in SBF solution [21, 27]

The EIS spectra of titanium after anodizing at 0.5 A/m² in 0.5-2 M H_3PO_4 (Fig. 7), exhibit a behavior typical of a metallic material covered by a porous film which is exposed to an electrolytic environment [6]. Two time constants are seen in the spectra: the first in the high-frequency part arises from the ohmic electrolyte resistance and the impedance resulting from the penetration of the electrolyte through a porous film, and the second in low-frequency part accounts for the processes at the substrate/electrolyte interface.

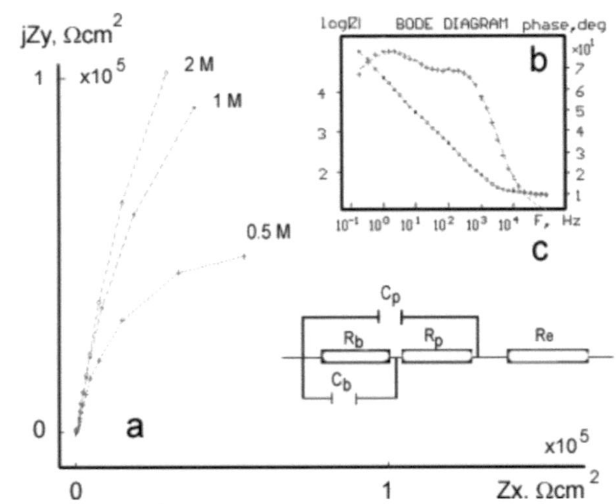

Fig. 7. Impedance spectra for titanium anodised in 2 M H_3PO_4 exposed to 0.9% NaCl solution a) Nyquist spectra, b) Bode diagrams and results oftheir fitting to c) equivalent circuit

The EIS data can be fitted to the equivalent circuit in Fig. 7c, which consists of a solution resistance R_s, the capacitance C_p of the barrier layer, the charge transfer resistance associated with the penetration of the electrolyte through the pores R_p; and the polarization resistance of the barrier R_b as well as the electrical double-layer capacitance at the substrate/electrolyte interface C_b. In the case of surface layer formed in 2 M H_3PO_4 the specimen is covered by a passive oxide film of higher impedance (Fig.7a). The significant increase of the resistance R_b and R_p values for the 2 M H_3PO_4 anodized samples, over those determined for the 0.5 M H_3PO_4 anodized titanium, confirm that the EIS results are complementary to those obtained by E_{cor} measurements and potentiodynamic polarization studies [11,31].

Titanium as metal very sensitive to the pre-treatment [33], due to the polishing, rinsing with water and drying, is usually covered by an air-formed oxide film and on immersion into acid solution shows potentials of the active-transition region [32-34]. However, in solutions of low pH may become active. In sulphate solution the anodic oxide film on titanium dissolves giving Ti^{3+} ions [11]. Typical activation behaviour and slow decrease in the open-circuit potential (-0.3 V SCE), is observed on immersing titanium into 1M HCl [35,36]. Titanium behaves differently in H_3PO_4 solutions. Although the values of E_{cor} ranging from -0.1V to -0.6V (SCE) indicate in Pourbaix diagram [29] that the oxide film should dissolve to Ti^{3+}, on immersion to H_3PO_4 solutions titanium shows the continuous shift of potential towards the anodic direction [32]. Such a tendency indicates that the rate of the anodic reactions is continually decreasing, as a result of the presence of an adsorbed, additional layer on the metal surface [37]. Thermodynamic data [11,33] for the potential $E \geq -0.8V$ (SCE), indicate that the following reactions on titanium are likely:

$$Ti_2O_3 + 6\,H^+ + 2e^- = 2\,Ti^{2+} + 3\,H_2O \quad E^o = -\,0.478\text{ V} - 0.177pH - 0.059\log(Ti^{2+}) \tag{1}$$

$$Ti_2O_3 + H_2O = 2\,TiO_2 + 2\,H^+ + 2\,e^- \quad E^o = -\,0.556\text{ V} - 0.059\text{ pH} \tag{2}$$

$$Ti_2O_3 + 3\,H_2O = 2\,TiO_2 \cdot H_2O + 2\,H^+ + 2\,e^- \quad E^o = -\,0.139V - 0.059\text{ pH} \tag{3}$$

$$2\,Ti(OH)_3 + H_2O = 2\,TiO_2 \cdot H_2O + 2\,H^+ + 2e^- \quad E^o = -0.091V - 0.059\text{ pH} \tag{4}$$

In acidic solutions within the cathodic region, the only oxide dissolution is reaction (1). This reaction determines the potential-current changes in active-passive region of titanium in 0.5 M and 3-4 M H_3PO_4, whereas due to the shift of the corrosion potential E_{cor} towards the anodic direction for titanium immersed in 1-2 M H_3PO_4 electrolyte, its direct oxidation proceeds according to reactions 3 and 4. These results indicate that, the layer of phosphates (Fig. 6) blocks the oxide dissolution. Insight into the adsorption of phosphates to TiO_2 surface revealed the hypothesis according to which they form covalent bonding to oxygen [38], or metal ions react with phosphate anions forming a gel of metal hydrophosphates [39]. Both proposed processes would lead to local increase of pH at the oxide surface and in consequence to the increase of concentration of dihydro-phosphate ions (E-pH diagram) [29]. Then, on phosphates covered titanium oxide electrode, the following gel like layer formation could proceed

$$2\,H_2PO_4^- + 2\,H_3O^+ + (n\text{-}2)\,H_2O \rightarrow 2\,H_3PO_4 \times n\,H_2O \tag{5}$$

This attribution agrees with Morligde's *et. al.* results on aluminum [40]. Both reactions: the phosphates adsorption and gel-like layer formation are non-faradaic, but are competitive towards the oxide dissolution (reaction 1) with regard to proton consumption. The advantageous effect of these two reactions on the anodizing, may be attributed to an inadequate supply of H^+ ions to keeping up with the demand for the reaction (1) of oxide dissolution. The increasing coverage of the anodized titanium surface by phosphates ions with the electrolyte concentration provides the evidence of a direct influence of electrolyte anions in suppressing the formation of dissolved titanium ions. According to potential/pH diagram for P-Ti-H_2O system [41], $H_3PO_4{\times}0.5H_2O$, the product of reaction 5, is stable thermodynamically in solutions of pH ranging to 3.

Thus, due to the applied anodizing conditions formation of either thin and porous oxide layer [11-23] or gel-like phosphates rich layer of $H_3PO_4{\times}0.5H_2O$ [11, 24-29, 42], covering thicker oxide layer on titanium can be obtained.

2.3 Bioactive layer

Apart from mechanical properties and biocompatibility, which make titanium and its alloys the materials of choice for various applications (artificial hip and knee joints, dental prosthetics, vascular stents, heart valves) also enhancement of bone formation is desired feature of a metallic implant developed through adequate surface treatments to obtain proper osseointegration.

Fast deposition of hydroxyapatite (HAp) coatings on titanium and its alloys Ti6Al4V and Ti6Al7Nb substrates anodised in H_3PO_4 was observed [21,23,25]. Anodizing in 0.5 M H_3PO_4, which produces phosphates enriched porous sub-surface layer on of titanium and its alloys Ti6Al4V and Ti6Al7N [22] or anodizing in 2 M H_3PO_4 which generates phosphates rich gel-like layer [31,42] may be used to enhance hydroxyapatite (HAp) deposition (Fig.8). For the latter anodic layers soaking the anodised substrates in simulated body fluid (SBF) resulted in the deposition of a uniform coating in 24 hours (Fig. 9). SEM and EDS investigations revealed that after 9 days thick coating consists of HAp globular of diameter varied from 100 to 300 nm aggregates. The Ca-O-P deposits merge in large clusters and they are seen in large numbers on both alloys, particularly on Ti6Al4V anodized in 2 M H_3PO_4.

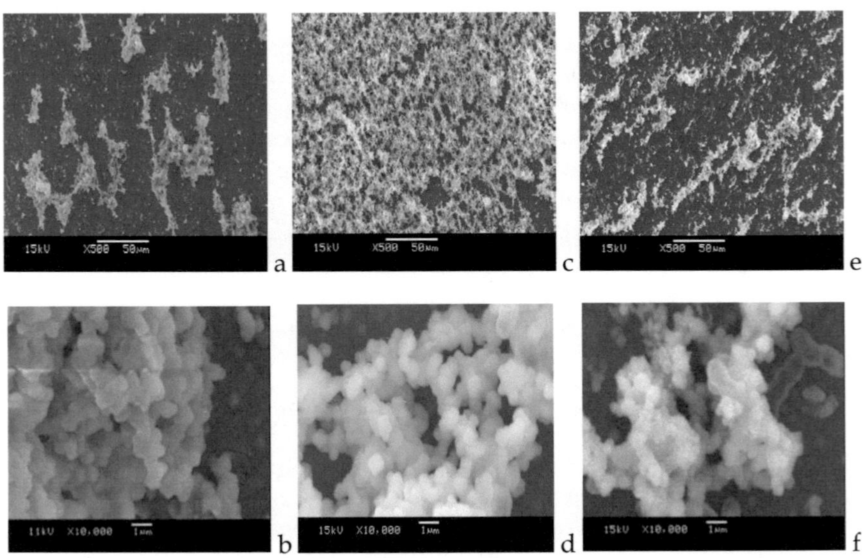

Fig. 8. SEM micrographs of titanium (a,b) and its alloys: Ti6Al4V (c,d) Ti6Al7Nb (e,f) surface anodized at 0.5 A/m^2 in 0.5 M H_3PO_4 after 24 h (abc) and 9 days (b,d,f) in SBF solution [21, 23]

SEM observations (Fig.8) and EDS microanalysis indicate the presence of deposits dispersed on the surface of anodised titanium and its implant alloys. However, deposits are are non-uniformly dispersed on a surface. Titanium and its two alloys anodised in 0.5M H_3PO_4 are covered with very thin oxide layer, which includes numerous and more scattered Ca-O-P deposits of diameter varied from 200 to 800 nm, suggesting the heterogeneous nucleation of Ca-O-P on TiO_2 covered surface. Althogh just after 24 hours deposits are seen on the surface of the 3 materials (more deposists on the Ti6Al4V alloy) the continuous films cover the whole surfaces after 9 days in SBF solution. At higher magnification it is seen that the film on titanium is formed of more flatter layer of deposits and broken layer of titanium oxides with titanium phosphates, whereas film on both alloys comprise small globules of Ca-O-P. The ratio of Ca/P ranging from 1.26 to 1.42 corresponds to non-steochiometric hydroxyapatite.

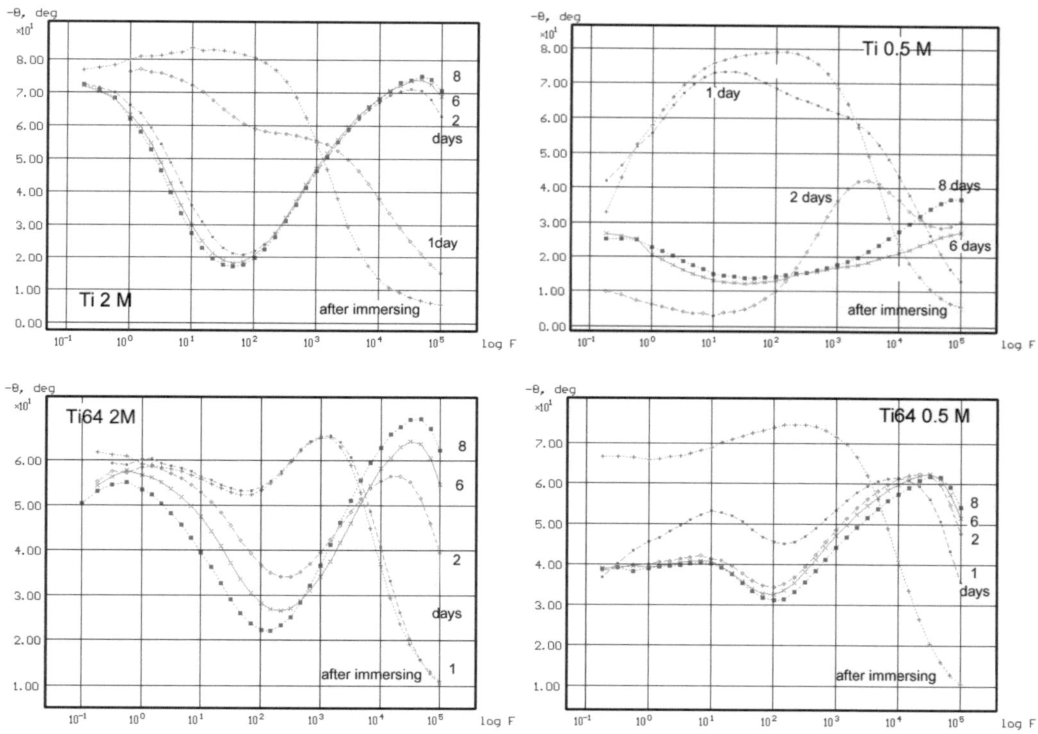

Fig. 9. EIS) spectra recorded during immersion of titanium and Ti6Al6V samples anodized in 0.5 M and 2 M H_3PO_4 in SBF solution [21,23]

Impedance (EIS) spectra (Fig. 9) recorded during immersion of the anodized titanium and the Ti6Al4V alloy in simulated body fluid (SBF) for titanium and Ti6Al4V alloy show changes in capacitance and structure of surface layers as well as differences between coatings on titanium and its alloy and confirm the SEM observations (Fig. 8). Titanium exhibits two-layered structure: the inner oxide layer is covered by an outer layer of more or less uniformly distributed various size Ca-P-O deposits. On contrary the Ti6Al4V alloy is coated by a more uniform and dense layer of deposits (Fig. 8 c,d) and much lower concentration of titanium oxide on a surface [25].

3. Nanostructural oxide layer formed in phosphate solutions

In the last 20 years anodizing has been also used as a method to form nanooxides on metal surfaces. Formation of self-organized titania nanotubes with high level organization of pores on large surfaces [43-48] became very useful technology applied to many purposes, i.e. to modification of surgical implant surfaces and to biomedical sensing. Titania nanotubes, just like barrier type titania, combine very well with osseous tissue and can be a perfect basis for osteoblasts in surgical implants. Studies focused on controlling the size and arrangement of pores [49-54], aiming at bone ingrowth and on use of titania nanotubes platform for biosensing, due to their capability to combine with e.g. enzymes, proteins or biological cells, brought promising results [55,56].

Formation of nanotubes at different polarization parameters in various electrolytes [57-60] as well as various scan rates during the very first seconds of anodizing, may help tailoring the oxides for effective implantation and improve their properties for biomedical and sensor applications. For the latter applications titania nanotubes require better ordering ie. controlled diameter achieved during improved oxide growth kinetics. For the last 5 years in several papers [61-64] it has been revealed that the value of polarisation determines the diameter of nanotubes. Every additional 5V of potential increases the nanotubes diameter of about 20nm, whereas the time of anodizing determines the length of the nanotube layer. Moreover, the low pH and organic aqueous solutions assure more regular shapes of nanotubes.

3.1 Nanotubes on titanium in phosphate solutions

Electrochemical oxidization of titanium can be carried out in electrolytes with or without HF additives [65-67]. Attempts to assess the optimal scan rate/fluoride concentration ratio for formation of structurally uniform nanotubes [60] revealed that 1M H_3PO_4+0.3% wt. HF is the most proper electrolyte for anodizing at $0.5V^{s-1}$. To study the effect of phosphates concentration, layers of titania nanotubes were produced in electrolytes of different phosphoric acid concentration. Their properties as the future coatings on titanium for medical uses were characterized by SEM/EDS observations and capacitance tests in simulated body fluids. Formation of oxide layers on titanium in phosphoric acid solutions with additions of fluoride ions [50] at 25°C, is usually carried out in two stages: the first stage potentiodynamic to the desired potential and the second stage, potentiostatic with fixed potential on electrodes for over 2 hours (Fig. 10).

Fig. 10 shows the behavior of titanium polarized from the OCP (Open Circuit Potential) to 20V with a sweep rate of $0.5Vs^{-1}$ in phosphoric acid solutions of different concentration (1M, 2M and 3M H_3PO_4) containing 0.4 wt.% HF. Flat polarization curve confirm passive behavior of titanium anodized in 2 and 3M H_3PO_4+0.4 wt.% HF, contrary to current transients recorded in 1M H_3PO_4+0.4 wt.% HF. The increase of current with potential in that region usually can be explained by the presence of some pores [9]. Polarization curves for more concentrated phosphate solutions show the presence of the anodic peaks, which can be ascribed to the oxygen evolution [6] followed by a broad passive region. By fixing the concentration of HF (fixing the dissolution rate) the decrease of current with potential indicates that oxide formation dominates over oxide dissolution at relatively higher field strengths or/and passive layer of phosphates is formed over nanotube titania in 2 and 3M H_3PO_4+0.4 wt.% HF solutions.

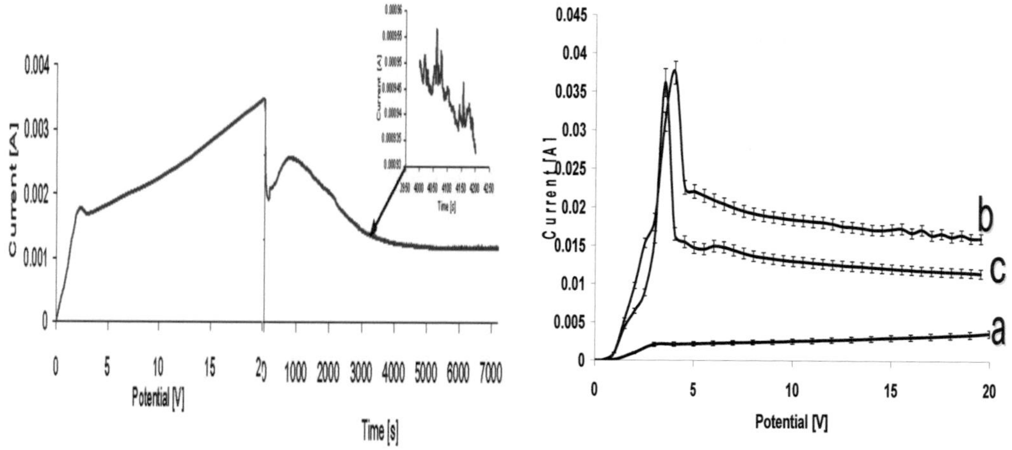

Fig. 10. Transients for potentiodynamic experiments recorded for titanium anodized to 20V (with scan rate 500mV/s) at various concentration of supporting electrolyte H_3PO_4, a) 1M, b) 2M, c) 3M with addition of 0.4% wt. HF [53]

Anodic titania nanotubes formed on titanium in 1-3 M H_3PO_4 with 0.4% wt. HF (Fig. 11) show the morphology of nanotubes on titanium which differ in diameter and the layer thickness due to electrolyte concentration.

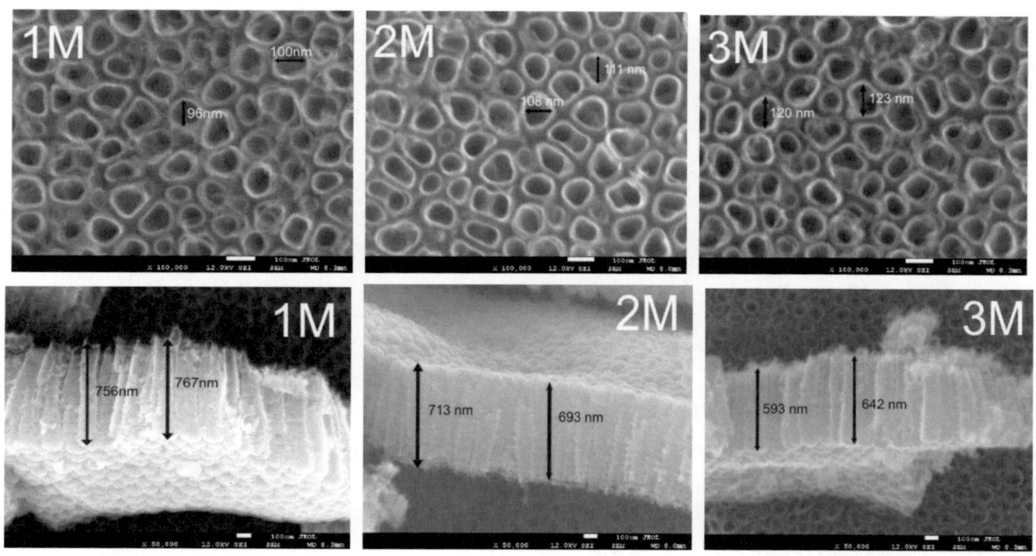

Fig. 11. SEM images of titania nanotubes formed anodically prepared at 20V for 2 h in aqueous solutions of H_3PO_4 ranging from 1 M to 3 M with 0.4% wt. HF (field emission JEOL 7600F) [53,54]

SEM observations (Fig. 11) confirmed formation of a highly organized nano-sized pores, ranging from 90 to 120 nm in all applied electrolytes. As apparent, the average nanotube diameter is slightly affected by the supporting electrolyte concentration. Also, the increase of the latter from 1M to 3M under fixed HF concentration results in significant decrease of nanotube layer thickness, from 760±35 nm to 590±35 nm, respectively.

H_3PO_4 Concentration	Diameter (+/- 10nm)	Thickness (+/- 35nm)
1M	100nm	750nm
2M	**110nm**	700nm
3M	120nm	590nm

Table 1. Diameters of nanotubes and the thickness of their layer on titanium formed in 1-3 M H_3PO_4 with 0.4% wt. HF

The XPS analysis (Fig. 12) revealed that the highest amount of fluorides in oxide surface layer was obtained in 1M H_3PO_4+0.3% wt. HF, but in this case the lowest amount of phosphates adsorbed above nanotubes was observed. Using higher concentrations of phosphoric acid 2-3 M H_3PO_4 Judging on the results of XPS analysis, the competition between fluorides and phosphates is observed during anodizing and the higher concentration of the latter is responsible for higher bioactivity of nanotubes formed in 2M H_3PO_4+0.4% wt. HF [53].

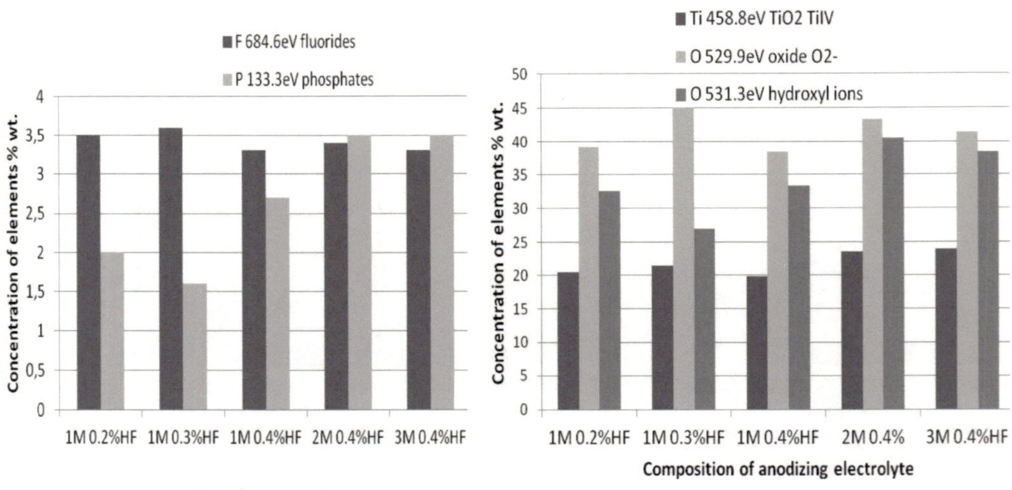

Fig. 12. Results of XPS analysis of nanotubes formed on titanium in 1-3 M H_3PO_4 solution containing different amount of fluorides, from 0.2% wt to 0.4% wt. HF VSW (Vacuum Systems Workshop, Ltd.) Kα Al (1486.6 eV) X-ray radiation working at power of 210 W (15 kV - voltage, 14 mA – emission current) [53].

The XPS spectra (Ti 2p, O 1s, P 2p and F 1s) revealed that nanotube layers consist of Ti, O, F and P species. The Ti 2p spectra for all samples showed only one doublet line. The position of the Ti 2p3/2 peak on the binding energy scale at 458.8 eV corresponded to titanium dioxide and Ti IV phosphates [16]. One type of phosphorus was revealed by the P 2p3/2 peak position at 133.3 eV associated to phosphate type species, indicating that species from the electrolyte are indeed adsorbed over the oxide film during anodizing. Oxygen was found to exist in two forms. The binding energies of the O 1s spectra corresponded to hydroxyl groups OH^- (531.3eV) and oxygen in oxides O^{2-} (529.9 eV). The presence of fluorine in the surface layer was confirmed by the F 1s spectra of binding energy 684.6 eV associated probably with Ti. As it is shown in Fig. 12 the concentrations of titanium corresponding to titanium dioxide and Ti IV phosphates are nearly the same in all tested samples, whereas the concentrations of the other elements vary with the composition of the anodizing electrolyte. Samples anodized in 1M H_3PO_4+0.3% wt. HF shows the highest amount of O^{2-} (asTiO_2) and fluorides, but the lowest amount of phosphates and hydroxyl ions. It means that previously recommended [6] conditions of uniform nanotubes formation on titanium implant materials favor titanium oxidation and enhance transport of fluorides in formed titania. The similarly high concentrations of relevant elements (Ti and O), together with the highest amount of phosphates of all controlled, are observed in samples anodized in 2-3M H_3PO_4+0.4% wt. HF. It indicates that the use of more concentrated phosphate electrolyte leads to the increase of phosphates adsorbed over the surface layer of nanotubes in competition with much smaller and more mobile fluorides.

3.1.1 Competition between phosphates and flurides

This is well known that at anodizing a competition between oxide formation and its dissolution exists, and that HF is the key factor, which causes the production of porous oxide layer. However, there is also a competition between phosphates and fluorides in the process of oxide nanotubes formation. These 2 anions differ in size, charge and rates of diffusion in oxides (Tabl. 2)

	Diffusion coefficient [mol/m^{2}s]	Van der Waals radius (pm)
Fluoride ion	**2.82 10^{-4}**	**147**
Phosphate ion	**0.88 10^{-9}**	**484**

Table 2. Values of diffusion coefficients and Van der Waals radius for fluoride and phosphate ions

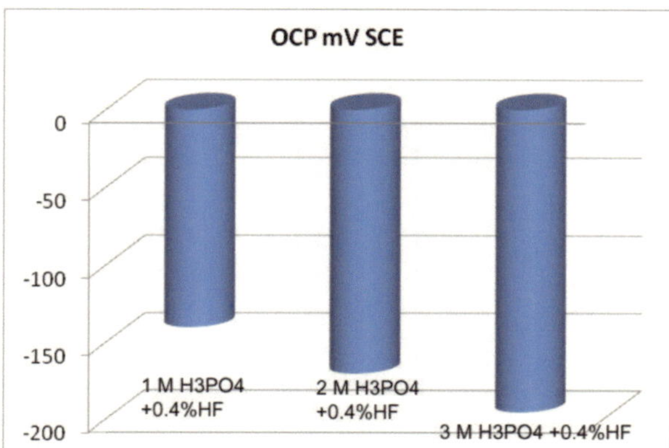

Fig. 13. Values of the OCP for nanotubes formed in 1M, 2M and 3M H_3PO_4 [unpublished results]

The effect of the phosphoric acid concentration in fluoride containing electrolytes on the properties of oxide nanotubes formed at anodizing was characterized by SEM/EDS observations and capacitance characteristics when immersed in simulated body fluids (SBF) in order to predict their behavior as the future coatings on titanium for biomaterial applications [53].

The values of the OCP (Fig. 13) for nanotubes formed in 1M, 2M and 3M H_3PO_4, each containing 0.4% wt. HF, measured at 25°C in SBF solution 1 h after anodizing are -0.140 V, -0.170V and -0.195V (SCE), respectively, indicate the observed earlier [68] decrease of the OCP of oxide produced in more concentrated phosphoric acid solution.

The impedance spectra for titanium anodized in 1-3M H_3PO_4+0.4% wt. HF were obtained at the OCP for frequency ranging from 10^5 to 0.18 Hz with ac amplitude 10 mV. The spectra recorded 1 hour after immersion in SBF solution show that variations in chemical composition of the surface layer over obtained nanotubes are confirmed by variations in capacitance characteristics.

The results of EIS tests (Fig.14) indicate nearly the same properties (similar impedance and –θ angle values) for ohmic resistance of the electrolyte and its penetration through nanotube films. However in the low frequency range, the impedance values are sensitive to the

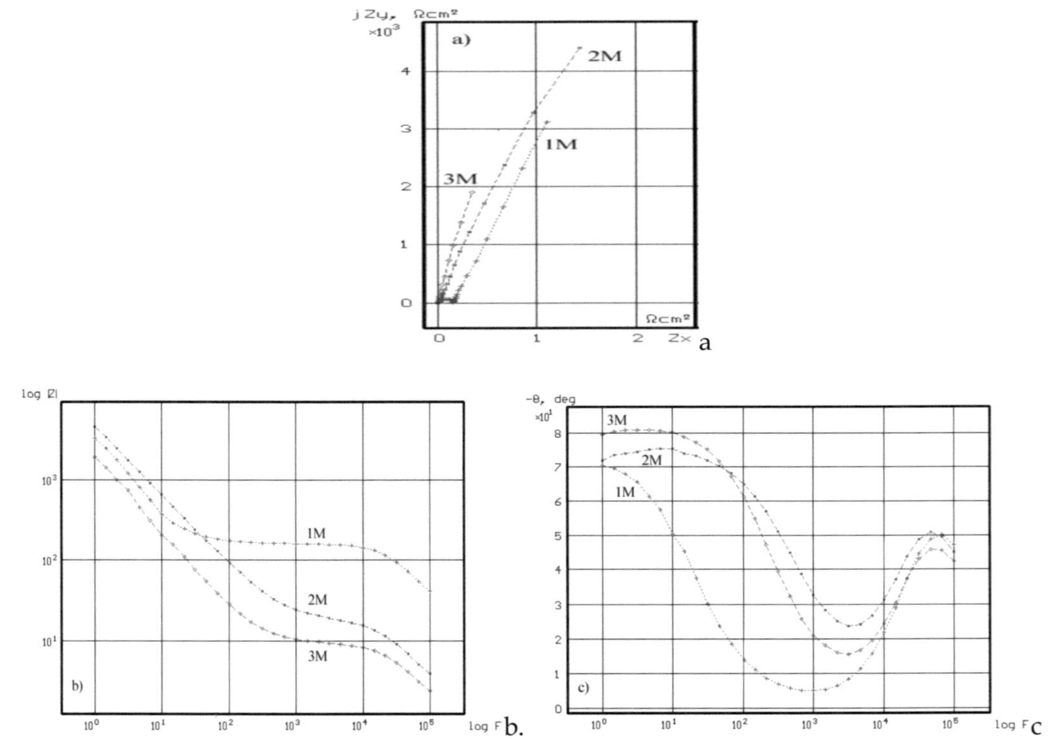

Fig. 14. EIS spectra (a- Nyquist, b,c- Bode spectra) for titania nanotubes formed on titanium in 1-3 M H_3PO_4 solutions with addition of 0.4% wt. HF recorded at the OCP in SBF solution at 25°C [53]

phosphate concentration in anodizing electrolyte, accounting for the processes at the nanotube layer/electrolyte interface which can be associated with deposited products. As the changes between spectra occurred during the first hour of exposure to SBF solution, one can assume that the deposition processes on nanotube layers formed in 2-3M H_3PO_4+0.4% wt. HF are quick.

The formation of porous metal oxides, ie. titania and alumina, is explained by a field-enhanced model [11,61,63,69] that depends on the ability of ions to diffuse through the metal oxide. Thus, due to the large size the incorporation of the phosphate ions is difficult, but the increased fluoride concentration in solution leads to its ability to migrate and intercalate into the oxide films during the anodizing [70]. The XPS results show (Fig. 12), that the increased fluoride concentration is accompanied by the decreased phosphates and hydroxyl ions in adsorbed layer over nanotubes [53]. It correlates very well with the results of titanium anodizing in electrolyte not containing fluorides, where the gel-like protective layer of phosphates was formed over the oxide [42].

3.2 Nanotubes on implant alloys in phosphate solutions

Titanium and its implant alloys, mainly ternary alloys of Ti-6Al-7Nb or, are widely used in biomedical implants and dental fields due to their unique mechanical, chemical properties, excellent corrosion resistance and biocompatibility [71-73].

Further improvement of the unique properties of nanotube anodic layers for medical applications, particularly for enhancement of bone in-growth [74] and biosensing [75] require not only the development of the formation method on two phase titanium alloys, but also providing the proper morphology and structure. Reported efforts to form anodic nanotube layers on Ti alloys such as Ti-6Al-7Nb, TiAl [76], or Ti45Nb [77] showed the formation of highly inhomogeneous surfaces due to selective dissolution of the less stable phase and/or different reaction rates of the different phases of the alloys.

Studies on development of nanotubes growth on the Ti6A4V [60] and Ti-6Al-7Nb alloys [54] were focused on varying the HF concentrations in the phosphoric acid media, in order to establish the pore size distribution and estimate the critical scan rate/concentration ratio for the initiation of nanopitting in compact oxide layer, which would be decisive for the formation of uniform nanotubes on both two-phase alloys.

Among several parameters influencing the quality of nanotubes formed anodically, such as potential, time of anodizing, fluoride ions concentration and scan rate of polarization, particularly the last two seem to be determiners for nanotubes structure and morphology. As an example to show the effect of fluoride ions concentration on the morphology of nanotubes on the implant alloy, the anodizing of the two phase ($\alpha+\beta$) Ti6Al7Nb alloy samples in 1 M H_3PO_4 containing 0.2%; 0.3% and 0.4 % wt. HF to 20V using scan rate 500mV/s and then holding them at that potential for further 2h in the same electrolyte, was performed. Nanotubes of diameter ranging from 50nm to 80nm, with thicker walls over β-phase grains than over α- phase grains, were obtained. During the formation process, which includes two stages: the first potentiodynamic and the second potentiostatic (20V), different electrochemical behaviour was observed in electrolytes of various fluoride concentration.

The implant alloy Ti6Al7Nb (Fig 15) of black α phase (hcp) and white β phase (bcc) irregular shape platelets forming variously oriented colonies, with the surface fraction of α and β phases 78% and 22%, respectively, was enriched with aluminum in oxides over α phase and enriched with niobium over β phase anodic nanotubes.

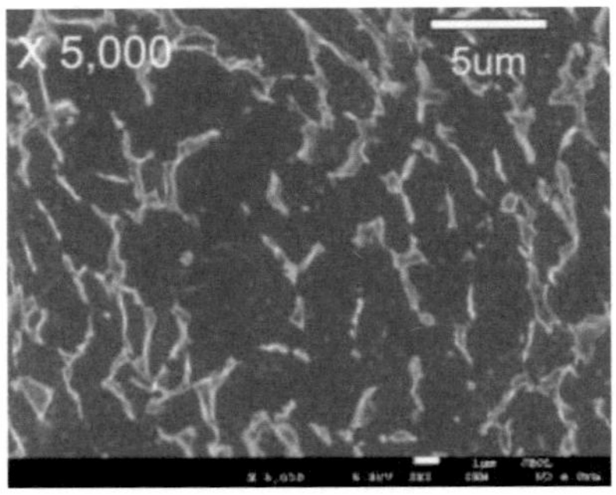

Fig. 15. Microstructure of the Ti-6Al-7Nb alloy [54]

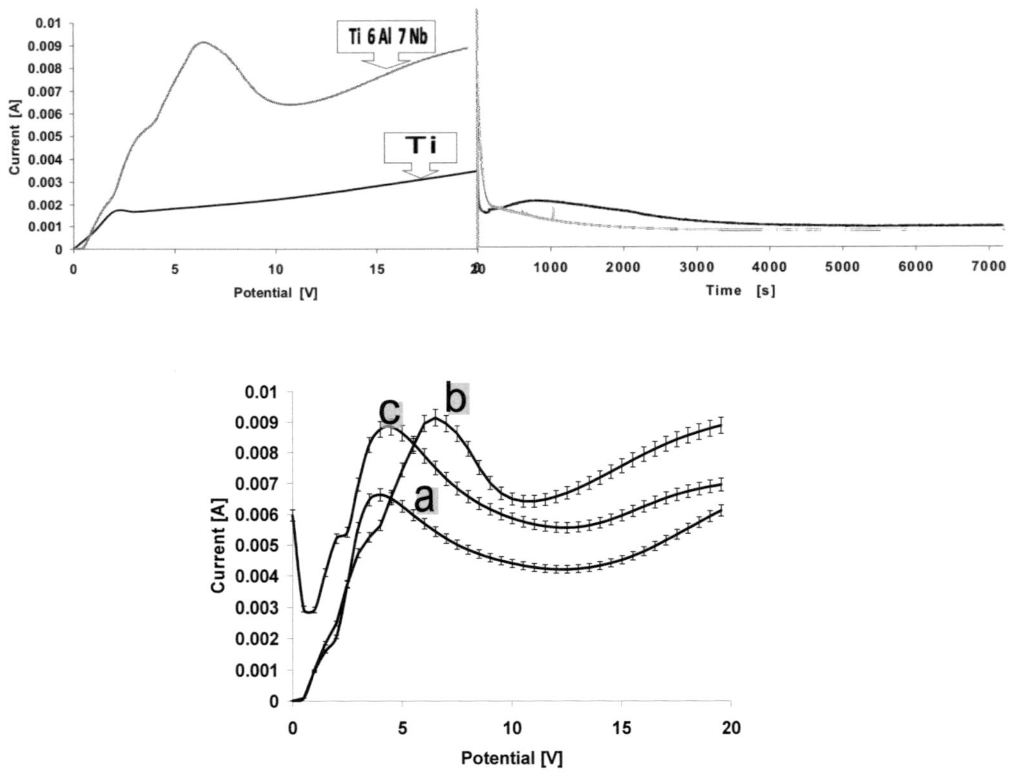

Fig. 16. Current transient for potentiodymic and potentiostatic stages recorded at anodizing the Ti-6Al-7Nb alloy and titanium (for comparison) at 20V for 2h in 1M H_3PO_4 containing 0.3% HF (scan rate in the potentiodynamic stage 500mV/s) and current transients recorded during potentiodynamic stage of anodizing of the Ti6Al7Nb alloy in 1 M H_3PO_4 with different fluoride concentration, a) 0.2%HF b) 0.3%HF c) 0.4%HF [54]

The typical current transients Fig. 16 recorded during the anodizing of the Ti-6Al-7Nb alloy in 1M H_3PO_4 containing 0.3 wt.% HF are similar to current transients observed during nanotube oxide layers formation on other alloys in other electrolytes [78]. As in previously described process the whole treatment consists of the potentiodynamic polarization from the OCP to 20V with a scan rate of $0.5Vs^{-1}$, followed by the potentiostatic polarization at 20V for further 2 h. However, contrary to constant current density increase observed at anodizing of pure Ti [49,50,79], during the potentiodynamic sweep to 20V at the alloy Ti-6Al-7Nb anodizing the current transients show 2 peaks: the first at about 2-3 V due to oxygen evolution and the second at about 4-6V linked probably to Al oxidation. In the potentiostatic stage of anodizing the current density for the alloy decreases until the end of the treatment, while in case of Ti a broad peak is seen at about 900 s of the anodizing (Fig. 16). According to [78,79] the broad peak, typically recorded in the potentiostatic stage of the process, indicates the dissolution of oxide before reaching final balance between both processes: oxide formation and oxide dissolution during nanotubes formation. Such the balance determines a steady-state oxide layer formation stage during anodizing of metals [11].

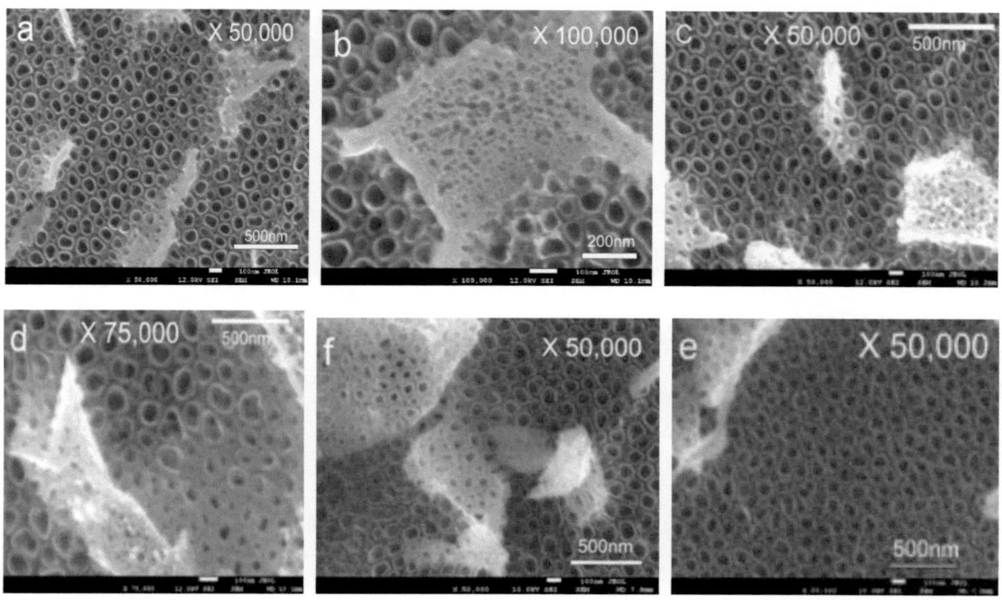

Fig. 17. SEM images of nanotubes produced on the Ti6Al7Nb (a ,c, e α-phase; b, d, f- β-phase) by anodization at 20V for 2h in 1M H_3PO_4 containing (a),(b) 0.2%HF, (c),(d) 0.3% HF, (e),(f) 0.4% wt. HF

Small pits on β phase grains and regular nanotubes on α-phase are observed in 0.2 wt.% HF (Fig. 17a,b). Irregular tubes on β-phase and regular tubes on α-phase grains are seen after anodising in 0.3 wt. % HF (Fig. 17c,d). Both phases are covered with regular nanotubes in case of samples anodised in 0.4 wt. % HF (Fig. 17e,f), but on β- phase nanotube walls are thicker than on α- phase. Fig. 16 illustrates the dissolved oxide over α-phase on Ti6Al4V alloy and bigger size of nanotubes over β- phase in more concentrated phosphoric acid solution.

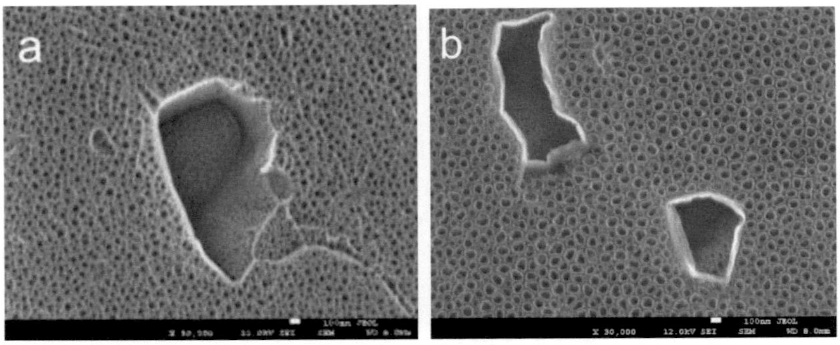

Fig. 18. SEM images of nanotubes produced on the Ti6Al4V by anodization at 20V for 2h in 1M H_3PO_4 containing (a) 0.2%HF, (b) 0.4% wt. HF

According to EDS analysis nanotubes formed on the Ti-6Al-7Nb alloy showed that those films are predominately TiO_2 with small amounts of Ti_2O_3, Al or Nb oxides (Table 3). Aluminium and niobium are present in their most stable oxidation states, Al_2O_3 and Nb_2O_3. The amount of alloying elements in the nanotube oxide layer was influenced by the underlying metal microstructure, where Nb was present in the β- phase and Al in the α-phase [80].

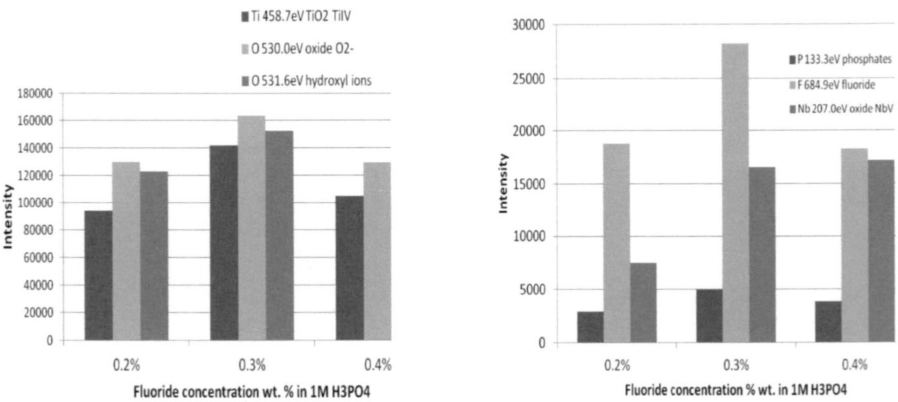

Fig. 19. Results of XPS examination of surface layer of nanotubes formed on the Ti6Al7Nb alloy in 1 M H_3PO_4 with different fluoride concentration

In the combined SEM and XPS examinations [54] (Fig. 17 and 19) the highest intensities for all controlled elements and groups: titanium oxide and titanium phosphates (458.7eV), oxides (530eV), hydroxyl ions (531.6eV), phosphates (133.3eV) and fluorides (648.6eV), clearly confirm that the most advantageous scan rate and electrolyte composition for the formation of uniform nanotube layer on the Ti6Al7Nb alloy, are $0.5Vs^{-1}$ during potentiodynamic stage of anodizing in 1M H_3PO_4 containing 0.3% wt. HF. Interesting is that also the intensity of niobium (207.3eV) in the most stable of the niobium oxides Nb_2O_5 [81], increases with fluoride concentration, but seems to reach the limit in these conditions for 0.3% wt. HF (Fig. 19). The highest current density (Fig. 14) is linked to the biggest nanotube diameters, as it was observed in case of pure titanium anodised in the same conditions [50].

Fluoride concentration	0.2%HF [weight %]		0.3%HF [weight %]		0.4%HF [weight %]		Compact oxide [weight %]
Phases	α	β	α	β	α	β	
Titanium	63.90	51.66	61.39	35.70	59.52	38.59	69.22
Oxygen	32.09	36.66	34.44	45.43	36.40	44.99	19.72
Aluminium	4.00	2.91	4.17	2.69	4.08	1.99	4.45
Niobium	---	8.77	---	16.18	---	14.43	6.53

Table 3. Results of EDS analysis of nanotube layers obtained by anodizing at 20V for 2h in 1M H_3PO_4 containing 0.2; 0.3; 0.4% wt. HF [53]

Due to chemical similarity of titanium and niobium [11, 82] electrochemical behaviour of the Ti-6Al-7Nb electrode should be qualitatively similar to that of the titanium and niobium electrodes in the potential range from -1 to 4V (SCE). Electrochemical oxidation of niobium electrode leads to formation of sub-oxides NbO and NbO_2 at the OCP, which partly transform into Nb_2O_5 oxide at 20V, according to the equations 6-7 [83,84]:

$$Nb + H_2O - 2e^- \rightarrow NbO + 2H^+ \quad (6)$$

$$NbO + H_2O - 2e^- \rightarrow NbO_2 + 2H^+ \quad (7)$$

$$2NbO_2 + H_2O - 2e^- \rightarrow Nb_2O_5 + 2H^+ \quad (8)$$

The dissolution process of niobium oxide (β-phase) (5) increases with increasing fluoride concentration [85], so the fluoride concentration is a crucial factor for nanotubes growth on Ti-6Al-7Nb. Structural and metallurgical aspects of the formation of self-organized anodic oxide nanotube layers on alloys are crucial for medical application to the advanced techniques of biological media immobilization which require morphologically uniform surface.

4. Titania layers formed in phosphate solutions for biosensing

The additional advantageous property of phosphate rich compact and nanotubular anodic oxide layers on titanium is its ability to attach enzymes, proteins or biological cells. To test such the possibility in order to apply anodic surfaces for H_2O_2 biosensing two electrodes were prepared: 1) the first electrode prepared by the electropolymerization of conducting polymer (PANI) on the surface of Ti/TiO_2 (compact) electrode [55], 2) the second electrode was prepared by using titania nanotubes on titanium as a platform of the 3rd generation biosensor [56]. In both cases the HRP (horseradish peroxide) enzyme was immobilized on the sensing surface. By using either cyclic voltammetry or amperometric modes the feasibility and electrochemical parameters for H_2O_2 monitoring on Ti/TiO_2 surface were checked in the simulated body fluid (SBF). Both electrodes were sensitive to H_2O_2, however the second electrode only in the presence of thionine as the mediator [56]. Two peaks seen on cyclic voltammograms (Fig. 18) for the Ti/TiO_2 (nanotube) electrode with immobilized HRP, indicate the sensitivity of the prepared platform to the presence of H_2O_2 in the analyte.

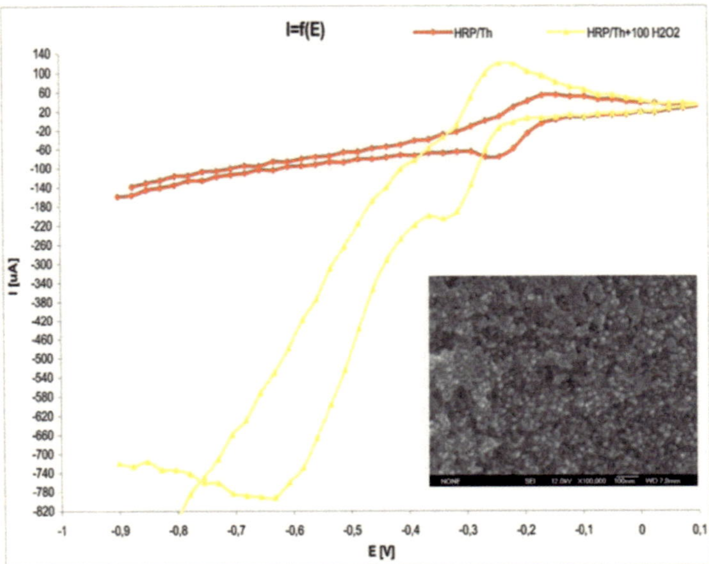

Fig. 20. SEM of Ti/TiO_2 (nanotube) covered with HRP and cyclic voltammograms for $Ti/TiO_2/HRP$ in 0.1 M PBS (pH 6.8) in the presence of H_2O_2, scan rate 100 mV/s [56]

5. Conclusions

Titanium surfaces can be modified by electrochemical treatment in the phosphoric acid solutions for better corrosion resistance, improved physicochemical and electrochemical properties and bioactivity. The formation of oxide layers enriched with phosphorus of 30-120nm thick can be formed in 0.5 M H_3PO_4 at both galvanostatic anodizing current density values varied in the range of 0.1-0.5 Am^{-2} and potentiostatically at up to 60V giving yellowish layers porous on the surface. Due to the presence of phosphates they are highly bioactive in comparison to oxides formed in other electrolytes and are covered with hydroxyapatite deposists after 9 days in the SBF solution.

Anodizing in more concentrated 2M solutions of the phosphoric acid generates a gel-like film with thickness of about 100nm on titanium. The film, containing a large number of phosphates nuclei, exhibits its effectiveness to activate titanium surface for biomimetic coating of calcium phosphate. The electrochemically treated titanium was able to form uniform Ca–P coatings on titanium after 48 hour immersions in the SBF solution. The treatment is a simple method to generate bioactive metal surfaces, besides other methods such as alkaline treatment applied to titanium implant materials.

Electrochemical treatment in the phosphoric acid solutions with the addition of 0.2-0.4% wt. HF allows to form on titanium and its implant alloys nano-sized pores (nanotubes) in more concentrated phosphoric acid solutions (1-3M). Their morphology, electrochemical properties and chemical composition are in close relation with the anodic polarization parameters and with the concentration of both ions: phosphates and fluorides. The highest amount of fluorides in surface layer is obtained when using 1M H_3PO_4+0.3% wt. HF, but in this case the lowest amount of phosphates adsorbed above nanotubes is observed. The use of higher concentrations of phosphoric acid (2-3M H_3PO_4 with 0.4% wt. HF assures the formation of nanotubes containing the high concentration of both bioactivity enhancing elements, fluorides and adsorbed phosphates. The obtained titania nanotubes show the significantly higher bioactivity in vitro during the first hour of immersion in SBF in comparison to barrier titanium oxide.

Depending on fluoride ion concentrations in anodizing electrolyte morphologically different nanotubular layers have been obtained on both phases of two titanium alloys: the Ti6Al4V and the Ti6Al7Nb alloys. Self organized nanotubes grow on both phases (α and α+β) in 1M H_3PO_4 containing 0.4% wt. HF, though smaller pore size and thicker wall tubes are obtained on the β phase. The electrochemical behavior of both phases of the alloys differs due to fluoride concentrations which is the key parameter in controlling their morphology. Uniform nanotubes are obtained in 2M H_3PO_4 containing 0.3% wt. HF at scan rate of polarization $0.5Vs^{-1}$ during potentiodynamic stage of anodizing. Such conditions assure the highest fluoride and phosphate concentrations in surface layer of nanotubes on titanium and nanotubes containing niobium oxide on the Ti6Al7Nb alloy. Both features promise a proper coating for improved osteoblast cell adhesion on artificial implants and for biosensing.

6. Acknowledgment

Funding from the Polish Ministry of Science and Higher Education under the N507 082 31/2009 project and by the National Centre of Research and Development under ERA-NET/MNT/TNTBIOSENS/1/2011 Project is gratefully acknowledged. My thanks go also to Professor Patrick Schmuki of University of Erlangen for inspiration on nanotubes and to my coworkers from the Biomedical Engineering Division at University of Zielona Góra.

7. References

[1] Trasatti S., Lodi G.: Electrodes of Conductive Metallic Oxides. ed. S. Trasatti, Chapt. B, Elsevier Amsterdam (1981) 521.

[2] Luckey H., Kubli F., Titanium Alloys in Surgical Implants. ASTM 796 (1981).

[3] Marciniak J.: Biomateriały (in Polish) Technical University of Upper Silesia Press (2002), ISBN 83-7335-031-4.

[4] Chrzanowski W., Szewczenko J., Tyrlik-Held J., Marciniak J., Zak J.: Influence of the anodic oxidation on the physicochemical properties of the Ti6Al4V ELI alloy. J Mater Proces Techn 162-163 (2005) 163-168.

[5] Hanawa T., Mamoru O.: Calcium phosphate naturally formed on titanium in electrolyte solution. Biomaterials 12 (1991) 767.

[6] de Assis S. L., Wolynec S., Costa I.: Corrosion characterization of titanium alloys by electrochemical techniques. Electrochimica Acta 51 (2006) 1815–1819.

[7] Krasicka-Cydzik E., Kowalski K., Głazowska I.: Bioactive surface layers formed electrochemically on titanium materials in phosphoric acid solution. 3rd Central Eur Conf. Krakow (2006) eds. K. Szaciłowski.

[8] Sul Y.T.: The significance of the surface properties of oxidized titanium to the bone response: special emphasis on potential biochemical bonding of oxidized titanium implant. Biomaterials 24 (2003) 3893-3907.

[9] Bylica J., Sieniawski J.: Titanium and its alloy. PWN, Warszawa (1985).

[10] Corbridge D.E.C.: Phosphorus. the 5th Edit. Elsevier, Amsterdam - Lausanne - New York-Oxford -Shannon -Tokyo (1995).

[11] Krasicka-Cydzik E.: Formation of thin anodic layers on titanium and its implant alloys. University of Zielona Gora Press, Zielona Góra (2003) ISBN: 83-89321-80-7.

[12] Krasicka-Cydzik E.: Electrochemical and Corrosion properties of Ti6Al4V in phosphoric acid solutions. Biomater. Eng. (1999) 7-8, 26-32.

[13] Krasicka-Cydzik E.: Electrochemical aspects of tailoring anodic layer properties on titanium alloys. Corrosion Protection XLII (1999) 48-52.

[14] Krasicka-Cydzik E.: Influence of phosphoric acid on the rate of titanium alloys anodizing. Arch. BMiAut., Poznań 20 (2000) 155.

[15] Krasicka-Cydzik E.: Impedance properties of anodic films formed in H_3PO_4 on selected titanium alloys. Mat. Eng. 7, 2 (2000) 5-11.

[16] Krasicka-Cydzik E.: Effect of polarization parameters on the selected properties of surface layer on Ti6Al4V alloy anodized in phosphoric acid solutions. Arch. TM i Aut. 21, 1 (2001) 171-178.

[17] Krasicka-Cydzik E.: Impedance examination of titanium and its selected implant alloys. Biomater. Eng. 14 (2001) 27-31.

[18] Krasicka-Cydzik E.: Formation and properties of anodic film on titanium in phosphoric acid solutions. Mater. Eng. Vol. IX (2002) 2, 9.

[19] Krasicka-Cydzik E.: Passivity of implant titanium alloys in phosphoric acid solution of high concentration. Corrosion Protection (2002) 282.

[20] Krasicka-Cydzik E.: Surface modification of phosphoric acid anodized implant titanium alloys. 7th World Biom. 2004, ISBN: 1-877040-19-3

[21] Krasicka-Cydzik E., Głazowska I., Michalski M.: Bioactivity of implant titanium alloys after anodizing in H_3PO_4. Biomater. Eng. (2004) 38-42.

[22] Krasicka-Cydzik E.: Method of anodic coating formation on titanium and its alloys. Polish Patent RP 185176, University of Technology Zielona Gora, LfC Ltd. 2003.

[23] Głazowska I., Krasicka-Cydzik E.: Impedance characteristics of anodized titanium in vitro. Biomater. Eng. (2005) 8, 47-53, 127—130.

[24] Krasicka-Cydzik E., Głazowska I., Michalski M.: Microscopic examination of anodic layers on implant titanium alloys after immersion in SBF solution. Biomater. Eng. (2005), 47-53, 130—133

[25] Krasicka-Cydzik E., Kowalski K., Głazowska I.: Electrochemical formation of bioactive surface layer on titanium. J Achiev Mater Manufact Engineering 18 (2006) 1-2, 147—150.

[26] Krasicka-Cydzik E., Głazowska I.: Influence of alloying elements on behavior of anodic layer in phosphoric acid solution. Biomater. Eng. (2007) 67—68, 29—31.

[27] Krasicka-Cydzik E., Głazowska I., Michalski M.: Hydroxyapatite coatings on titanium and its alloys anodised in H_3PO_4. EUROMAT 2005, European Congress on Advanced Materials and Processes. Prague, Czech Rep, 2005.

[28] Unal I., Phosphate Adsorption On Titanium Oxide Studied By Some Electron Spectroscopy, Diploma , Universite de Geneva, Sept. 1999

[29] Pourbaix M., Atlas of Electrochemical Equilibria in Aqueous Solutions, National Association of Engineers, Houston, 1974

[30] Kurosaki M., Seo M., Corrosion behavior of iron thin film in deaerated phosphate solutions by an electrochemical quartz crystal microbalance, Corrosion Science 45 (2003) 2597.
[31] Krasicka-Cydzik E.: Gel-like layer development during formation of thin anodic films on titanium in phosphoric acid solutions, Corrosion Science Vol. 46 (2004) 2487–2502.
[32] Krasicka-Cydzik E.: Studies on transition of titanium from active into passive state in phosphoric acid solutions, in: Passivation of metals and semiconductors, and properties of thin oxide layers. ed. P. Marcus, V. Maurice .- Amsterdam: Elsevier (2006) 193–198.
[33] Kelly E.J., in: J. O'Bockris, B.E. Conway, R.E. White (Eds.), Modern Aspects of Electrochemistry, Plenum Press, New York, London, 1982, pp. 332-336
[34] Torresi, R.M. Camara O.R., De Pauli C.P., Influence of the hydrogen evolution reaction on the anodic titanium oxide film properties, Electrochim. Acta 32 (9) (1987) 1357.
[35] Kelly E.J., Anodic Dissolution and Passivation of Titanium in Acidic Media, J. Electrochem. Soc. 126 (12) (1979) 2064
[36] Makrides A.C, Kinetics of Redox Reactions on Passive Electrodes, J. Electrochem. Soc. 111 (4) (1964) 392
[37] Felhosi I., Telegdi J., Palinkas G., Kalman E., Kinetics of self-assembled layer formation on iron, Electrochem. Acta 47 (2003) 2335-2340.
[38] Healy K.E., Ducheyne P., Hydration and preferential molecular adsorption on titanium in vitro, Biomaterials. 13 (1992) 553.
[39] Wagh A.S.,. Yeong S.Y, Chemically bonded phosphate ceramics: I, a dissolution model of formation, J. Am. Ceram. Soc. 86 (11) (2003) 1838-1855.
[40] Morlidge J.R. P. Skeldon, G.E. Thompson, H. Habazaki, K. Shimizu, G.C. Wood, Gel formation and the efficiency of anodic film growth on aluminium Electrochim. Acta, 44 (1999) 2423.
[41] Roine A., HSC Ver. 2. 03, Outkumpu Research Oy, Pori, Finland, 1994
[42] Krasicka-Cydzik E.: Method of phosphate layer formation on titanium and its alloys, Polish Patent, PL 203453, Univ. Zielona Gora, 2009.
[43] Grimes C. A., Mor G. K.: TiO_2 Nanotube Arrays. Springer 2009, ISBN 978-1-4419-0067-8.
[44] Bavykin D.V., Walsh F. C.: Titanate and Titania Nanotubes, RSC (2010).
[45] Ghicov A., Tsuchiya H., Macak J. M., Schmuki P.: Titanium oxide nanotubes prepared in phosphate electrolytes. Electrochem. Comm. 7 (2005) 505-509.
[46] Tsuchiya H., Macak J. M., Taveira L., Balaur E., Ghicov A., Sirotna K., Schmuki P.: Self-organized TiO2 nanotubes prepared in ammonium fluoride containing acetic acid electrolytes. Electrochem. Comm. 7 (2005) 576-580.
[47] Tsuchiya H., Macak J. M., Taveira L., Schmuki P.: Fabrication and characterization of smooth high aspect ratio zirconia nanotubes. Chem. Phys. Letters 410 (2005) 188-191.
[48] Zhao J., Wang X., Chen R., Li L.,:Synthesis of thin films of barium titanate and barium strontium titanate nanotubes on titanium substrates. Materials Letters 59 (2005) 2329 - 2332.

[49] Krasicka-Cydzik E., Głazowska I., Kaczmarek A., Białas-Heltowski K.: Influence of floride ions concentration on growth of anodic sel-aligned layer of TiO2 nanotubes. Biomater. Eng. R. 11,77-80 (2008) 46.

[50] Lukacova H., Plesingerova B., Vojtko M., Ban G., Electrochemical Treatment of Ti6Al4V, Acta Metallurgica Slovaca, 16, (2010) 3, 186-193.

[51] Krasicka-Cydzik E., Kowalski K., Kaczmarek A.: Anodic and nanostructural layers on titanium and its alloys for medical applications. Mater. Eng. 5 (2009) 132.

[52] Krasicka-Cydzik E., Głazowska I., Kaczmarek A., Klekiel T., Kowalski K.: Nanostructural oxide layer formed by anodizing on titanium and its implant alloy with niobium. Biomater. Eng. 85 (2009) 325.

[53] Krasicka-Cydzik E., Kowalski K., Kaczmarek A., Glazowska I., Bialas Heltowski K.: Competition between phosphates and fluorides at anodic formation of titania nanotubes on titanium. Surface and Interface Analysis 42, 6-7 (2010) 471 – 474.

[54] Kaczmarek A., Klekiel T., Krasicka-Cydzik E.: Fluoride concentration effect on the anodic growth of self aligned oxide nanotube array on Ti6Al7Nb alloy. Surface and Interface Analysis, 42, 6-7 (2010) 510.

[55] Machnik M., Głazowska I., Krasicka-Cydzik E.: Investigation for Ti/TiO_2 electrode used as a platform for H_2O_2 biosensing. Mater. Eng. 5 (2009) 363-365.

[56] Łoin J., Kaczmarek A., Krasicka-Cydzik E.: Attempt to elaborate platform of the IIIrd generation biosensor for H_2O_2 on the surface of Ti covered with titania nanotubes. Biomedical Eng. Vol. 16, no 2 (2010) 54 – 56.

[57] Beranek R., Hildebrand H., Schmuki P., Self-Organized Porous Titanium Oxide Prepared in H2SO4/HF Electrolytes, Electrochem. and Solid-St. Lett. 2003: 6, B12

[58] Zhao J., Wang X., Chen R., Li L., Fabrication of titanium oxide nanotube arrays by anodic oxidation, Solid State Commun. 2005, 134, 705-710.

[59] Macak J.M., Sirotna K., Schmuki P., *Self-organized porous titanium oxide* prepared in Na2SO4/NaF electrolytes, Electrochim. Acta 50, 2005, 629.

[60] Krasicka-Cydzik E., Glazowska I., Kaczmarek A., Bialas-Heltowski K., Influence of scan rate on formation of anodic TiO2 nanotubes, Engineering of Biomaterials, 2008, 11, 77-80.

[61] Gong D.,. Grimes C. A., Varghese, O. K Chen Z., Hu W., Dickey E. C., Titanium oxide nanotube arrays prepared by anodic oxidation Journal Mater. Res. 2001, 16, 3331-3334

[62] Ghicov A., Tsuchiya H., Macak J. M., Schmuki P., Titanium oxide nanotubes prepared in phosphate electrolytes, Electrochem. Comm. 2005, 7, 505–509

[63] Yasuda K., Schmuki P., Control of morphology and composition of self-organized zirconium titanate nanotubes formed in (NH4)2SO4/NH4F electrolytes, Electrochim. Acta, 2007 52, 4053–4061

[64] Qiu J., Yu W., Gao X., X. Li, Sol–gel assisted ZnO nanorod array template to synthesize TiO2 nanotube arrays, Nanotechnology, 2007, 18, 295604

[65] Zwilling V., V, Aucouturier M, Darque-Ceretti E. Anodic oxidation of titanium and TA6V alloy in chromic media. An electrochemical approach, Electrochim. Acta 1999; 45921

[66] Yang B., Uchida M. Kim H.M., Zhang X.,Kokubo T.: Preparation of bioactive titanium metal via anodic oxidation treatment, Biomaterials 2004; 25, 1003

[67] Sul Y.-T., Johansson C. B, Jeong Y, Albrektsson T. The electrochemical oxide growth behaviour on titanium in acid and alkaline electrolytes, ed. Eng. Phys. 2001; 23, 329]

[68] Kodama A., Bauer S., Komatsu A., Asoh H., Ono S., Schmuki P., Bioactivation of titanium surfaces using coatings of TiO2 nanotubes rapidly pre-loaded with synthetic hydroxyapatite, Acta Biomaterialia 2009; 5, 2322–2330.

[69] Heusler K.E., The influence of electrolyte composition on the forfmation and dissolution of passivating films, Corrosion Science, 1989, 29, 131

[70] De-Sheng K, The influence of fluoride on the physicochemical properties of anodic oxide films formed on titanium surfaces, Langmuir 2008, 24, 5324-5331

[71] Gonzalez J.E.G,. Mirza-Rosca J.C, *Study of corrosion behaviour of titanium and some of its alloys for biomedical and dental implant applications*, J. Electroanal. Chem. 1999; 471, 109.

[72] Williams J. D.F, in Williams (Ed.) Biocompatibility of clinical implant materials, CRC Press, Boca Roton, FL, 1981, 45.

[73] Fonseca C., Barbosa M.A, Corrosion behaviour of titanium in biofluids containing H2O2 studied by electrochemical impedance spectroscopy, Corrosion Science (2001); 43, 547-559

[74] Velten D., Eisenbarth E., Schanne N., Breme J, Biocompatible Nb2O5 thin films prepared by means of the sol–gel process, Journal of Materials Science: Materials in Medicine 2004, 15, 457–461

[75] Xiao P., Garcia B.B., Guo Q, Liu D., Cao G., TiO2 nanotube array fabricated by anodization in different electrolytes for biosensing, Electrochemistry Communications 9, (2007), 2441-2447

[76] Tsuchiya H.; Berger, S., Macak, J. M.; Ghicov, A., Schmuki, P., Self-organized porous and tubular oxide layers on TiAl alloys, Electrochemistry Communications (2007); 9, 2397-2402.

[77] Zorn G., Lesman A., Gotman I. Oxide formation on low modulus Ti45Nb alloy by anodic versus thermal oxidation, Surface & Coatings Technology 2006; 210, 612-618

[78] Macak J. M., Tsuchiya H., Taveira L., Ghicov A, Patrik Schmuki P., Self-organized nanotubular oxide layers on Ti-6Al-7Nb and Ti-6Al-4V formed by anodization in NH4F solutions, Journal of Biomedical Materials Research Part A, 2008, 75A, 4, 928–933.

[79] Macak J.M., Hildebrand H., Marten-Jahns U., Schmuki P, Mechanistic aspects and growth of large diameter self-organized TiO2 nanotubes, J. Electroanal. Chem. (2008); 40 621, 254-266

[80] Metikoš-Hukovič M. A Kwokal, J Piljac, The influence of niobium and vanadium on passivity of titanium-based implants in physiological solution, Biomaterials 2003; 24, 3765-3775

[81] Halbritter J., On the Oxidation and on the Superconductivity of Niobium, Applied Physics, 1987, A 43, 1–28

[82] I. Sieber I., Hildebrand H., Friedrich A., Schmuki P., Formation of self-organized niobium porous oxide on niobium, Electrochem. Comm. 2005; 7, 97–100

[83] Freitas M.B, Bulhoes L., Breakdown and Crystallization Processes in Niobium Oxide-Films in Oxalic-Acid Solution, J. Appl. Electrochem, 1997; 27, 612-615

[84] Heusler K.E., Schultze M., Electron-transfer reactions at semiconducting anodic niobium oxide films, Electrochemica Acta, 1975, 20, 237-244
[85] El-Mahdy G.A, Formation and dissolution behaviour of niobium oxide in phosphoric acid solutions, Thin Solid Films, 1997; 307, 141-147.

9

Surface Modification Techniques for Biomedical Grade of Titanium Alloys: Oxidation, Carburization and Ion Implantation Processes

S. Izman[1], Mohammed Rafiq Abdul-Kadir[1], Mahmood Anwar[1], E.M. Nazim[1], R. Rosliza[2], A. Shah[1] and M.A. Hassan[1]

[1]Universiti Teknologi Malaysia, UTM Skudai, Johor,
[2]TATi University College, Malaysia, Kemaman, Terengganu, Malaysia

1. Introduction

Titanium and titanium alloys are widely used in a variety of engineering applications, where the combination of mechanical and chemical properties is of crucial importance. Aerospace, chemical and automotive industries as well as the medical device manufacturers also benefited from the outstanding properties of titanium alloys. The wide spread of its uses in biomedical implants is mainly due to their well-established corrosion resistance and biocompatibility. However, not all titanium and its alloys can meet all of the clinical requirements for biomedical implants. For instance, it is reported that bare titanium-vanadium alloy has traces of vanadium ion release after long period exposure with body fluid (López et al., 2010). Excessive metal ions release into the body fluid and causes toxicity problems to the host body. A new group of titanium alloy such as Ti-Nb and Ti-Zr based are recently introduced in the market to overcome the toxicity of titanium-vanadium based alloy (Gutiérrez *et al.*, 2008). Although, these alloys have a high strength to weight ratio and good corrosion resistance and biocompatible, but it suffers from poor tribological properties which limits their usefulness to a certain extent especially when they are applied to joint movements. Wear debris generated from these articulation joints can induce inflammation problem and toxic effect to the human body. In biomedical point of view, post implantation is very crucial stage where the interaction between the implanted material surface and the biological environment in human body is critically evaluated. Either in the short or long run, the toxic effect becomes an issue to host body. Hence, the implant material surface has a strong role in the responses to the biological environment. In order to improve the biological and tribological properties of implant materials, surface modification is often required (Huang *et al.*, 2006, Kumar *et al.*, 2010b). This chapter embarks on the commonly used implant biomaterials, followed by general overview on the surface modification techniques for treating titanium alloy. The basic principles of oxidation, carburization and ion implantation methods and their developments are discussed in the following sections.

2. Implant biomaterials

Biomedical implant is defined as an artificial organ used for restoring the functionality of a damaged natural organ or tissue of the body (Liu *et al.*, 2004). In other words it is expected to perform the functions of the natural organ or tissue without adverse effect to other body parts (Andrew *et al.*, 2004). Fig.1 shows a typical hip and knee joint implants replacement. Biomedical engineering is a new discipline where engineering principles and design concepts are applied to improve healthcare diagnosis, monitoring and therapy by solving medical and biological science related problems. At least three different terminologies that always are being referred to biomedical implant materials, i.e. biocompatibility, biodegradable and biomaterials. Biocompatibility is defined as the immune rejection or inflammatory responses of the surrounding tissue systems to the presence of a foreign object in the body. Whereas biodegradable material means the implant material can easily decompose in the body. Their presence in the body is temporary and usually they degrade as a function of time, temperature or pressure. Biomaterials must possess biocompatibility and sometimes biodegradable properties. A typical example of simultaneously possess biocompatibility and biodegradable properties is drug delivery capsule where their presence to release drugs inside the body over a specific time without causing any toxic effect to the surrounding tissues (Hollinger, 2006). There are varieties of biomaterials such as metallic, ceramics and polymers that have been used as biomedical devices. Metallic biomaterials can be grouped as steels, cobalt and titanium based alloys. Among non-metallic polymeric based biomaterials are polyethylene terephthalate, polytetrafluoroethylene, ultrahigh molecular weight polyethylene (UHMWPE) and lactide-co-glycolide. While titania (TiO_2), titanium carbide (TiC), titanium nitride (TiC), bioglass, hydroxyapatite (HA), silicon carbide (SiC) are typical examples of ceramic biomaterials. All biomaterials must be free from cytotoxicity. Plain steel would corrode easily in the body and become toxic. On the

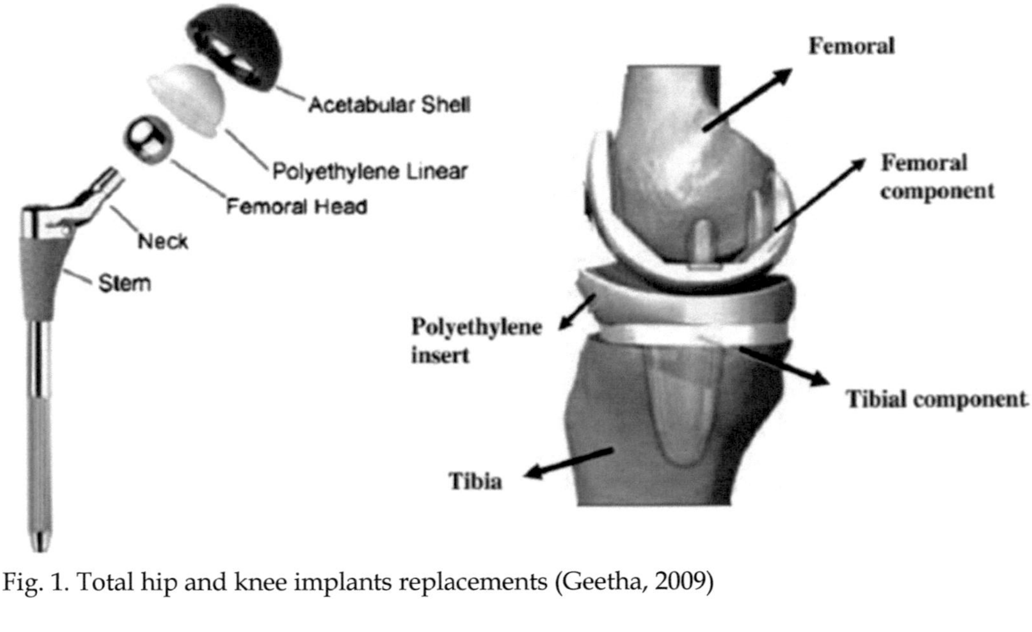

Fig. 1. Total hip and knee implants replacements (Geetha, 2009)

other hand, biomedical grade 316L stainless steel possesses higher biocompatibility properties and it can be used as implant as well as for surgical devices. However, due to its heavy weight, this group of materials have been gradually replaced by other lighter biomaterials such as titanium alloys. Co based alloys is another metallic biomaterials but its high elastic modulus compared to bone causes stress shielding effect especially in load bearing applications (Kumar *et al.*, 2010b). Stress shielding effect is the reduction of bone density due to the removal of normal stress from bone by an implant. Among metallic materials, titanium and its alloys are considered as the most convincing materials in medical applications nowadays because they exhibit superior corrosion resistance and tissue acceptance when compared to stainless steels and Co-based alloys. Table 1 shows the various metallic biomaterials used for medical applications.

Material designation	Common or trade name	ASTM Standard	ISO Standard
Steel biomaterial :			
Fe-18Cr-14Ni-2.5Mo	316L Stainless Steel	ASTM F 138	ISO 5832-1
Fe-18Cr-12.5Ni-2.5Mo,Cast	316L Stainless Steel	ASTM F 745	-
Fe-21Cr-10Ni-3.5Mn-2.5Mo	REX 734	ASTM F 1586	ISO 5832-9
Fe-22Cr-12.5Ni-5Mn-2.5Mo	XM-19	ASTM F 1314	-
Fe-23Mn -21Cr-1Mo-1N	108	F-04.12.35	-
Cobalt base biomaterials:			
Co-28Cr-6Mo Casting alloy	Cast CoCrMo	ASTM F 75	ISO 5832-4
Co-28Cr-6Mo Wrought alloy#1	Wrought CoCrMo, Alloy 1	ASTM F 1537	ISO 5832-12
Co-28Cr-6Mo Wrought alloy#2	Wrought CoCrMo, Alloy 2	ASTM F 1537	ISO 5832-12
Co-28Cr-6Mo Wrought alloy#3	Wrought CoCrMo,GADS	ASTM F 1537	-
Co-20Cr-15W-10Ni-1.5Mn	L-605	ASTM F 90	ISO 5832-5
Co-20Ni-20Cr-5Fe-3.5Mo-3.5W-2Ti	Syncoben	ASTM F 563	ISO 5832-8
Co-19Cr-7Ni -14Fe-7Mo-1.5W-2Ti	Grade 2 "Phynox"	ASTM F 1058	ISO 5832-7
Co-20Cr-15Ni -15Fe-7Mo-2Mn	Grade 1 "Elgiloy"	ASTM F 1058	ISO 5832-7
Co-35Ni-20Cr-10Mo	35N	ASTM F 562	ISO 5832-6
Titanium base biomaterials:			
Ti Cp-1	CP-1(α)	ASTM F 67	ISO 5832-2
Ti CP-2	CP-2(α)	ASTM F 67	ISO 5832-2
Ti-3Al-2.5V	Ti-3Al-2.5W(α/β)	ASTM F 2146	-
Ti-5Al-2.5Fe	Tikrutan (α/β)	-	ISO 5832-10
Ti-6Al-4V	Ti 364(α/β)	ASTM F 1472	ISO 5832-3
Ti-6Al-4V ELI	Ti 364ELI(α/β)	ASTM F 136	ISO 5832-3
Ti-6Al-7Nb	Ti 367(α/β)	ASTM F 1295	ISO 5832-11
Ti-15Mo	Ti-15Mo (Metastable Beta)	ASTM F 2066	-
Ti-12Mo-6Zr-2Fe	"TMZF" (Metastable Beta)	ASTM F 1813	-
Ti-11.5Mo-6Zr-4.5Sn	"Beta 3" (Metastable Beta)	-	-
Ti-15Mo-5Zr-3Al	Ti-15Mo-5Zr-3Al(Metastable Beta)	-	ISO 5832-14
Ti-13Nb-13Zr	Ti-13Nb-13Zr (Metastable Beta)	ASTM F 1713	-
Ti-35Nb-7Zr-5Ta	"TiOsteum" (Metastable Beta)	-	-
Ti-45Nb	Ti-45Nb (Metastable Beta)	ASTM B 348	-

Table 1. Metallic biomaterials for medical and surgical implants (Courtesy from ATI, Allvac, USA)

3. Overview of surface modification techniques

Since all biomedical devices subject to extremely high clinical requirements, a thorough surface modification process is needed prior to implantation process into the human body. The main reasons to carry out various surface modification processes on implant materials for biomedical applications can be summarized as follows:

i. Clean implant material surface from contaminations prior to implantation
ii. Increase bioactivity, cell growth and tissue attachments after implantation
iii. Increase hardness of implant to reduce wear rate especially in articulation joint applications
iv. Introduce passive layer to prevent excessive ion release into body environment
v. Promote antibacterial effect
vi. Increase fatigue strength of implants

The proper surface modification techniques keep the excellent bulk attributes of titanium alloys, such as good fatigue strength, formability, machinability and relatively low modulus. It also improves specific surface properties required by different clinical requirements. Table 2 summarizes the typical surface modification schemes used to treat titanium and its alloys for implant.

Surface modification methods	Modified layer	Objective
Mechanical methods • Machining • Grinding • Polishing • Blasting	Rough or smooth surface formed by subtraction process	Produce specific surface topographies; clean and roughen surface; improve adhesion in bonding
Chemical methods • Chemical treatment • Acidic treatment	<10 nm of surface oxide layer	Remove oxide scales and contamination
• Alkaline treatment	~1 μm of sodium titanate gel	Improve biocompatibility, bioactivity or bone conductivity
• Hydrogen peroxide treatment	~5 nm of dense inner oxide and porous outer layer	Improving biocompatibility, bioactivity or bone conductivity
Sol-gel	~10 μm of thin film, such as calcium phosphate, TiO_2 and silica	Improve biocompatibility, bioactivity or bone conductivity
Anodic oxidation	~10 nm to 40 μm of TiO_2 layer, adsorption and incorporation of electrolyte anions	Produce specific surface topographies; improved corrosion resistance; improve biocompatibility, bioactivity or bone conductivity

CVD (Chemical Vapour Deposition)	~1 µm of TiN, TiC, TiCN, diamond and diamond-like carbon thin film	Improve wear resistance, corrosion resistance and blood compatibility
Biochemical methods	Modification through silanized titania, photochemistry, self-assembled monolayers, protein-resistance, etc.	Induce specific cell and tissue response by means of surface-immobilized peptides, proteins, or growth factors
Physical methods • Thermal spray • Flame spray • HVOF • DGUN	~30 to 200 µm of coatings, such as titanium, HA, calcium silicate, Al_2O_3, ZrO_2, TiO_2	Improve wear resistance, corrosion resistance and biological properties
PVD (Physical Vapour Deposition) • Evaporation • Ion plating • Sputtering	~1 µm of TiN, TiC, TiCN, diamond and diamond-like carbon thin film	Improve wear resistance, corrosion resistance and blood compatibility
Ion implantation and deposition • Beam-line ion implantation • PIII	~10 nm of surface modified layer and/or ~µm of thin film	Modify surface composition; improve wear, corrosion resistance, and biocompatibility
Glow discharge plasma treatment	~1 nm to ~100 nm of surface modified layer	Clean, sterilize, oxide, nitride surface; remove native oxide layer

Table 2. Summary of surface modification methods used for titanium and its alloys implants (Liu *et al.*, 2004)

Among the popular surface modification methods are mechanical, chemical, sol-gel, oxidation, carburization and ion implantation. The last three methods will be discussed in detailed in later sections. Mechanical surface treatments include machining, grinding, and blasting. These methods were discussed in depth elsewhere (Lausmaa *et al.*, 2001). The main goal of mechanical modification is to obtain particular surface roughness and topographies on implant surface. In general, mechanical surface treatments lead to rough structures which finally increase the surface area of implant. This condition is considered more favourable for the implant because it facilitates biomineralization process to take place (Sobieszczyk, 2010a). Surface roughness enhances cell attachment, proliferation and differentiation of osteogenic cells and is the key factor for the osseous integration of metallic implants. Among the mechanical methods, blasting is the most popular technique for achieving desired surface roughness on titanium. The common abrasive particles used as the blasting media are silicon carbide (SiC), alumina (Al_2O_3), biphasic calcium phosphates (BCP), hydroxyapatite and ß-Tricalcium phosphate (Citeau *et al.*, 2005). One of the disadvantages of blasting is that it may lead to surface contamination and local inflammatory reactions of surrounding tissues as a result of dissolution of abrasive particles into the host bone (Gbureck *et al.*, 2003). A range of surface roughness (Ra = 0.5-1.5 μm) shows stronger bone response after the implantation compared to implants with smoother or rougher surface (Sobieszczyk, 2010b). This observation was in contrast with the findings by Fini *et al.* (2003), that rougher surface show encouraging results. Their results were confirmed in the vivo experiments using titanium implants having roughness of 16.5 - 21.4 μm inserted in the cortical and trabecular bone of goats.

Chemical methods include acid treatment, alkali treatment, sol-gel, oxidation, chemical vapour deposition (CVD), and biochemical modification. Following discussion is limited to the first four chemical methods. Since oxidation method itself is a big field, this technique is separately discussed in section 4.

Acid treatment is a popular surface treatment method to clean substrate surface by means of removing oxide and contamination. A mixed acids solution is frequently used for this purpose (Nanci *et al.*, 1998). It is also noted that TiO_2 is the dominant oxide layers formed on the substrate due to high affinity of titanium to react with O_2. These oxides need to be removed prior to other surface treatments such as HA coating, thermal oxidation or carburization and ion implantation. A recommended standard solution for acid treatment is composed of HNO_3 and HF (ratio of 10 to 1 by volume) in distilled water. Hydrofluoric acid has natural tendency of quickly attack TiO_2 in the acid solution and forms soluble titanium fluorides and hydrogen. This acid solution also can be used to minimize the formation of free hydrogen that prevent surface embrittlement occurs due to inclusion of hydrogen in titanium (American Society for Testing and Materials, 1997). A group of researchers investigated the decontamination efficiency to the Ti surface using three acids, $Na_2S_2O_8$, H_2SO_4, and HCl (Takeuchi *et al.*, 2003). They found that HCl was the most effective decontamination agent among these three due to the capability to dissolve titanium salts easily without weakening Ti surfaces.

Alkali treatment can be simply defined as simple surface modification by alkali solution such as NaOH or KOH to form bioactive porous layer on substrate materials. Later, this method followed by thermal treatment to dehydrate and transform amorphous structure into porous crystalline. The combined treatment is called Alkali Heat Treatment (AHT). The alkali treatment process is started by immersing titanium alloy in a 5–10 M NaOH or KOH solution for 24 hr (Kim *et al.*, 1996). After that specimens have to rinse with distilled water followed by ultrasonic cleaning. It is then dried in an oven. Finally heat treatment is carried out by heating the specimens around 600–800 ^{o}C for 1 hr. The heat treatment is performed at very low pressure for avoiding oxidation of titanium at high temperature. The porous surface formed on treated titanium surface disclosed the formation of sodium titanate hydrogel on the titanium substrate. It was observed that after thermal treatment, a large quantity of crystalline sodium titanate with rutile and anatase precipitated. Bioactive bone-like apatite was obtained on the surface after soaking the treated titanium in simulated body fluid (SBF) for 4 weeks. They found that bone-like apatite layer which is bioactive can be formed on other surfaces such as bioglass, hydroxyapatite and glass–ceramic by using this method. It is noted that bioglass, hydroxyapatite and glass–ceramic all are the examples of bioactive ceramics. Recently, a group of researchers investigated the effect of substrate surface roughness on alkali treated CP-Ti for apatite formation after immerging in SBF solution for seven days (Ravelingien *et al.*, 2010). They found that apatite formation increased with the moderate surface roughness. However, very smooth surface (< 0.5 μm) causes sudden decrease in apatite formation.

Sol-gel consists of two terms, sol and gel. A sol can be defined as a colloidial suspension of very small solid particles in a continuous liquid. Gel can be defined as a substance that contains a continuous solid skeleton enveloping a continuous liquid phase (Brinker, 1990). The sol–gel process consists of five main steps: (1) hydrolysis and polycondensation; (2) gelation; (3) aging; (4) drying; (5) densification and crystallization (Piveteau *et al.*, 2001). Two different techniques usually used to carry out the sol–gel process: (i) spin coating technique and (ii) dip coating technique. In spin coating technique the specimens are spun to spread the coating solution on the substrate using centrifugal force where in dip coating specimens are dipped or submerged in the solution. The sol–gel process is popular for depositing thin (<10 μm) ceramic coatings (Liu *et al.*, 2004). In the biomedical area, the sol–gel process is considered new field. Sol-gel method capable of producing various types of coatings on titanium and titanium alloys for biomedical applications. Examples of these coatings are titanium oxide (TiO_2), calcium phosphate (CaP), and TiO_2–CaP composite. Sol–gel technique also has been applied for some silica-based coatings. It has a great potential to replace plasma spray for synthesizing the composite hydroxyapatite/titania coating on the titanium substrate with high adhesion and good bioactivity (Kim *et al.*, 2004). It is reported that plasma spray method results in chemical inhomogeneity and low crystallinity of HA coating on titanium alloys (Wang *et al.*, 2011). In contrast, sol-gel technique produces high crystalline HA microstructure and better chemical homogeneity due to ability to mix the calcium and phosphorus precursors at molecular-level. They also found that atomic diffusion accelerated when increasing the calcining temperature or prolong the calcining time. Other advantages of sol-gel method in comparison with other conventional thin layer oxidation processes are: i) low densification temperature, ii) better control of the homogeneity, chemical composition and crystalline structure of the thin coating, iii) Cost effective and less complicated equipment.

4. Oxidation

Oxidation is a chemical reaction between metal and oxygen. This reaction occurs naturally. However, this reaction can be started with exciting the atoms by providing external energy. In simple way, an oxidation is defined as a chemical reaction by the interaction of metal with oxygen to form an oxide. The oxidation behaviour of a metal depends on various factors and the reaction mechanism usually quite complex. The phenomena started with adsorption of oxygen molecules from the atmosphere, and then followed by nucleation of oxides, formation of a thin oxide layer, finally growth to a thicker scale. During the growth process, nodule formation and scale spallation may also take place (Khanna, 2004) . The total chemical reaction for the formation of oxide (M_aO_b) by oxidation between metal (M) and oxygen gas (O_2) the can be written as:

$$aM + (b/2)\,O_2 = M_aO_b \quad (1)$$

The mechanism of oxidation process is illustrated in Fig. 2. The initial step started by the adsorption of gas on the clean metal surface during the metal-oxygen reaction. As the reaction proceeds, oxygen may dissolve in the metal forming an oxide on the surface either as a film or as oxide nuclei. The gas adsorption and initial oxide formation both are functions of various factors: (i) surface orientation, (ii) crystal defects at the surface, (iii) surface preparation, and (iv) impurities in both the metal and the gas. The oxides formed on surface separates the metal and the gas and sometimes act as a barrier for further oxide formation. This barrier oxide is called protective oxide layer. The oxide can be continuous film or porous structure. Oxides can also be liquid or volatile at high temperature. In general, the reaction mechanism for a specific metal will be a function of several factors: (i) pre-treatment and surface preparation of the metal, (ii) temperature, (iii) gas composition, (iv) pressure and (v) required time of reaction (Kofstad, 1988). The oxidation mechanism can be generalised both at room temperature as well as at high temperature. The basic difference between oxidation at room temperature and high temperature is the reaction rate. At room temperature reaction rate is very slow where at high temperature the rate is accelerated. There are various types of oxidation for surface modification of biomedical grade titanium alloy such as (i) Thermal oxidation, (ii) Anodic oxidation, (iii) Micro-arc oxidation (MAO). These techniques are discussed separately in 4.2, 4.3 and 4.4 respectively.

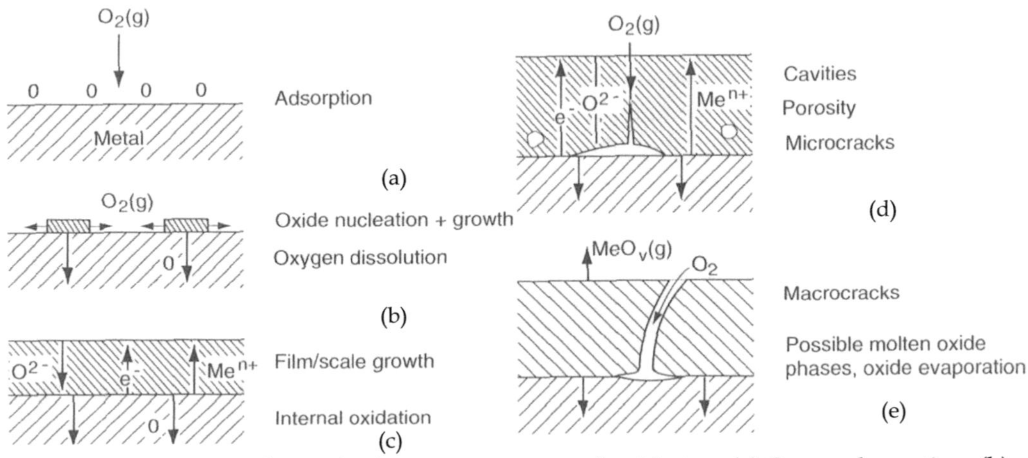

Fig. 2. Scale Formation during high temperature metal oxidation: (a) O_2 gas absorption, (b) O_2 dissolution, (c) Thin oxide film formation, (d) Oxide layer growth, (e) Thick oxide layer (Kofstad, 1988)

4.1 Mechanism of oxidation based on thermodynamic point of view

In oxidation, the chemical reaction between a metal (M) and the oxygen gas (O_2) can be written as:

$$M\ (s) + O_2\ (g) = MO_2\ (s) \tag{2}$$

In thermodynamic point of view, if oxygen potential in the environment is greater than the oxygen partial pressure in equilibrium with the oxide then an oxide will form on the surface of that metal. This equilibrium oxygen pressure is determined from the standard free energy of formation of the oxide. This equilibrium oxygen pressure is also called the dissociation pressure of the oxide in equilibrium with the metal. From equation 2, the standard free energy of the oxidation can be written as,

$$\Delta G^{o} = -RT \ln \left(\alpha_{MO2} / \alpha_{M} . P_{(O2)} \right) \tag{3}$$

Where,

ΔG^o = Gibbs free energy
R = Universal gas constant
T= Absolute temperature
α_{MO2} and α_M = Activities of the oxide and the metal respectiely
and α_M (element activity in alloy) = $\gamma_M . X_M$;
γ_M = the activity coefficient of metal in the alloy ;
X_M = mole fraction of metal in the alloy;
$P_{(O2)}$ = Partial pressure of the oxygen gas.
If a value coefficient γ_M is not available, ideal behaviour is assumed and γ_M is assigned the value of unity. Assuming element activity of the solid constituents is unit, i.e. the metal and oxide, the equation 3 becomes

$$\Delta G^o = - RT \ln P_{(O2)} \quad (4)$$

Or,

$$P(O_2) = \exp(\Delta G^o / RT) \quad (5)$$

Therefore, equation 5 can be used to determine the partial pressure of oxygen required for any metal to form oxide at any temperature from the standard free formation energy. Standard free energy for the formation of oxides is a function of temperature. This can be obtained from the Ellingham/Richardson diagrams which is mentioned elsewhere (Khanna, 2004).

4.2 Thermal oxidation

Thermal oxidation occurs when metals or alloys are heated in a highly oxidizing atmosphere such as air or in the presence of oxygen. It is one of the cost- effective surface modification methods to deliberately generate a barrier oxide layer on titanium alloy. Thermal oxidation treatment aims for obtaining a ceramic coating, mainly focussed on rutile structure. Particularly, oxidation at temperature above 200 °C promotes the development of a crystalline oxide film. Many researchers reported that the thermally formed oxide layer enables increment in hardness, wear resistance and corrosion resistance of titanium and its alloy (Borgioli *et al.*, 2005, Kumar *et al.*, 2009). This protective oxide layer also reduce ion release inside body fluid and thus helps the body from metal toxicity (López *et al.*, 2010). During thermal oxidation process, titanium can easily reacts with air due to its affinity to oxygen. Three types of oxides structure can be produced through this method, which are rutile, brookite or anatase structures. Among the three, rutile structure is more preferable for several reasons. Rutile structure is more inert to bacterial attack (Bloyce *et al.*, 1998), has high hardness and low friction coefficient that can reduce wear as compared to the other two structures (Krishna *et al.*, 2005).

Many researchers investigate thermal oxidation method to solve ion release issues through increasing the corrosion resistance of titanium alloy. A group of researchers studied the chemical composition of oxide layer produced by thermal oxidation on vanadium free TiNb and TiZr based alloys (López *et al.*, 2001). Their aim was to reduce ion release as well as improving corrosion resistance for better biocompatibility. They reported that Ti, Al and Zr based oxide dominate the surface where small amount of Nb based oxide formed. They also explored further to study the chemical composition of the deeper oxide layer surface and found that rutile structure dominates in oxide layer of TiZr based alloy (López *et al.*, 2002). Another group of workers investigated the oxide structure to develop thicker oxide layer for improving the corrosion resistance as well as biocompatibility (Morant *et al.*, 2003). They found that oxide layer was compact and uniform with the granular structure in TiNb based alloy and longitudinal groove structure in TiZr based alloy. It was investigated the corrosion- wear responses by increasing hardness through thermal oxidation of CP-Ti and Ti-6Al-4V (Dearnley *et al.*, 2004). They found that corrosion-wear resistance improved by oxidation where surface of oxidized Ti-6Al-4V is harder than CP-Ti.

Some investigators study the effect of oxidation time on surface roughness of oxide layer through this method to improve surface structure for better corrosion resistance (Gutiérrez *et al.*, 2006). It is noted that higher surface roughness will provide better cell adhesion. They observed that surface roughness increases with the increase of oxidation time. A group of researchers investigated in depth chemical composition of oxide layer to understand the diffusion of elements in the substrate during oxidation (Gutiérrez *et al.*, 2008). Their motivation was to produce thick oxide layer for reducing ion release as well as corrosion protection. They observed that TiZr based alloy showed thicker oxide layer than TiNb based alloy but less homogenous. Another group of researchers studied extensively to optimize the oxidation temperature and time (Kumar *et al.*, 2010b). Their objectives were to produce well adherence oxide layer with rutile structure for improving corrosion resistance and biocompatibility. They observed that best corrosion resistance achieved by oxidation at 650 ^{o}C for 24 hr and the hardness increased threefold at 650 ^{o}C for 48 hr compare to bare metal.

Excessive wear rate is another issue that limit the usage of titanium alloy in various articulation applications. To address the wear resistance issue, another group of researchers investigated the effect of oxidation and temperature on hardness of the oxide layer formed through thermal oxidation on CP-Ti (Yan *et al.*, 2004). They found that thickness of the oxide layer increases with increasing temperature or time and hardness also increases accordingly. Another team of researchers studied the effect of temperature on adhesion and hardness of the oxide layer through thermal oxidation (Rastkar *et al.*, 2005). Their aim was to improve sliding wear resistance by providing hard surface on TiAl based alloy. They observed that higher temperature oxide layer is non-adherent where lower temperature produced adherent oxide layer and also hardness increases with the oxidation temperature increment. Other group of workers also investigated the effect of oxidation time on oxide layer produced through this method (Guleryuz *et al.*, 2005). They wanted to evaluate the dry sliding wear performance on Ti-6Al-4V by providing hard surface. They observed that hardness and surface roughness increases with the increase of oxidation time and these hard oxide layers show significant improvement in dry sliding wear resistance. Another team of researchers investigated the effect of oxidation time and temperature to developed well adherent rutile based oxide surface in order to improve wear resistance (Biswas *et al.*, 2009). It is also noted that rutile structure provides higher hardness compared to anatase structure. They observed that hardness is proportional to oxidation time as well as temperature. However, higher temperature shows significant increase of hardness compared to higher oxidation time.

An appropriate articulating implant should possess the modulus of elasticity close to the bone. Otherwise, this could lead to stress shielding effect which is a loss of bone density. A team of researchers made an effort to address the stress shielding issue of TiNb and TiZr based implant alloys (Munuera *et al.*, 2007). They studied the surface structural properties by evaluating the nanoscale elastic properties of oxide layers at various oxidation times. They found that most cases the Young moduli of the oxide layer are lower than 65 GPA and in some cases it is almost near to bone i.e. 20 GPA. In other study, the nanomechanical properties of oxide scale (hardness and Young modulus) was also investigated (Cáceres *et al.*, 2008). They noticed that TiZr based alloy shows increment in hardness and Young modulus after thermal oxidation. However, TiNb shows reduction in hardness and Young modulus at prolonged oxidation duration which is near to bone. Several researchers also investigated the effective oxidation parameters to produce rutile structure through thermal oxidation. As mentioned earlier, rutile structure is more preferable compared to other structures due to better resistant to bacterial attack and also having higher hardness. A group of researchers carried out thermal oxidation process on Ti–6Al–7Nb, Ti–13Nb–13Zr, and Ti–15Zr–4Nb at 750 °C for 24 hours (López *et al.*, 2003). They reported that TiNbZr based alloys present a thicker scale with rutile structure. Other group of researchers studied the effect of producing rutile structure on AISI 316L coated with titanium (Krishna *et al.*, 2005). They found that the presence of rutile structure improves the hardness and corrosion resistance. Another group of workers investigated the effect of thermal oxidation temperature on the Commercial Pure Titanium (CP-Ti) (Kumar *et al.*, 2010a). They reported that rutile structure can be obtained at 800 °C after continuous heating for 24 hours. Another group of researchers carried out experiments to investigate the effects of different pickling times as well as temperature on the adhesion strength of oxide layer formed on the Ti-6Al-4V after oxidation process (S. Izman *et al.*, 2011a). It was revealed that the thickness of oxide layer increases with pickling time but the adhesion strengths become lower. It was also found that the adhesion strength of oxide layer formed on Ti substrate surface increases with the increase of temperature while the thickness of the oxide layer decreased within 40 °C pickling temperature. Izman *et al* took an attempt to evaluate the effect of thermal oxidation temperature on surface morphology and structure of the Ti13Nb13Zr biomedical material (S. Izman *et al.*, 2011c). It is noted that all thermally oxidized samples exhibit the presence of oxides without spallation regardless of the thermal oxidation temperatures. Surface morphology of oxidized substrates changes from smooth to nodular particles-like shape when the oxidation temperature increases from low to high. Rutile structure dominants the surface area when the substrate is thermally oxidized at 850 °C. In summary, thermal oxidation is a simple and low cost method to produce protective oxide layer with rutile structure on titanium alloys. Studies show that the Young modulus of rutile structure is near to that of the bone (less than 65 GPa) and has antibacterial effect, better corrosion and wear resistance. Despite these encouraging properties, limited works have been reported on the adhesion strength of oxide layer formed on the titanium substrate.

4.3 Anodic oxidation

Anodic oxidation is an electrochemical reaction which is a combined phenomenon of diffusion between oxygen and metal ion. In this technique, the metal ion is driven by an electric field. This phenomenon leads to oxide layer formation on the surface of anode (Liu *et al.*, 2004). Anodic oxidation can be used for producing different types of protective oxide layer on different metals. Common electrolytes used in the process are various diluted acids such as H_2SO_4, H_3PO_4, acetic acid, etc. The main advantage of anodic oxidation compared to other oxidation methods is their ability to form bioactive oxide film on surface of titanium and its alloy. Anodic oxidation increases thickness of the oxide layer for reducing ion release as well as improving corrosion protection. By varying the anodic oxidation parameters, such as current, process temperature, electrolyte composition, and anode potential, the oxide film properties i.e. chemical or structural can be changed.

Principal reactions cause oxidations at the anode are as follows:
At the Ti/Ti oxide interfaces:
$Ti \leftrightarrow Ti^{2+} + 2e^-$
At the Ti oxide/electrolyte interfaces:
$2H_2O \leftrightarrow 2O^{2-} + 4H^+$ (oxygen ions react with Ti to form oxide) ,
$2H_2O \leftrightarrow O_2$ (gas) $+ 4H^+ + 4e^-$ (O_2 gas evolves or stick at anode surface).
At both interfaces:
$Ti^{2+} + 2O^{2-} \leftrightarrow TiO_2 + 2e^-$

In anodic oxidation, a linear correlation exists between the oxide film thickness and applied voltage. If the final oxide thickness is d and the applied voltage is U, then the relationship is $d = \alpha U$. α is a constant and its typical range is 1.5–3 nmV^{-1}. Ishizawa and Ogino *et al.* is the pioneer in developing Ca and P contained oxide layer through anodizing titanium in β-glycerophosphate sodium and calcium acetate contained electrolyte (Ishizawa and Ogino, 1995). They further proceeded exploring and able to transform it into hydroxyapatite by applying hydrothermal treatment. The results showed that the electrolyte possessed some impurities (e.g. sodium). These impurities decreased oxide layer's strength. A group of researchers reported that desirable cellular behaviour such as cell growth, cell attachment, etc. can be obtained from the thin HA layer on the surface of CP-Ti which was produced by anodization and subsequently followed by hydrothermal treatment (Takebe *et al.*, 2000). It is observed that cellular attachment and spreading are affected by this thin HA layer on the CP-Ti surface. It is also revealed a thin HA layer on titanium surface shows more osteoconductive behaviour to cell attachment as compared to bare CP-Ti. Other group of workers investigated new electrolytes consists of calcium glycerophosphate and calcium acetate for producing anodic oxide films that consist of Ca and P on titanium implants (Zhu *et al.*, 2001). The anodic oxide film of titanium obtained using this method is highly crystalline with porous structure and rich in Ca and P. The recommended optimum conditions are: (i) 350 V as final voltage, (ii) 70 A m^{-2} as current density, and (iii) concentrations of the calcium glycerophosphate (0.02M) and calcium acetate (0.15 M). Ca and P ratio near to 1.67 was achieved using this recommended condition. Positive biological response also observed from the properties of that anodic oxide layer surface. Yang *et al.* reported that using anodic oxidation in H_2SO_4 solution united with consequent heat treatment is an efficient method for obtaining titanium alloy with bioactive surface (Yang *et al.*, 2004). They also observed that the porous structure Titania of anatase and/or rutile phase covered on the surface after anodic oxidation. It was interesting to observe that apatite can be formed on titanium alloy by anodic oxidation in simulated body fluid. The initial time for apatite formation was inversely proportional to the quantity of rutile or anatase phase (Liu *et al.*, 2004). Apatite cannot be formed without spark discharge on the surface although anatase was produced on anodically oxidized titanium. Hence, a combination of anodic oxidation with heat treatment is required for the apatite formation on titanium in SBF without spark discharge treatment. Heat treatment induces apatite formation in SBF since the amount of anatase and/or rutile increases by the heat treatment.

This also indicates that prior to the formation of apatite on the surfaces, a titanium oxide with three-dimensional micro-porous structure may be essential. It is also noted that surfaces can be bioactive by containing Ca and/or P which leads to osteoinduction of new bones. Wojciech (2011) investigated the effective anodic voltage for producing better corrosion resistance bioactive oxide layer containing Ca and P on TiZr based alloy through anodic oxidation method. He found that lower anodic voltage produced highest corrosion resistance. However, higher anodic voltage provides bioactive oxide layer rich in Ca and P on TiZr based alloy which also increase the corrosion resistance. In summary, anodic oxidation is a simple and effective method of surface modification for providing better bioactive surface of titanium alloys which also homogenous and highly crystalline. However, bioactive apatite formation on titanium alloy through this method required post-treatment such as hydrothermal heat treatment. The oxide film produced by anodic oxidation method shows various properties such as better biocompatibility, corrosion resistance, osteoconductive, etc. These properties rely on the microstructure and composition of the materials as well as anodic oxidation parameters, such as current, temperature, anode potential and electrolyte composition.

4.4 Micro-arc oxidation (MAO)

Another name of micro-arc oxidation is anodic spark oxidation or plasma electrolytic oxidation (PEO). Micro arc oxidation is an electrochemical surface modification process for producing oxide coatings on metals such as Al, Ti, Mg, Ta, W, Zn, and Zr and their alloys (Liu *et al.*, 2010). According to Yerokhin *et al.* (1999), MAO can be defined as a complex plasma-enhanced physico-chemical process which involved micro-arc discharge, diffusion and plasma chemical reactions. Basically, it is a new type of anodic oxidation technique, but the difference between MAO and the conventional anodic oxidation is it employs higher potentials to discharges and the resulting plasma modifies the structure of oxide layer. This process can be used to grow crystalline oxide coating with thickness range from ten to hundreds μm. The coating thickness depends on process parameters such as current density, process time, electrolyte temperature, applied voltage, electrolyte composition, alloy composition (Dunleavy *et al.*, 2009). A large number of short-lived sparks (micro arc discharges) produced in MAO process is a result of localized electrical breakdown of the growing coating. These discharges play the key role in the coating growth mechanism as they deposit 'craters' on the free surface of the growing coating. In MAO process, the anode is immersed in electrolyte which is an aqueous solution. The anode is made from valve metals. Valve metals usually refer to Ti, Al, Mg, Ta, W, Zn, and Zr due to their usage as a cathode to emit electron in electronic valve. They are also known as 'thermionic valve' materials in early days. However, disputes on the right definition of these terms have been remained among researchers since Al is not a suitable material for high temperature resistance to emit electron. In MAO, an unequal alternating voltage between the anode and cathode initiates an electrical discharge at the anode. The typical voltage range for anode and cathode is from 150 to 1000V and from 0 to 100V respectively. Temperature and local pressure in the discharge channels are among the parameters that affect the MAO coating qualities such as high strength, well adhesion, high micro-hardness, and wear resistance (Liu *et al.*, 2010). Since MAO can provide high hardness and a continuous barrier, this coating is suitable for protection against wear, corrosion or heat as well as electrical insulation (Curran and Clyne, 2005). General characteristics of these coatings are porous, firm adhesion to substrates and the pores are homogenously distributed on the coating's surface with nanostructure grains (Kim *et al.*, 2002). Due to superior corrosion resistance, thermal stability, photocatalytic activity, wear resistance and CO sensing properties makes MAO coatings as a popular research area (Shin *et al.*, 2006, Jin *et al.*, 2008). MAO has been popular in the biomedical community since Ishizawa *et al.* pioneered the technique to biomedical titanium implants (Ishizawa and Ogino, 1995). Biomimetic deposition of apatite is possible on Ca and P-containing MAO coatings (Song et al., 2004). Zhao *et al.* found that the MAO coatings benefit osteoblast adhesion (Zhao *et al.*, 2007). They compared the adhesion performance of MAO coatings on various modified smooth and rough surfaces. Other researchers investigated the effect of variations in the electrolyte compositions to produce different kinds of nanostructured composite coatings under this method (Kim et al., 2007, Yao *et al.*, 2008) . Cimenoglu *et al.* investigated the MAO coating on Ti6Al7Nb and found that oxide layer shows grainy appearance rather than porous and contained calcium titanate precipitates, HA and rutile structure (Cimenoglu *et al.*, 2011). In summary, MAO is a potential method for producing porous nanostructured coatings on Ti and its alloys which promote best osteoblast cell adhesion. This technique has been spreading into the field of orthopaedic and dental implant materials.

5. Carburization of titanium alloy

Poor tribological properties limit the usefulness of titanium alloy in many engineering applications (Bloyce *et al.*, 1994). Moreover, not all titanium and its alloys can meet all of the clinical requirements. In order to improve the biological, chemical, and mechanical properties, surface modification is often performed (Huang *et al.*, 2006, Kumar *et al.*, 2010b). Till now various surface modification techniques by thermo-chemical process have been studied and applied for improving wear resistance of titanium alloys. These are carburizing, nitriding and oxidation (Biswas *et al.*, 2009, Tsuji et al., 2009b, Savaloni et al., 2010). Among them, carburization technique is one of the methods that can be used to form hard ceramic coating on titanium alloys. The main objective of carburization is to provide hard surface on titanium and its alloys for increasing wear resistance in articulation application since titanium carbide is one of the potential biocompatible carbide layers (Bharathy *et al.*, 2010). It is also one of the cost-effective surface modification methods to deliberately generate a carbide layer on titanium alloy. Many researchers reported that the carbide layer enables to increase hardness, wear resistance and corrosion resistance to titanium and its alloy (Kim *et al.*, 2003). Sintered solid titanium carbide is a very important non-oxide ceramics that widely used in the fields of wear resistance tools and materials due to its high melting point (3170 ^{o}C), low density, high hardness (2500 ~3000HV), superior chemical and thermal stability, and outstanding wear resistance (Courant *et al.*, 2005). Apart from sintering, titanium carbide layer can be created by other surface modification methods, such as plasma carburizing process, thermal carburization or high-temperature synthesis, carburization by laser melting, gas-solid reaction or gas carburization and sol-gel process (Lee, 1997, Yin *et al.*, 2005, Cochepin *et al.*, 2007, Luo *et al.*, 2011). Among these methods, thermal carburization process is considered as the simplest and the most cost effective. Typically, one of the main obstacles for TiC coating is the high affinity of titanium to oxygen which leads to form TiO_2 easily on the surface. To overcome this issue , vacuum carburization or inert gas environment is introduced to remove O_2 contents in carburization chamber (Wu *et al.*, 1997). Another common problem related to carburization is non uniform hardness profile across the carburized layers due to variation of carbon concentration in the surface region (Saleh *et al.*, 2010). The discussion of this chapter starts with the basic mechanism of carburization followed by three popular carburization methods, i.e. thermal, gas and laser melting.

5.1 Basic mechanism of carburization

Carburization is a process widely used method to harden the surface and enhance the properties of components that made from metal. Carburizing consists of absorption and diffusion of carbon into solid metal alloys by heating at high temperature. Historically, the carburizing process is generally done at elevated temperatures with a carbon medium that can supply adequate quantity of atomic carbon for absorption and diffusion into the steel (Luo *et al.*, 2009). The carbon medium that use for carburizing process can be solid (charcoal), molten salt (cyanide), a gaseous or plasma medium (Prabhudev, 1998). There are three methods of carburizing process, i.e. solid carburizing, liquid carburizing, and gas carburizing. All these three methods have their own compounds medium that is used for the carbon supply during the process. In solid carburizing process, carburizing compound such as charcoal or graphite powder is used for its medium. In the liquid carburizing method, molten cyanide is used for carbon enrichment. Lastly, for the gas carburizing method, hydrocarbon gas or plasma is used as the source of the carburizing medium.

During carburizing, the atomic carbon is liberated from carbonaceous medium due to decomposition of carbon monoxide into carbon dioxide and atomic carbon as given below:

$$2CO \rightarrow CO_2 + C_{at} \qquad (6)$$

Then, the carbon atom from carburizing medium is transferred to the surface of the metal. These metal surfaces will absorb the carbon and diffuse deep into it. Thus, this phenomena results increase in hardness of the substrate materials surface.

5.2 Thermal carburization

Thermal carburization process is considered the earliest carburization technique and it is a kind of solid carburization. Generally solid particle such as charcoal, graphite powder, etc is used as a carbon source to surround titanium substrate during carburization process. Titanium can easily react with oxygen in ambient environment and form a thin passive layer of TiO_2 on the outer surface with thickness range of 3 to 7 nm (Liu *et al.*, 2004) . This passive layer becomes a barrier for carbon atom diffusion into the titanium surface. Since titanium is highly affinity to oxygen, an inert or vacuum environment is preferable for conducting carburization process. Argon gas is commonly used as a medium to remove oxygen in tube or muffle furnace heating chamber from pre-oxidizing the titanium substrate surface. The quality of carburized layer largely depends on the carburizing temperature, soaking time, source of carbon (type and particle size) and the absence of oxygen level in the chamber. The carburizing parameters may have significant effects on the thickness, adhesion, density and chemical composition of carburized layer formed on the titanium substrate. Studies on titanium carbide powder synthesis by carbothermal method in argon environment requires high temperature in the range of 1700–2100 °C (Weimer, 1997) and long reaction time (10–24 h) (Gotoh *et al.*, 2001). Other workers tried to synthesize TiC powder at lower temperatures and shorter time with success. For instance, Lee *et al.* studied the chemical kinetics at various temperatures (1100 to 1400 °C) for synthesising TiC from CP-Ti alloy and graphite powders (Lee and Thadhani, 1997). They found that Ti with compacted graphite powder shows highly activated state of reactants which reduce activation energies by 4-6 times, undergo a solid state diffusion reaction. They also concluded that increasing temperature will increase the rate of heat released. This released heat generates localized melting of unreacted Ti and initiate a combustion reaction. It has been reported that the carburizing rate of titanium dioxide, TiO_2 into TiC can be accelerated by using the finest and homogenous carbon powder (Maitre *et al.*, 2000). Sen *et al* produced fine and homogeneous TiC powders by carbothermal reduction of titania/charcoal in a vacuum furnace at different reaction temperatures from 1100 °C to 1550 °C (Sen *et al.*, 2010). They observed that reaction temperature increases, uniform crystal grain arises with the liberation of much CO and higher temperature (at 1550 °C) produced large amount of TiC. They also noticed that as reaction temperature increased, formation of the compounds was in sequenced as Ti_4O_7, Ti_3O_5, Ti_2O_3, TiC_xO_{1-x} and TiC. Hardly found researchers study TiC formation on titanium solid substrate. Izman *et al* initiated the study to investigate the effects of different carburizing times on the adhesion strength of carbide layer formed on the Ti-6Al-7Nb (S. Izman *et al.*, 2011b). Prior to carburization process, all samples were treated to remove residual stress and oxide scales by annealing and pickling processes respectively. Hard wood charcoal powder was used as a medium. The carburizing process was carried out under normal atmospheric condition. They found that a mixture of oxide and carbide layers formed on the substrate and the thickness of these layers increases with carburizing time. It was also revealed that the longer carburizing time provides the strongest adhesion strength and TiC as the dominant layer. Porous structure of TiC was observed and this structure is believed able to facilitate the osteoblast cell growth on implant. In summary, thermal carburization is a simple and cost effective method to produced TiC for increasing the wear resistance properties of titanium and its alloys. However, the technique has not been explored rigorously this far. Issues regarding carbide grain growth, carbon particle agglomeration, non-uniform carbide particle shapes and large amounts of unreacted TiO_2 and carbon in the substrate are still under on-going research.

5.3 Gas carburization

The main difference between thermal and gas carburization process is the carbon source medium. Instead of solid, hydrocarbon gas is used as a carbon source and carburization process takes place either under gaseous or plasma condition. This process is typically performed using plasma or flowing hydrocarbon gas over the Ti and Ti alloy substrate at high temperature in a inert gas or vacuum furnace. Gas carburizing also have various categories such as hydrocarbon gas carburizing (using methane or ethane), plasma carburizing, etc. The advantage of gas carburizing over solid carburising is faster processing time but this method is costlier compared to solid carburization (Robert *et al.*, 1994). Due to high affinity to oxygen, plasma carburizing method has difficulties in carburizing the Ti alloys because thin protective titanium oxide film easily forms on its surface which cause in obstruction of the carbon diffusion (Okamoto *et al.*, 2001). Kim *et al* carburized Ti6Al4V at 900 °C and 250MPa pressure using CH_4–Ar–H_2 plasma for 6 hrs to increase the wear resistance (Kim *et al.*, 2003). Hardness of titanium alloy was improved significantly from 400HV to 1600HV with the carburized layer thickness of about 150 μm along the surface.

They revealed that fine and homogeneous dispersion of hard carbide particles such as TiC and V_4C_3 found in the carburized layer able to improve wear resistance as well as fatigue life for more than two folds. Tsuji *et al.* carburized Ti6Al4V at 600 °C in a Ar gas conditioned furnace using CH_4-H_2 plasma for 1 hr for improving hardness from 400 HV to 600 HV (Tsuji *et al.*, 2009a). They also investigated the effects of combining plasma-carburizing and deep rolling on the notched surface microstructure and morphology, micro-hardness and notch fatigue life of Ti-6Al-4V alloy specimen in a laboratory at an ambient temperature. They reported that the notch root area of plasma-carburized specimen's surface roughness has been significantly improved by deep-rolling. This method effectively introduces compressive residual stress and work hardening in the substrate. Plasma-carburization with subsequent deep-rolling largely enhances the notch fatigue strength of specimen in comparison with untreated specimen. The developed compressive residual stress and work hardening zone influence the initial crack growth rate of deep-rolled carburized specimen. The thickness of this zone is approximately 350 μm depth from the surface. However, the crack rapidly propagates toward the inside after it passes through this zone. They concluded that plasma-carburizing process combined with deep-rolling effectively improves the notch fatigue properties of Ti-6Al-4V alloy. Another researcher made an effort to investigate the plasticity effect on titanium alloy after being treated under gas carburization. Luo *et al.* carburized Ti6Al4V at 1050 °C in a vacuum furnace using C_2H_2 gas for 4 hrs for improving the of hardness from 350 HV to 778 HV (Luo *et al.*, 2011). TiC or also called titanium cermets were successfully formed on the surface. It was reported that the plasticity of the titanium cermets was slightly lower (10.86%) than original titanium bare material. This indicates that the carburized titanium has significantly improved in fracture toughness as compared to typical ceramics material. They concluded that carburization is a way to produce titanium cermets efficiently which consists of hard surface, high toughness and plasticity. All these properties make titanium carbide as a potential candidate for artificial articulation material. In summary, the primary objective of gas carburizing is to produce carburized layer on the substrate in order to increase wear resistance property of titanium alloys. However, improvement in hardness introduces other issues such as reduction in plasticity and fatigue strength in the titanium substrate.

5.4 Carburization by laser melting

Laser carburizing technique is developed from laser surface hardening of steel. In a simple way, laser carburization can be defined as a process of using laser as a source of high energy to perform carburization. There are various types of laser carburizing methods where the categories are based on laser source, such as Neodymium Yttrium Lithium Fluoride (Nd:YLF), Neodymium Yttrium Aluminium Garnet (Nd:YAG), Titanium Sapphire (Ti:Sapphire), CO_2 laser, etc. Laser carburizing process involves carbon diffusion into the metal substrate using laser irradiation. The typical source of carbon is graphite powder. Other type of powder such as TiC is also being used in laser melting technique to form carburized layer on titanium based materials. Fig. 3 shows a schematic diagram of laser melting working principle. This process involved heating of specimen through continuous or pulse wave laser irradiation, rapid melting, intermixing or diffusion of carbon particle, and rapid solidification of the pre-deposited alloying elements on substrate to form an alloyed zone or carburized layer.

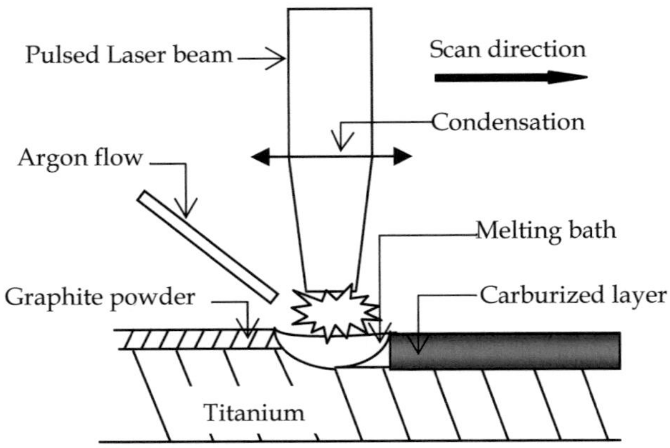

Fig. 3. Schematic diagram of typical pulsed laser carburization set up.

Investigations on laser carburizing technique were extended from steel to α-Titanium (Fouilland *et al.*, 1997), commercial pure titanium (Courant et al., 2005) , and biomedical grade titanium (Sampedro *et al.*, 2011). Laser melting carburization produces thick coating ranged between one and several hundred micrometers depending on the irradiation conditions. Other carburizing methods are more suitable for producing thin film coating. Another advantage of this method compared to other techniques is that it's capability of coating complex substrate geometry and shape such as notches or grooves where through other methods very difficult to reach these inaccessible areas. Wide heat affected zone is a general issue for thermal or plasma method heating which leading to shape distortion. On the other hand, laser carburizing method is free from these disadvantages since an accurate focused heating on the work piece can be controlled easily. Other commonly controlled laser processing parameters are laser power (W), scanning speed (mm/min), pulse/deposition time (ms), laser frequency (Hz) and overlapping factor (%). The effects of these variables are investigated in terms of changes to the hardness, compositions, heat affected zone, pores, cracks and microstructure of the carburized zone. For instance, a group of researchers investigated the effect of processing time on the TiC microstructure formation on titanium alloy using Nd-YAG laser (Courant *et al.*, 2005). They observed that the time ratio has a significant effect on the carburized microstructure. A lower time ratio caused an increase in pulse power leading to form a thick layer of melted zone with rich in carbon but free from graphite formation. In contrast, higher time ratio produces large amount of graphite formation in the melted zone which can act as a solid lubrication. This phenomenon shows the potential to reduce abrasive wear rate and hence increase the tribological performance of articulation implants. One group of researchers compared the effect of process parameters (laser power and scanning speed) on solidification of TiC microstructure using two different laser sources on Ti-6Al-4V substrate (Saleh *et al.*, 2010). They found that TiC appears either in the form of dendrites or as particles located inside the grains and at the grain boundaries. This resulted significant increment in microhardness of the surface after carburizing process. They concluded that both Nd-YAG and the CO_2 lasers able to produce macroscopically homogeneous microstructures of carburized layers. However, the former laser produces deeper carburized layer compared to the later. Recently, another group of workers studied pulse wave laser method (Nd-YAG laser) to form TiC layer on CP-Ti. They investigated the effect of process parameters (irradiated energy per length and pulse duration) on the microstructure as well as hardness of the substrate surface. It is noted that the microhardness of the surface increased 3-5 times higher than the base metal substrate when increasing the pulse duration. It is also observed that the microhardness of microstructure reduced by decreasing the irradiated energy per unit length of the material where irradiated energy can be reduced by increasing the process travel speed (Hamedi *et al.*, 2011). In a summary, laser carburization is a potential route of strong surface hardening method with a short process time to increase the wear resistance property of the titanium alloys without affecting its bulk properties. This method provides hardest carbide layer compared to other two carburization techniques.

6. Ion implantation and deposition

Ion implantation or ion beam processing is a procedure in which ions of a material are accelerated in an electric field and bombarded into the solid substrate surface. Various ions such as oxygen, nitrogen, carbon, etc. can be implanted on any substrate material for a coating purpose to modify the substrate surface. When carbon is implanted on substrate material then the effect of the surface modification is similar to carburization. Similarly, this method also can be applicable for nitridation as well as oxidation. Two common types of ion implantation process are (i) Conventional beam line ion implantation and (ii) Plasma immersion ion implantation (PIII) method. The basic difference between the beam line ion implantation and plasma immersion ion implantation method is the target function. In beam line ion implantation, the target is totally isolated from the ion beam generation. In contrast, the target is an active part of the ion generation through bias voltage in PIII system (Savaloni *et al.*, 2010). Fig. 4 shows the two typical types of ion implantation systems. The ion implantation phenomena started with the acceleration of ions and it directed towards a substrate (titanium in the present case) which is called target. The energy of the ions is usually in the range of several kilo electronvolt to few mega electronvolt. This level of energy could cause significant changes in the surface by the ions penetration. However, the energy of ions is selected carefully to avoid deep penetration inside the substrate. Therefore, the surface modifications are limited to the near-surface region and a depth of 1 μm from the surface is normal (Rautray *et al.*, 2011). In other words, bulk material properties will not be affected by the ion implantation process.

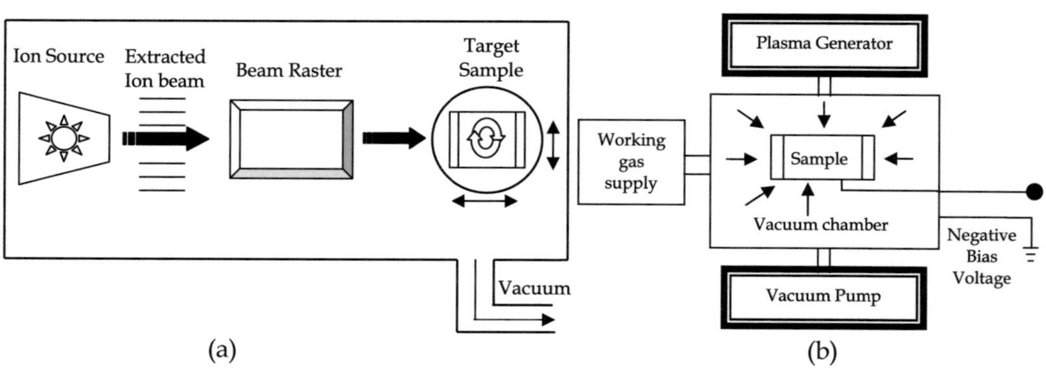

Fig. 4. Schematic diagram of (a) beam line ion implantation system and (b) PIII ion implantation system

For instance, in carbon ion implantation process, the implanted carbon ions are limited either to form titanium carbides or carbon atoms with C–C bonds near the surface. This may result in improvement of the mechanical properties as well as biocompatibility of titanium alloys. However, studies on the issue regarding corrosion resistance of TiC formation by ion implantation carburization are still underway. It is reported that a very high carbon ion dose of implantation (10^{18} cm^{-2}) will reduce surface hardness of the titanium substrate (Viviente *et al.*, 1999). The reaction between excess carbon and titanium produces mixed layer of graphite (C–C bonds) and TiC which cause reduction in hardness. In other study, a moderate dose of ion implantation from 5×10^{15} to 1×10^{17} cm^{-2} able to create nanocrystalline titanium carbide (TiC) layer which hardness of more than two folds on Ti–6Al–4V alloy substrate (Liu *et al.*, 2004). Liu et al. also reported that the tribological properties of titanium alloys are significantly improved at ion implant doses of over 4×10^{17} cm^{-2}, producing friction coefficients of 0.2–0.3. Ion implantation method is free from some disadvantages of plasma process such as thick coating, different phases of mixed crystalline and low crystallinity which leads to delamination problem (Rautray *et al.*, 2011). Effects of ion implantation process on wear resistance also have been studied by various researchers. Williams *et al.* investigated carbon ion implantation effect on the wear resistance of Ti–6Al–4V alloys in a corrosive environment with the composition of 0.9% NaCl or 0.9% NaCl + 10% serum (Williams, 1985). Two-stage carbon ion implantation: 2.5×10^{16} cm^{-2} at 35 kV followed by 1.6×10^{17} cm^{-2} at 50 kV were carried out for the test. They revealed that ion implanted sample shows reduction in corrosion current by a factor of 100 compared to that untreated samples. A group of researchers investigated Ti–6Al–4V alloy's corrosion resistance after 80 kV, 3×10^{17} cm^{-2} carbon ion implantation (Zhang *et al.*, 1991). They carried out examinations using electrochemical methods in two media: 0.5 M H_2SO_4 and (HCl + NaCl) solution (pH = 0.1) at 25 °C. In both solutions, ion implanted samples show higher corrosion potential (E_{cor}) than unimplanted samples. They also reported that the increment in the surface corrosion resistance was due to a durable solid passive layer formation. Other group of researchers experimented various carbon doses on the titanium alloy for evaluating corrosion resistance of TiC formation at energy of 100 keV. in 0.9% NaCl solution at a temperature of 37 °C (Krupa *et al.*, 1999). They revealed that the corrosion resistance of titanium alloy improved significantly by producing a continuous solid nanocrystalline TiC layer when applying 1×10^{17} C^+ cm^{-2} of carbon dose or more. Another group of workers studied the formation of TiC on titanium alloy using PIII method by varying the deposition times (Baba *et al.*, 2007). They concluded that the formation of TiC through ion implantation on titanium alloy depends on amount of carbon ion implantation which is proportional to ion implantation process time. Corrosion resistance on biomedical grade titanium alloy can also be improved by nitrogen ion implantation. It was reported that increasing of N^+ flux will influence the corrosion potential, corrosion current and passive current. These changes lead to initial increase in the corrosion resistance of the titanium alloy (Savaloni *et al.*, 2010). Other group of researchers investigated the effect of process temperature and implantation time on the corrosion properties of Ti-6Al-4V. It was found that prolonged implantation times do not contribute to a major changes in corrosion resistance where process temperature does (Silva *et al.*, 2010). They also reported that the best corrosion resistance achieved at 760 °C with 2 hr processing time. Previous studies on PIII method basically focused on single non-metallic ion implantation to improve tribological properties such as hardness as well as corrosion resistance. However, in the recent development, it shows that the research interests in this method have been expanded to include implanting both metal and non-metallic ion simultaneously on titanium alloy. The main driving force of introducing this dual implantation method is to address the clinical and tribological issues concurrently. For example, Ca and Mg ion implanted into titanium alloy for increasing the bone integration (Kang *et al.*, 2011). Ag and N ion have been used to have dual effects on titanium alloy (Li *et al.*, 2011). Ag provides antibacterial effect and nitride layer (TiN) formed on the titanium surface increases wear and corrosion resistance. The ion implantation sequence in this dual method also has impact on the deposited particle size and distribution. In a summary, ion implantation method is more suitable for wear and corrosion resistance application. However, recent research trend on ion implantation shows the focus is not only on tribological issues but also on the effect in clinical aspects. Therefore, various metallic ions implantation on titanium alloy appear to be a future prospective research area.

7. Conclusions

Various surface modification methods used for improving properties of biomedical grade titanium and its alloys are discussed in this chapter. There are at least six (6) different methods available in the current practice. These are mechanical, chemical, physical, sol-gel, carburization and ion implantation. Oxidation and carburization methods are discussed in detail while the discussions on other methods are in brief.

Oxidation method modifies the titanium surface into various types of oxides. The main objective is to produce porous oxide structure for promoting cell growth and cell attachment. There are cases where corrosion and wear resistance are also improved by applying this technique. The recent trend shows that the oxide layer formed on the titanium substrate serves as a basis for growing hydroxyapatite layer to increase bioactivity.

Carburization is mainly used to improve wear resistance by increasing titanium surface hardness via thermal, gas and laser melting methods. Hardness of titanium carbide layer formed through these methods varies from 1.5 to 5 times as compared to bare material. Higher hardness of carbide layer assists to increase wear and corrosion resistance of implant surface.

Ion implantation method provides better wear and corrosion resistance than other thermal surface modification techniques. In the recent trend, ion implantation technique is found to provide dual effects concurrently such as wear resistance and antibacterial effect.

Generally, it is observed that the overall trends of surface modification methods seem to shift from the use of conventional source (chemical, induction heater and gas) to the application of advanced technology (electrolyte based, laser, plasma and ion). This could be due to the low efficiency of conventional methods that require longer time and huge amount of energy. The works on surface modifications also appear to expand from focusing on tribological issues such as wear resistance, corrosion resistance and hardness of modified layer to clinical issues such as cell growth, cell attachment and antibacterial effects. These developments demand newer technologies in the future for providing solutions of dual issues simultaneously, i.e. tribological and clinical.

8. Acknowledgements

Authors would like to express highest gratitude to Ministry of Higher Education (MOHE)and Ministry of Science, Technology and Innovation (MOSTI) for providing grant to conduct this study via vote numbers Q.J130000.7124.02H60, 78611 and 79374. Authors also would like to thank the Faculty of Mechanical Engineering, UTM for providing their facilities to carry out this study.

9. Nomenclature

BCP	Biphasic Calcium Phosphates
CVD	Chemical Vapour Deposition
CP	Commercial Pure
DGUN	Detonation Gun
HVOF	High Velocity Oxygen Fuel spraying
HA	Hydroxyapatite
LSA	Laser Surface Alloying
MAO	Micro Arc Oxidation
Nd:YAG	Neodymium Yttrium Aluminium Garnet
Nd:YLF	Neodymium Yttrium Lithium Fluoride
PIII	Plasma Immersion Ion Implantation
SBF	Simulated Body Fluid

10. References

American Society for Testing and Materials (1997). *ASTM standard B600,* Philadelphia, American Society for Testing and Materials.

Andrew, W. B., Chandrasekaran, Margam (2004). *Service characteristics of biomedical materials and implants,* London, Imperial College Press.

Baba, K., Hatada, R., Flege, S., Kraft, G. & Ensinger, W. (2007). Formation of thin carbide films of titanium and tantalum by methane plasma immersion ion implantation. *Nuclear Instruments and Methods in Physics Research Section B: Beam Interactions with Materials and Atoms,* vol.257, pp.746-749

Bharathy, P. V., Nataraj, D., Chu, P. K., Wang, H., Yang, Q., Kiran, M. S. R. N., Silvestre-Albero, J. & Mangalaraj, D. (2010). Effect of titanium incorporation on the structural, mechanical and biocompatible properties of DLC thin films prepared by reactive-biased target ion beam deposition method. *Applied Surface Science,* vol.257, pp.143-150

Biswas, A. & Dutta Majumdar, J. (2009). Surface characterization and mechanical property evaluation of thermally oxidized Ti-6Al-4V. *Materials Characterization,* vol.60, pp.513-518

Bloyce, A., Morton,P. , Bell,T. (1994). *ASM Handbook,* OH, ASM International

Bloyce, A., Qi, P. Y., Dong, H. & Bell, T. (1998). Surface modification of titanium alloys for combined improvements in corrosion and wear resistance. *Surface and Coatings Technology,* vol.107, pp.125-132

Borgioli, F., Galvanetto, E., Iozzelli, F. & Pradelli, G. (2005). Improvement of wear resistance of Ti-6Al-4V alloy by means of thermal oxidation. *Materials Letters,* vol.59, pp.2159-2162

Brinker, C. J. S., G.W. (1990). *The Physics and Chemistry of Sol–Gel Processing,* San Diego, American Press.

Cáceres, D., Munuera, C., Ocal, C., Jiménez, J. A., Gutiérrez, A. & López, M. F. (2008). Nanomechanical properties of surface-modified titanium alloys for biomedical applications. *Acta Biomaterialia,* vol.4, pp.1545-1552

Cimenoglu, H., Gunyuz, M., Kose, G. T., Baydogan, M., Uğurlu, F. & Sener, C. (2011). Micro-arc oxidation of Ti6Al4V and Ti6Al7Nb alloys for biomedical applications. *Materials Characterization,* vol.62, pp.304-311

Citeau, A., Guicheux, J., Vinatier, C., Layrolle, P., Nguyen, T.P., Pilet, P., Daculsi, G. (2005). In vitro biological effects of titanium rough surface obtained by calcium phosphate grid blasting. *Biomaterials,* vol.26, pp.157-165

Cochepin, B., Gauthier, V., Vrel, D. & Dubois, S. (2007). Crystal growth of TiC grains during SHS reactions. *Journal of Crystal Growth,* vol.304, pp.481-486

Courant, B., Hantzpergue, J. J., Avril, L. & Benayoun, S. (2005). Structure and hardness of titanium surfaces carburized by pulsed laser melting with graphite addition. *Journal of Materials Processing Technology,* vol.160, pp.374-381

Curran, J. A., Clyne, T. W. (2005). Thermo-physical properties of plasma electrolytic oxide coatings on aluminium. *Surface and Coatings Technology,* vol.199, pp.168-176

Dearnley, P. A., Dahm, K. L. & Çimenoğlu, H. (2004). The corrosion–wear behaviour of thermally oxidised CP-Ti and Ti–6Al–4V. *Wear,* vol.256, pp.469-479

Dunleavy, C. S., Golosnoy, I. O., Curran, J. A. & Clyne, T. W. (2009). Characterisation of discharge events during plasma electrolytic oxidation. *Surface and Coatings Technology,* vol.203, pp.3410-3419

Fini, M., Savarino, L., Aldini, N.N., Martini, L., Giaveresi, G., Rizzi, G., Martini, D., Ruggeri, A., Giunti, A., Giardino, R. (2003). Biomechanical and histomorphometric investigation on two morphologically differing titanium surfaces with and without frluorohydroxyapatite coating: an experimental study in sheep tibiae. *Biomaterials,* vol.24, pp.3183-3192

Fouilland, P. L., Ettaqi, S., Benayoun, S. & Hantzpergue, J. J. (1997). Structural and mechanical characterization of Ti/TiC cermet coatings synthesized by laser melting. *Surface and Coatings Technology,* vol.88, pp.204-211

Gbureck, U. M., A. Probst, J. Thull, R. (2003). Tribochemical structuring and coating of implant metal surfaces with titanium oxide and hydroxyapatite layers. *Materials Science and Engineering C,* vol.23, pp.461-465

Geetha, M. S., A.K.R. Asokamani and Gogia, A.K. (2009). Ti Based Biomaterials: the Ultimate Choice for Orthopaedic Implants. *Progress in Materials Science,* vol.54, pp.397–425

Gotoh, Y., Fujimura, K., Koike, M., Ohkoshi, Y., Nagura, M., Akamatsu, K. & Deki, S. (2001). Synthesis of titanium carbide from a composite of TiO2 nanoparticles/methyl cellulose by carbothermal reduction. *Materials Research Bulletin,* vol.36, pp.2263-2275

Guleryuz, H. & Cimenoglu, H. (2005). Surface modification of a Ti–6Al–4V alloy by thermal oxidation. *Surface and Coatings Technology,* vol.192, pp.164-170

Gutiérrez, A., Munuera, C., López, M. F., Jiménez, J. A., Morant, C., Matzelle, T., Kruse, N. & Ocal, C. (2006). Surface microstructure of the oxide protective layers grown on vanadium-free Ti alloys for use in biomedical applications. *Surface Science,* vol.600, pp.3780-3784

Gutiérrez, A., Paszti, F., Climent, A., Jimenez, J. A. & López, M. F. (2008). Comparative study of the oxide scale thermally grown on titanium alloys by ion beam analysis techniques and scanning electron microscopy. *Journal of Materials Research* vol.23, pp.2245-2253

Hamedi, M. J., Torkamany, M. J. & Sabbaghzadeh, J. (2011). Effect of pulsed laser parameters on in-situ TiC synthesis in laser surface treatment. *Optics and Lasers in Engineering,* vol.49, pp.557-563

Han, Y., Hong, S. H. & Xu, K. W. (2002). Porous nanocrystalline titania films by plasma electrolytic oxidation. *Surface and Coatings Technology,* vol.154, pp.314-318

Hollinger, S. A. G. A. J. O. 2006. *An introduction to biomaterials,* Florida, Tailor & Francis.

Huang, H., Winchester, K. J., Suvorova, A., Lawn, B. R., Liu, Y., Hu, X. Z., Dell, J. M. & Faraone, L. (2006). Effect of deposition conditions on mechanical properties of low-temperature PECVD silicon nitride films. *Materials Science and Engineering: A,* vol.435-436, pp.453-459

Ishizawa, H., Ogino, M. (1995). Characterization of thin hydroxyapatite layers formed on anodic titanium oxide films containing Ca and P by hydrothermal treatment. *Journal of Biomedical Materials Research Part: A,* vol.29, pp. 1071–1079

Jin, F., Chu, P. K., Wang, K., Zhao, J., Huang, A. & Tong, H. (2008). Thermal stability of titania films prepared on titanium by micro-arc oxidation. *Materials Science and Engineering: A,* vol.476, pp.78-82

Kang, B.-S., Sul, Y.-T., Jeong, Y., Byon, E., Kim, J.-K., Cho, S., Oh, S.-J. & Albrektsson, T. (2011). Metal plasma immersion ion implantation and deposition (MePIIID) on screw-shaped titanium implant: The effects of ion source, ion dose and acceleration voltage on surface chemistry and morphology. *Medical Engineering & Physics* vol.33, pp.730-738

Khanna, A. S. (2004). *Introduction to High Temperature Oxidation and Corrosion,* California, ASM International.

Kim, H. M. M., F. Kokubo, T. Nakamura,T. (1996). Preparation of bioactive Ti and its alloys via simple chemical surface treatment. *Journal of Biomedical Materials Research Part A,* vol. 32, pp. 409–417

Kim, H. W., Koh,Y.H., Li, L.H., Lee,S., Kim, H.E. (2004). Hydroxyapatite coating on titanium substrate with titania Buffet layer processed by sol-gel method. *Biomaterials,* vol.25, pp.2533-2538

Kim, M. S., Ryu, J.J., Sung, Y.M. (2007). One-step approach for nano-crystalline hydroxyapatite coating on titanium via micro-arc oxidation. *Electrochemical Communication,* vol.9, pp.1886–1891

Kim, T.-S., Park, Y.-G. & Wey, M.-Y. (2003). Characterization of Ti-6Al-4V alloy modified by plasma carburizing process. *Materials Science and Engineering A,* vol.361, pp.275-280

Kofstad, P. 1988. *High Temperature Corrosion,* New York, Elsevier Applied Science Publisher Ltd.

Krishna, D. S. R. & Sun, Y. (2005). Thermally oxidised rutile-TiO2 coating on stainless steel for tribological properties and corrosion resistance enhancement. *Applied Surface Science,* vol.252, pp.1107-1116

Krupa, D., Jezierska, E., Baszkiewicz, J., Wierzchoń, T., Barcz, A., Gawlik, G., Jagielski, J., Sobczak, J. W., Biliński, A. & Larisch, B. (1999). Effect of carbon ion implantation on the structure and corrosion resistance of OT-4-0 titanium alloy. *Surface and Coatings Technology,* vol.114, pp.250-259

Kumar, S., Narayanan, T. S. N. S., Raman, S. G. S. & Seshadri, S. K. (2009). Thermal oxidation of CP-Ti: Evaluation of characteristics and corrosion resistance as a function of treatment time. *Materials Science and Engineering: C,* vol.29, pp.1942-1949

Kumar, S., Narayanan, T. S. N. S., Raman, S. G. S. & Seshadri, S. K. (2010a). Thermal oxidation of CP Ti -- An electrochemical and structural characterization. *Materials Characterization,* vol.61, pp.589-597

Kumar, S., Sankara Narayanan, T. S. N., Ganesh Sundara Raman, S. & Seshadri, S. K. (2010b). Thermal oxidation of Ti6Al4V alloy: Microstructural and electrochemical characterization. *Materials Chemistry and Physics,* vol.119, pp.337-346

Lausmaa, J., Brunette , D.M., Tengvall , P. Textor , M. Thomsen , P. 2001. *Titanium in Medicine,* Berlin, Springer.

Lee, J. H., Thadhani, N. N. (1997). Reaction synthesis mechanism in dynamically densified Ti + C powder compacts. *Scripta Materialia,* vol.37, pp.1979-1985

Li, J., Qiao, Y., Ding, Z. & Liu, X. (2011). Microstructure and properties of Ag/N dual ions implanted titanium. *Surface and Coatings Technology,* vol.205, pp.5430-5436

Liu, X., Chu, P. K. & Ding, C. (2004). Surface modification of titanium, titanium alloys, and related materials for biomedical applications. *Materials Science and Engineering: R: Reports,* vol.47, pp.49-121

Liu, X., Chu, P. K. & Ding, C. (2010). Surface nano-functionalization of biomaterials. *Materials Science and Engineering: R: Reports,* vol.70, pp.275-302

López, M. F., Gutiérrez, A. & Jiménez, J. A. (2001). Surface characterization of new non-toxic titanium alloys for use as biomaterials. *Surface Science,* vol.482-485, Part 1, pp.300-305

López, M. F., Gutiérrez, A., Jiménez, J. A., Martinesi, M., Stio, M. & Treves, C. (2010). Thermal oxidation of vanadium-free Ti alloys: An X-ray photoelectron spectroscopy study. *Materials Science and Engineering: C,* vol.30, pp.465-471

López, M. F., Jiménez, J. A. & Gutiérrez, A. (2003). Corrosion study of surface-modified vanadium-free titanium alloys. *Electrochimica Acta,* vol.48, pp.1395-1401

López, M. F., Soriano, L., Palomares, F. J., Nchez-Agudo, M. S., Fuentes, G. G., Gutierrez, A. & Nez3, J. A. J. (2002). Soft x-ray absorption spectroscopy study of oxide layers on titanium alloys. *Surface and Interface Analysis* vol.33, pp.570-576

Luo, Y., Ge, S.-R. & Jin, Z.-M. (2009). Wettability modification for biosurface of titanium alloy by means of sequential carburization. *Journal of Bionic Engineering,* vol.6, pp.219-223

Luo, Y., Jiang, H., Cheng, G. & Liu, H. (2011). Effect of carburization on the mechanical properties of biomedical grade titanium alloys. *Journal of Bionic Engineering,* vol.8, pp.86-89

Maitre, A., Tetard, D. & Lefort, P. (2000). Role of some technological parameters during carburizing titanium dioxide. *Journal of the European Ceramic Society,* vol.20, pp.15-22

Morant, C., López, M. F., Gutiérrez, A. & Jiménez, J. A. (2003). AFM and SEM characterization of non-toxic vanadium-free Ti alloys used as biomaterials. *Applied Surface Science,* vol.220, pp.79-87

Munuera, C., Matzelle, T. R., Kruse, N., López, M. F., Gutiérrez, A., Jiménez, J. A. & Ocal, C. (2007). Surface elastic properties of Ti alloys modified for medical implants: A force spectroscopy study. *Acta Biomaterialia,* vol.3, pp.113-119

Nanci, A. W., J.D. Peru, L.Brunet,P. Sharma,V. Zalzal,S. Mckee, M.D. (1998). *Journal of Biomedical Materials Research,* vol.40, pp.324

Okamoto, Z., Hoshika, H., Yakushiji, M. (2001). *Heat Treatment,* vol.40, pp.88

Piveteau, L.-D., Brunette, D.M., Tengvall,P. , Textor,M., Thomsen, P. 2001. *Titanium in Medicine,* Berlin, Springer.

Prabhudev, K. H. 1988. *Handbook of Heat Treatment of Steels*, New Delhi, Tata McGraw-Hill Publishing Company Ltd.

Rastkar, A. R. & Bell, T. (2005). Characterization and tribological performance of oxide layers on a gamma based titanium aluminide. *Wear,* vol.258, pp.1616-1624

Rautray, T. R., Narayanan, R. & Kim, K. H. (2011). Ion implantation of titanium based biomaterials. *Progress in Materials Science,* vol.56, pp.1137-1177

Ravelingien, M., Hervent, A.-S., Mullens, S., Luyten, J., Vervaet, C. & Remon, J. P. (2010). Influence of surface topography and pore architecture of alkali-treated titanium on in vitro apatite deposition. *Applied Surface Science,* vol.256, pp.3693-3697

Robert, H. T., Dell, K. A. , Leo, A. 1994. *Manufacturing Processes Reference Guide*, New York, Industrial Press Inc.

S. Izman, Abdul-Kadir, M. R., Anwar, M., Nazim, E. M., L.Y.Kuan & E.K.Khor (2011a). Effect Of Pickling Process On Adhesion Strength Of Ti Oxide Layer On Titanium Alloy Substrate. *Advanced Materials Research,* vol.146-147, pp.1621-1630

S. Izman, Abdul-Kadir, M. R., Anwar, M., Nazim, E. M., Nalisa, A. & M.Konneh (2011b). Effect of carburization process on adhesion strength of Ti carbide layer on titanium alloy substrate. *Advanced Materials Research,* vol.197-198, pp.219-224

S. Izman, Shah, A., Abdul-Kadir, M. R., Nazim, E. M., Anwar, M., Hassan, M. A. & Safari, H. (2011c). Effect Of Thermal Oxidation Temperature On Rutile Structure Formation Of Biomedical TiZrNb Alloy *Advanced Materials Research,*

Saleh, A. F., Abboud, J. H. & Benyounis, K. Y. (2010). Surface carburizing of Ti-6Al-4V alloy by laser melting. *Optics and Lasers in Engineering,* vol.48, pp.257-267

Sampedro, J., Pérez, I., Carcel, B., Ramos, J. A. & Amigó, V. (2011). Laser Cladding of TiC for Better Titanium Components. *Physics Procedia,* vol.12, Part A, pp.313-322

Savaloni, H., Khojier, K. & Torabi, S. (2010). Influence of N+ ion implantation on the corrosion and nano-structure of Ti samples. *Corrosion Science,* vol.52, pp.1263-1267

Sen, W., Sun, H., Yang, B., Xu, B., Ma, W., Liu, D. & Dai, Y. (2010). Preparation of titanium carbide powders by carbothermal reduction of titania/charcoal at vacuum condition. *International Journal of Refractory Metals and Hard Materials,* vol.28, pp.628-632

Shin, Y.-K., Chae, W.-S., Song, Y.-W. & Sung, Y.-M. (2006). Formation of titania photocatalyst films by microarc oxidation of Ti and Ti–6Al–4V alloys. *Electrochemistry Communications,* vol.8, pp.465-470

Silva, G., Ueda, M., Otani, C., Mello, C. B. & Lepienski, C. M. (2010). Improvements of the surface properties of Ti6Al4V by plasma based ion implantation at high temperatures. *Surface and Coatings Technology,* vol.204, pp.3018-3021

Sobieszczyk, S. (2010a). Surface modifications of Ti and its alloys. *Advances in Materials Sciences,* vol.10, pp.29-42

Sobieszczyk, S. (2010b). Surface Modifications of Ti and its Alloys. *Advances in Materials Sciences,* vol.10, pp.29-42

Song, W. H., Jun,Y.K., Han,Y., Hong, S.H. (2004). Biomimetic apatite coatings on micro-arc oxidized titania. *Biomaterials,* vol.25, pp.3341–3349

Takebe, J., Itoh, S., Okada, J., Ishibashi, K. (2000). Anodic oxidation and hydrothermal treatment of titanium results in a surface that causes increased attachment and altered cytoskeletal morphology of rat bone marrow stromal cells in vitro. *Journal of Biomedical Materials Research,* vol.51, pp. 398–407

Takeuchi, M. A., Y. Yoshida,Y. Nakayama,Y. Okazaki,M. Kagawa,Y. (2003). Acid pretreatment of titanium implants. *Biomaterials,* vol.24, pp.1821-1827

Tsuji, N., Tanaka, S. & Takasugi, T. (2009a). Effect of combined plasma-carburizing and deep-rolling on notch fatigue property of Ti-6Al-4V alloy. *Materials Science and Engineering: A,* vol.499, pp.482-488

Tsuji, N., Tanaka, S. & Takasugi, T. (2009b). Effects of combined plasma-carburizing and shot-peening on fatigue and wear properties of Ti-6Al-4V alloy. *Surface and Coatings Technology,* vol.203, pp.1400-1405

Viviente, J. L., Garcia, A., Loinaz, A.,Alonso, F., on˜Ate, J.I. (1999). Carbon layers formed on steel and Ti alloys after ion implantation of C+ at very high doses. *Vacuum,* vol.52, pp.141-146

Wang, D. G., Chen, C. Z., Ma, J. & He, T. (2011). Microstructure evolution of sol–gel HA films. *Applied Surface Science,* vol.257, pp.2592-2598

Weimer, A. W. 1997. *Carbide, nitride and boride materials-synthesis and processing,* London, Chapman & Hall.

Williams, J. M., Buchanan, R.A., Rigney, E.D. (1985). *In:* Proceedings of the ASM conference on applications of ion plating and ion implantation to materials, Atlanta,GA, pp. 141.

Wojciech, S. (2011). Preliminary investigations on the anodic oxidation of Ti–13Nb–13Zr alloy in a solution containing calcium and phosphorus. *Electrochimica Acta,* vol.56, pp.9831-9837

Wu, S. K., Lee, C. Y. & Lin, H. C. (1997). A study of vacuum carburization of an equiatomic TiNi shape memory alloy. *Scripta Materialia,* vol.37, pp.837-842

Yan, W. & Wang, X. X. (2004). Surface hardening of titanium by thermal oxidation. *Journal of Materials Science,* vol.39, pp.5583-5585

Yang, B., Uchida, M., Kim, H.-M., Zhang, X. & Kokubo, T. (2004). Preparation of bioactive titanium metal via anodic oxidation treatment. *Biomaterials,* vol.25, pp.1003-1010

Yao, Z. P., Gao, H.H, Jiang, .Z.H., Wang,F.P. (2008). Structure and properties of ZrO_2 ceramic coatings on AZ91D Mg alloy by plasma electrolytic oxidation. *Journal of American Ceramic Society,* vol.91, pp.555–558

Yerokhin, A. L., Nie,X., Leyland, A., Matthews,A., Dowey,S.J. (1999). Plasma electrolysis for surface engineering. *Surface and Coatings Technology,* vol.122, pp.73-93

Yin, X., Gotman, I., Klinger, L. & Gutmanas, E. Y. (2005). Formation of titanium carbide on graphite via powder immersion reaction assisted coating. *Materials Science and Engineering A,* vol.396, pp.107-114

Zhang, D., Yu, W., Wang, Z. & Dong, R. (1991). In: Corrosion control 7th APCCC China, China, pp. 793.

Zhao, B. H., Lee, I.S., Han,I.H., Park,J.C.,Chung, S.M. (2007). *Current Applied Physics,* vol.7S1, pp.e6-e10

Zhu, X., Kim, K.-H. & Jeong, Y. (2001). Anodic oxide films containing Ca and P of titanium biomaterial. *Biomaterials,* vol.22, pp.2199-2206